THE SCIENTIFIC JOURNAL

THE
SCIENTIFIC
JOURNAL

Authorship and the Politics of Knowledge
in the Nineteenth Century

Alex Csiszar

The University of Chicago Press
Chicago and London

The University of Chicago Press, Chicago 60637
The University of Chicago Press, Ltd., London
© 2018 by The University of Chicago

Published 2018
Paperback edition 2020
Printed in the United States of America

29 28 27 26 25 24 23 22 21 20   1 2 3 4 5

ISBN-13: 978-0-226-55323-8 (cloth)
ISBN-13: 978-0-226-75250-1 (paper)
ISBN-13: 978-0-226-55337-5 (e-book)
DOI: https://doi.org/10.7208/chicago/9780226553238.001.0001

Library of Congress Cataloging- in- Publication Data

Names: Csiszar, Alex, author.
Title: The scientific journal : authorship and the politics of knowledge in the
    nineteenth century / Alex Csiszar.
Description: Chicago ; London : The University of Chicago Press, 2018. |
    Includes bibliographical references and index.
Identifiers: LCCN 2017038012 | ISBN 9780226553238 (cloth : alk. paper) |
    ISBN 9780226553375 (e-book)
Subjects: LCSH: Science—Periodicals—Publishing—History—19th century. |
    Communication in science. | Technical writing—History—19th century.
Classification: LCC Q223.C85 2018 | DDC 501.4—dc23
LC record available at https://lccn.loc.gov/2017038012

♾ This paper meets the requirements of ANSI/NISO Z39.48-1992
(Permanence of Paper).

To my mother and the memory of my father

. . . & weekly scientific journals had their news sheets attached, with its accidents & offences, its police, its Waylayings & elopements, & the chase there anent, & crime, cons &c. &c. & did not these in time convey their influence even to philosophers?

—WILLIAM JARDINE, "Reality, a Tale of Physical Science, Founded on Facts" (ca. 1844)

CONTENTS

FIGURES

INTRODUCTION

"Broken Pieces of Fact"

We expect a lot of different things of the scientific journal. Not only is it the medium through which scientists are supposed to make knowledge claims public, but the journal literature is also alleged to be a relatively robust archive of humanity's scientific knowledge — it has variously been called scientists' "bible," "canon," and "heritage." Threatening the integrity of this archive has been likened to "sewage thrown into the pure stream of science." Today, vigilantes police the borders of the scientific literature, hoping to purge it of defective papers and to expel illegitimate publications.[1]

We expect journals ideally to be public and open to all — they embody the vision of science as public knowledge, even if what that means is often a subject of vigorous debate. Many advocates of a more open science have focused their efforts on expanding access to the scientific literature, from the British chemist and Marxist John Desmond Bernal's attempt to reform the publishing system in the 1940s to Aaron Swartz's "Guerilla Open Access Manifesto" and architects of databases of pirated papers.[2] At the same time, however, scientific papers are written to be read almost exclusively by small cadres of initiates.

The journal literature is also used to identify who counts as a legitimate scientific practitioner and who is a qualified expert in relevant fields. When questions arise as to what scientific consensus is on some matter of concern, governmental bodies, the public, and even scientists routinely look to the reputable journal literature dealing with that subject. The paper has become a base unit for sizing up careers, with publication lists a significant factor in decisions about hiring, tenure, and grants.[3]

Not very long ago this arrangement appeared to many observers to be a natural feature of how modern science is supposed to work. That has now changed. We are in the midst of an intense period of experimentation in genres, formats, and practices of evaluation in science.[4] As the hegemony of the scientific journal comes unbound, we can now ask how this format and genre became such a dominant institution in the first place. How did it ever become possible to suppose that scientific journals could be both permanent archive and breaking news, both a public repository and the exclusive dominion of experts, both a complete record and a painstakingly vetted selection? And what are the stakes in deciding that this no longer makes sense?

This book sets out to explore these problems by investigating two linked questions. First, how did so much epistemic weight come to be loaded into this one format over all others? Second, how did the public legitimacy of the scientific enterprise become so closely associated with the scientific literature? One quick way to state my argument is that you cannot really answer one of these questions without answering the other one too.

Let me explain. The scientific journal is not now—indeed, has never been—the main way in which scientists communicate among themselves or with others about the natural world. Scientists have always used a wide spectrum of media and formats to accomplish these things, including letters, telephone conversations, databases, conferences, laboratories, offices, classrooms, and (in the past few decades) email, video, and other networked platforms. But even if we agreed that what really matters are the conversations within laboratories, correspondence, or wine-fueled receptions at conferences, it remains the case that acts of submitting work to a journal, of refereeing, of a paper's being accepted or rejected by a more or less prestigious publication, provide a great deal of fodder for such conversations. In this way the rituals associated with the scientific literature structure the everyday life of science in ways that go well beyond the formal machinery itself. The value ascribed to publishing research papers, and the general expectations about the form that those papers ought to take, influence the kinds of projects researchers choose to pursue, the modes of collaboration that they routinely engage in, and the kinds of information sharing that research communities demand. Imagine for a moment that the making of an academic scientific career required scientists to publish not papers but a longer book that synthesized a field of information based on their own and others' research — one that might require the research assistance of a large team of assistants behind the scenes. All of these conditions on the making of scientific knowledge would certainly take quite a different form. Formats and genres have epistemic consequences.

Likewise, although scientific journals are supposed to be crucial for making knowledge public, scientific practitioners do not normally use them as a direct means of communicating that knowledge to non-practitioners. Rather, the scientific literature plays a prominent role in public representations of scientific expertise and objective judgment. We are told that the careful scrutiny exercised by the scientific community of what makes it into print means that the scientific literature is or ought to be a faithful representation of legitimate scientific opinion. Conversely, questioning the integrity of this process has become a favored strategy among those aiming to cast doubt on the integrity of whole scientific fields, from biomedicine to climatology. The scientific literature has also become a symbol of the public and democratic nature of scientific life. Observers call foul when researchers bypass the journal literature for reasons of industrial or military secrecy, or in favor of filing a patent. Imagine if the participants in a given scientific field decided to keep all of their discussions and ideas under tight lock and key, perhaps putting out a carefully crafted statement about the research consensus every year or two. Such a field would likely have a hard time convincing a modern democratic state to provide them with much financial support, or to take their findings seriously into account in policy discussions. Formats and genres have political consequences too.

The idea that problems of ordering knowledge are connected with questions of politics is not a very original one in the history of science. But this perspective has been largely absent from debates over the changing media landscape of science and of the future of scholarly publishing. In order to understand how we got into our current situation and how we might move forward, I think we need to stop carrying on as if problems of scholarly publishing are a matter simply of improving the means by which experts communicate with one another and in doing so reap professional rewards. This book shows that moments when the norms and forms of expert communication have been most in doubt are precisely those moments when scientific practitioners have sought — or been forced — to renegotiate their public status within a wider political landscape.[5] The litany of scientific crises we are now said to be living through — of replication, of peer review, of scholarly publishing — cannot be separated from political questions regarding the nature and status of scientific expertise in modern democracies.[6] Since the nineteenth century, the apparatus of specialized publishing has been an intersection point where expert cultures of credibility have overlapped, uneasily, with public criteria of accountability. This is what I mean when I say that the question about the epistemic weight of the scientific journal cannot be disconnected from the problem of the public legitimacy of science.

The modern scientific journal is largely an invention of the nineteenth century. Periodicals comprising natural philosophical news go back more than a century earlier, and in the twentieth century the forms and meanings that scientific journals could take continued to evolve. But it was in the hundred years following the French Revolution that scientific practitioners began to perceive the existence of a single, cohesive-enough genre — the scientific paper — as the chief avenue by which to make the results of their research into public knowledge claims. Although versions of this transition took place in much of the scientific world, it followed a remarkable course in France and Britain. In both these countries a single elite institution — the Royal Academy of Sciences and the Royal Society of London, respectively — had long claimed a privileged position as an authoritative audience for scientific claims. To see how these elite institutions first resisted, then embraced, and finally transformed this format to suit new ends is to see the modern scientific journal take shape. This book is about how this happened, and how scientific journals subsequently took on many of the traits and meanings with which we have come to associate them. Before saying more, let me list what I take these to be.

By the early twentieth century, most scientific journals were supposed to be made up largely of papers that were original contributions to knowledge: their central claims were not to be speculative opinions nor synthetic reviews of others' work.[7] The latter were signed by authors who took primary credit and responsibility for their contents, but they were also expected to highlight their reliance on other authors through citations. Journals might be published either by a society or as for-profit ventures (or both), but authors did not normally receive payment for their contributions (although payment for papers remained possible, and was relatively common, for example, in German periodicals). Of course, the scientific journal never referred to any truly unified format. Even before the advent of electronic publishing, journals varied widely in their size, frequency, and submission and acceptance procedures, and in the nature of their contents. Articles varied not only in their length, from short "letters" to more extended memoirs, but in the genre expectations of diverse research fields. Despite such variations, it became widely accepted that scientific journals—and in many respects scholarly journals more generally[8]—constituted a special class of publications that could be demarcated from other forms of periodical literature. In this view, the scientific literature had become a corpus of texts whose integrity demanded protection. Editors or experts called on to judge whether a paper ought to be published were imagined as doing their duty not only to a periodical's reputation and prestige but to science as a whole. (By the early twentieth century some periodicals employed referee systems, but the idea that this was a *sine qua non* of scholarly journals gained ground only later in the twentieth century.)

Through the early nineteenth century, many would have found the preceding description quite puzzling. Journals by no means constituted a uniquely appropriate medium for announcing a new claim to discovery, and if the category "scientific journal" existed at all, it did not correspond to the foregoing description. When learned journals first emerged in the seventeenth century, they took after newspapers and gazettes, by design ephemeral and often in disrepute. Aimed at readers who were members of the Republic of Letters, they included book reviews, reports of goings-on, and digests and excerpts of correspondence. While some became important venues for circulating discovery claims, it was not clear that the broad learned public with which they were associated was the appropriate one for esoteric or technical questions of natural philosophy.

Before the nineteenth century, the institutions most intimately associated with the evaluation of scientific knowledge were the learned societies and academies scattered throughout major European cities. Many of these specialized in making judgments on matters of science and technology in service to their respective monarch, from whom they derived their authority. Although their members were often participants in the Republic of Letters, these institutions generally kept their distance from journals. What they did publish were grand, weighty tomes that contained polished memoirs and which appeared at infrequent intervals. Even the most famous and arguably the first natural philosophical journal, the *Philosophical Transactions*, was reformulated along these lines when it finally became an official publication of the Royal Society of London in the mid-eighteenth century.

Journals that focused specifically on natural philosophy or fields of natural science remained relatively rare until the late eighteenth century. Their spread was spurred in part by entrepreneurs looking to develop new markets for their publications, but often also by their editors' commitment that scientific topics ought to be made accessible to broader publics. Their launch was often accompanied by statements that science (or some branch of it) was no longer simply an esoteric pastime of elites but a body of knowledge with far-reaching consequences for everyday life.[9]

These new publications hewed to contemporary ideas of what journals and magazines were for: they diffused information, provided a locus for correspondence, and digested and reprinted news of discoveries. One London publication put the distinction as follows: "the permanent records of Science are chiefly preserved in the Transactions of learned Societies; and are principally confined to the labours of their Members only. The monthly publications, edited by individuals, furnish an account of what may be regarded as the News of Philosophy."[10] Following the political transformations brought by the French Revolution, however, journals began to offer alternative models of sociability to the academies, with editors often arguing that public opinion—which they claimed to represent—was the only legitimate judge of truth, whether in politics, culture, or natural knowledge.

At first, powerful societies and academies resisted the influence of these journals. They tried to keep journalists out of their meetings, and they insisted that their own publications were made of wholly different stuff, unsullied by commercial exigencies. In 1817 the French physicist Jean-Baptiste Biot explained that such ephemeral publications could not compete with "grand academic collections where the slow, but continual progress of the human mind are deposited, and which are destined to last as long as there is civilization on the earth."[11] But protests such as Biot's were themselves symptoms of the increasing pressure that commercial journals were putting on elite institutions. Not only did the new journals often excerpt or print texts that had originally been submitted to academies and societies, but independent journals were increasingly publishing original papers, submitted directly by authors who chose to bypass them entirely.

By the 1830s the leaders of academies and societies changed their strategy. Instead of keeping journals at arm's length, they decided to publish journals of their own, modeled on their commercial competitors. This was a controversial move. The intrusion of journals into elite scientific institutions brought the

Figure 0.1 Literary genius parceled out in standardized morsels by the machinery of the press "like cakes on the counter of a pastry shop." Jean-Jacques Grandville, *Un autre monde* (Paris: H. Fournier, 1844), 272. NC1499 G66 A42 1844. Houghton Library, Harvard University.

market logic of the publishing business to natural philosophy in new ways. Biot worried that encouraging authors to abandon longer books or polished memoirs in favor of publishing in journals was turning academies into advertising bureaus. Moreover, because such short notes did not afford much space for evidence in support of an author's claims, conviction might come to rest dangerously on arguments from authority rather than from facts.[12] Worries over the role that the periodicity of the press played in shaping and distorting knowledge tapped into broader cultural concerns about distorting creative expression by forcing it into short serialized formats. In 1844 the French caricaturist Jean-Jacques Grandville depicted the press as a machine that cranked out literary genius in a continuous flow that could be sliced into uniform portions as if in a pastry shop (Figure 0.1).

But new commitments to the role of public opinion in conferring political legitimacy, and the increasingly central role of the periodical press in imagining this public, convinced even Biot that academies needed to reimagine the grounds of their authority to make claims about the natural world. As the power to certify scientific knowledge diffused out into the marketplace of the press, academies and societies invested journals with functions designed to confer reward and to demarcate true practitioners from others. What emerged over the rest of the nineteenth century was an assemblage of ideas and attributes drawn from other genres, including the learned journals of the Enlightenment, magazines and popular miscellanies, academic memoirs and transactions, journalists' accounts of scientific meetings, and even patent specifications.

As journals became not only purveyors of scientific news but also archives of discovery, it became more common to conceive of science as a series of discrete discovery events localized in time and connected with an individual author. This raised other tricky questions about the status of collective knowledge. Confidence in the intelligibility and accessibility of nature had long derived support from the metaphor of nature as a book. Although the trope had many variations, the Book of Nature was generally single-authored and bounded (with a logical beginning and end), and treated subjects in a cohesive, comprehensive manner.[13] But, as James Clerk Maxwell playfully suggested in 1856, little of this followed if nature were imagined instead as a *magazine.* Many subsequent observers, including Hermann von Helmholtz and Lord Rayleigh, routinely connected worries over the problem of synthesizing isolated facts with their reflections on the problem of synthesizing the growing expanse of scientific papers.[14]

As late as 1867, the Cambridge physiologist Michael Foster could refer to the short papers published in periodicals as "specimens of those broken pieces of fact, which every scientific worker throws out to the world, hoping that on them, some time or other, some truth may come to land."[15] But over the course of the century, claims that certain forms of print publishing constituted the only legitimate means of publicizing scientific knowledge progressively became more credible, and at the same time specialized serials came to form the solid core of science in print.

By 1902 an older Michael Foster had come to a new view. The habits of authors such as Charles Darwin, who published new findings in grand books, were now "out of place and even dangerous." Rather, "the writings of [T. H.]

Huxley furnish an example of the more common mode of publication adopted by men of science. Nearly all his important contributions to science were published in periodicals." It was through his papers that posterity would "judge of Huxley's worth as an investigator."[16]

Indeed, the consolidation of the scientific paper had as its correlate the emergence of the scientific author as a distinctive identity. For very long, many if not most practitioners of natural philosophy might indeed have been authors, but there was no special regime of scientific authorship.[17] During the early nineteenth century, the kinds of rewards and responsibilities that could be associated with writing about science were as varied as the genres and formats—including single-authored books, manuscripts, or periodicals genres—in which science might be found. In Britain, becoming too closely identified with the sort of professional authorship associated with scientific journals could even risk one's social standing among elites. But by the end of that same century, it was widely acknowledged that a career and reputation in science depended largely on one's unpaid contributions to particular kinds of periodicals. Although this identity had been formed in the crucible of debates over property in ideas that included the development of modern patent law, as well as international copyright, being held a scientific author generally came to be seen as separate from those other regimes of propriety and property, and in some cases even incompatible with them. While many might write about science, only certain writers were scientific authors.

In 1967, the Institute for Scientific Information, a company based in Philadelphia that was responsible for the Science Citation Index, produced an animated history of scientific communication as part of a film marketing its wares to scientists and librarians (Figure 0.2).[18] It opened with early natural philosophers enjoying drinks in a tavern, communicating through "face-to-face contact." Later, such raucous exchanges were supplemented by communication at a distance via correspondence. As the number of correspondents grew, the printing press came in handy as a time-saver for prolific philosophical correspondents, but the real breakthrough occurred when an enterprising individual bound several such letters together to form a journal. From then on, the scientific enterprise, depicted as a tree made of periodicals, grew uncontrollably. "The number of journals continued to grow, to hundreds, to thousands, to tens of thousands, from most of the countries of the world."

For over a century it has been common to associate the scientific journal

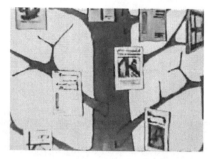

Figure 0.2 *Putting Scientific Information to Work*, a promotional video released by the Institute for Scientific Information in 1967, opened with an animated history of scientific communication. *Left to right, top to bottom*: Early men of science communicated via face-to-face contact, but this eventually gave way to written correspondence. The printing press made possible the heroic invention of the scientific journal, and soon journals were being founded as if they were growing on trees. By the twentieth century, scientists were drowning in journals, leading to the proliferation of indexes and classification schemes. The Science Citation Index promised to bring information overload under control. Courtesy of the Chemical Heritage Foundation.

with the invention of modern experimental science as a progressive enterprise. In 1917, for example, a pair of zoologists wrote that it was likely "that science could not have made the advance that it has but for the recognition of the periodical as the most convenient and efficient method of encouraging research." The physicist John Ziman posited in 1969 that "the invention of a mechanism for the systematic publication of fragments of scientific work may well have been the key event in the history of modern science."[19]

In the 1960s Ziman and others became interested in exploring what they understood to be a social history of science, and the idea that the spread of scientific papers might serve as a proxy for the development of science was hard to resist. The most consequential contributions to this project came from the sociologist Robert K. Merton and his students. Early in his career in the 1930s, when totalitarian states seemed to threaten both democracy and science, Merton had laid a theoretical foundation for studying science by outlining a set of behavioral norms that governed its social structure. Insofar as practitioners adhered to these norms — universalism, communism, organized skepticism, and disinterestedness — scientists could work together to extend knowledge. It was only in the late 1950s, however, that Merton decided to launch an empirical program in the sociology of science. He realized that to observe these rather abstract norms in action, he would need to get more concrete: just how were they instantiated in actual practice? Looking to instances of reward and censure in science, most paths seemed to lead to the scientific journal. Publishing practices seemed to be the nerve center where issues of reward, responsibility, and status merged on an everyday basis. Journals made it possible to reconcile the seeming contradiction between the norm of communism (specifically, the requirement that scientists share their discoveries openly) and the norm that scientists should be original: by publishing new claims, one received credit for doing so, but at the same time turned it into public (hence universal) property. Science was largely self-policing, and the scientific literature was largely where that happened.[20]

Sometimes observers linked the origins of the scientific journal to the tantalizing idea that the technology of printing had made possible the rise of Western civilization itself. Merton was inspired by historian Elizabeth Eisenstein's scholarship on the transformative role of the printing press to note that "printing thus provided a technological basis" for certain aspects of the ethos of science. But for Merton it was only when Henry Oldenburg, a German scholar who had moved to London and later became the first secretary of the

Royal Society of London, harnessed that technology into a new institution by creating the *Philosophical Transactions* in 1665, that the shift from secrecy to open science really took off.[21]

Subsequent historians and sociologists have largely eschewed this heroic narrative of the journal, but versions of it have become lodged in the historical stories passed down by practicing scientists and journalists.[22] Publishers have used it to justify their economic models, and it has also made it into ethics guides aimed at new researchers. Oldenburg has become both heroic inventor and media pioneer: "The solution to the problem of making new discoveries available to others while assuring their authors credit was worked out by Henry Oldenburg . . . Out of these arrangements emerged both the modern scientific journal and the practice of peer review."[23] Once the basic pattern was set, Oldenburg's creation simply needed to be reproduced wherever experimental science came to be cultivated, and as science grew such periodicals simply became more specialized. In the memorable words of the historian Derek de Solla Price, "Science encapsulated in papers became a sort of conspiracy that made knowledge run faster than people."[24]

One problem with these ideas is that they suppose that the format and uses of journals and papers have remained more or less constant throughout their existence. But this is not true. The *Philosophical Transactions* itself exemplifies the problem: few publications have taken on as diverse a set of formats and meanings as this publication has over its long history. For most of the period covered by this book, contemporaries did not even consider it to be a "journal" at all. The idea that there is essential continuity between early modern journals and their modern counterparts has encouraged us to project back onto these earlier epochs twentieth-century sensibilities about what journals are for, the meaning of credit, the character of public knowledge, and even the nature of trust in science.

Scholars objected to the heroic narrative in several ways, casting doubt on the idea that the journal literature was a useful source for understanding much about the social nature of knowledge. Beginning in the late 1970s, social scientists began to invade laboratory spaces to study science in the making, exploring aspects of scientific exchange that usually don't make it into print. The sociologist Harry Collins argued that all knowledge includes a tacit element, often inseparable from practice and not usually captured in documents.[25] The historian Steven Shapin argued that interpersonal trust was as central to the everyday life of science as bureaucratic systems of accountability, and that the criteria for adjudicating trustworthiness depended on broader social and cultural resources. Along with Simon Schaffer, Shapin argued more generally that the conventional boundaries defining who counts as an expert depend on problems of politics and knowledge that could not be disentangled. In this reading, not only would a focus on the scientific journal miss what is importantly social about science, but it would take for granted precisely what required investigation: by what means was a legitimate boundary between science and society established and maintained?[26]

Other historians did continue to study print genres of scientific communication, but they rejected any limitation of their study to those texts that were normally thought of as constituting the scientific literature. Eager to challenge entrenched ideas that science was the sole purview of recognized experts and elites, these scholars sought out the diverse social groups who were not only interested in but contributed to the investigation of nature. To do so, they focused on formats, media, and venues that fell outside the conventional limits of expert science, including popular scientific bestsellers, encyclopedias, textbooks, travel narratives, public lectures, and sites of oral communication such as soirées. This work has arguably been the richest and most productive area in studies of nineteenth-century science for the past two decades.[27]

But this book takes a different tack. Rather than challenging boundaries between elite scientific groups and other scientific publics, I am interested in tracing how these boundary conventions developed historically: how did it happen that certain formats and genres became relegated to popularization and education, while one format in particular was elevated to become a defining marker of scientific expertise?[28] This is to bring the focus of Shapin and Schaffer on the legitimacy of science to the historical study of genres and formats. To put it the other way around, it is to bring the sensibilities of book history to the politics of scientific knowledge.[29]

Ironically, because studies of the history of science in print in the past two decades have focused precisely on those formats that were previously marginalized, publications associated with elite and professional science have largely escaped historical scrutiny.[30] We have developed thick histories of so-called popular science publications that are too easily opposed to an ahistorical and unrealistic caricature of elite genres of intellectual life. One of the problems that has dogged the historiography of popular science is that the history of expertise, by which the popular is often defined in opposition, has gone largely unexamined.[31] The picture that has resulted is of a wildly diverse and dynamic world of scientific communication developing around a static core of specialized journals. As we bring to light the surprising social diversity of those who engaged in scientific inquiry, we should not neglect the processes through which their marginalization came to appear natural to later observers.

To put the point another way, I am interested in transformations in the imagined public of science. But this use of "public" is somewhat different from the sense it is routinely given in history of science scholarship. We often speak as if the public is a rough synonym for all those who are *not* scientific practitioners, as that which lies beyond the inner sanctum of scientific expertise. This has led to blind spots of historical interpretation, especially for the nineteenth century, when "the public" was a crucial category in debates over the changing basis of the legitimacy of expert communities in general. We might say that there is as much public *inside* science as outside it.[32]

Understanding the history of "the public" of science requires that we study how diverse conceptions of the public coexisted and also transformed one another. Instead of delineating the existence of previously overlooked social worlds of science by unearthing forgotten sites of scientific exchange, I explore ways in which the media practices of some of these alternative social worlds set the terms for the emergence of modern public-oriented scientific expertise. I follow transformations in practices of validation and judgment associated with expert science via the incorporation of genres, formats, and values that were first associated with those on the periphery. To put it another way, the historical significance of the emergence of new formats and genres which their creators intended for bringing scientific knowledge to new audiences is not simply whether they expanded access to knowledge. Rather, it is that the structures of power governing the scientific enterprise were ultimately reconfigured in the image of those alternative publics.

More generally, the history of the public is not simply a history of the social groups engaged in scientific exchanges, but also of individuals' beliefs about the nature of the groups—and modes of judgment—that are relevant to the legitimation of knowledge claims.[33] Thomas Broman has outlined one way in which this approach to the history of the scientific public might bear fruit. He posits that the ideal of critical judgment that came to be attributed to public opinion during the Enlightenment played a key role in establishing journals as legitimate loci for the exchange of scientific information and critique among German physiologists at the end of the eighteenth century.

Broman goes on to hypothesize that the appeal to universal consensus embodied in this Enlightenment conception of public legitimacy was not simply discarded as the sciences expanded and specialized in the nineteenth century but was rather a key constituent of emerging conceptions of the scientific expert.[34] Even as scientific practitioners came in practice to be associated with ever-more-rarified professional groups, their ability to speak credibly about nature continued to depend on the legitimacy of the universal public first associated with Enlightenment periodicals.

This paradox is precisely what this book explores: how did public opinion become a legitimating resource not only for general Enlightenment culture but for the more specialized domain of natural scientific inquiry? And how did this conception of public legitimacy persist in the emergence of the scientific expert? Broman develops his argument by following transformations in genres of periodical publishing in the German lands, from Enlightenment journals to more specialized, and ultimately professionalized, ones. But the shape of the story in Britain and France is different. There, the rise and transformation of scientific periodicals cannot be understood without taking into account the existence of elite, central institutions that had long claimed to embody legitimate authority and judgment in natural philosophy. In this situation, the rise of the new journals tended to appear as a challenge to these institutions' status as privileged audiences and appraisers of scientific claims. The emergence and spread of commercial journals focused on natural knowledge was all the more problematic in the context of the perceived spread of hack writers supposedly more interested in profiting from their writing than in disseminating truth. In the German lands, which did not see the rise of a vast and potentially subversive literary culture to the extent that Britain and France did, and where many leading authors and editors of journals were civil servants connected to universities, the rising importance of the press in scientific life did not spark controversy in quite the same way.

Given how central the scientific journal became to both everyday scientific life and public representations of scientific knowledge in the twentieth century, it seems surprising that no book has yet been written about its emergence. But if you consider the diversity and the vast extent of the subject, this becomes more understandable.[35] Historians who have considered the topic have usually limited their study either by focusing on a single publication (or publisher or editor),[36] or by undertaking general surveys.[37] Neither of these strategies, however, is likely to tell us very much about how the category of the scientific journal itself came into being, or how the idea of the scientific literature took on such importance. This book casts its net wider than individual case studies, but it is far from a comprehensive survey. (The vast majority of scientific periodicals, including many of historical importance, are not even mentioned in this book.) My strategy has instead been to focus each chapter on a series of episodes that illuminate a particular aspect of the changing role that periodicals were taken to play in the scientific enterprise. What unifies these episodes is that each of them involves some contest over the changing relationship between elite scientific institutions and the periodical press.

This strategy makes Britain and France a natural setting to focus on. In these nations, the complex, changing relationship between centrally located societies or academies and the periodical press provides a particularly illuminating vantage from which to trace changing beliefs about who is able to speak with authority about nature, and how this ought to be done. I do not intend to suggest by this choice that other nations and sites are of secondary importance. We are not yet at a place where a global history of the scientific journal can be written, and when we are, it will have to take into account not only Germany and the rest of Europe, but also East Asia, the Americas, other parts of the British Empire, and more. With that said, several of the characteristics associated with the scientific literature that this book focuses on do seem to have arisen in Britain and France as early as or earlier than in Germany. Despite rapid expansion of scientific and technical periodicals throughout the nineteenth century, German publications in the early twentieth century often worked differently from those based in Western Europe and the United States. These differences came to a head in the late 1920s, when library associations and universities, faced with ballooning subscription costs, began to protest authorial habits and publishing practices in Germany. Their litany of complaints included the excessive length of many papers, which often resembled historical treatises on a subject as much as they did original contributions to knowledge. Instead of being fixed, subscription prices were often determined by the number of pages published, and authors often received payment by the page, which combined to lead to mammoth annual volumes including supplemental issues that could include whole dissertations. Librarians in the United States, France, and Britain threatened partial boycotts of German serials unless German editors learned to edit their publications more rigorously, putting limits to their annual size, and unless authors learned to write scientific papers properly, singling out their original contributions and cutting out the rest. The confrontation dragged on for several years, reaching its apex when the heads of the German publishing firms Springer and Verlag Chemie traveled to the United States to plead their case to university librarians. Ferdinand Springer even took the step of translating an American guide to medical writing to be distributed to authors. He summed up its message as follows: "Say what you know, briefly and distinctly."[38]

By choosing to follow transformations in scientific publishing as they played out in more than one nation, I have tried to avoid the rash generalizations that historians sometimes make about what are often quite localized cultures of publishing, authorship, and discovery. I have no doubt that some such generalizations remain, however, especially since the choice to consider multiple sites has also led me to highlight evidence from the largest metropolises. I look forward to subsequent research that will deepen this picture not only within Britain and France but across the globe. Of course, these two countries by no means constitute independent cases; as the historical narratives that follow will show, national and scientific rivalries fostered a great deal of mutual influence in the publishing experiments pursued on either side of the Channel.

Finally, it is important not to overstate the importance of this specific format to the circulation of scientific information or to downplay the richness and complexity of the sundry modes of communication that remained central to science throughout the nineteenth century. One of the paradoxes I explore is that the scientific paper achieved privileged status in scientific life even as the actual channels through which science was communicated remained extremely diverse. The scientific journal that emerged as a dominant format in the later nineteenth century was an aggregation of ideas and functions drawn from the broader landscape of institutions and communications media that made up the social world of scientific practitioners. Indeed, each chapter is focused as much on other genres, formats, or sites that contributed to defining the meaning and character of the modern scientific journal—either by incorporation, opposition, or (more often) both—as it is on journals themselves. These include scientific meetings, newspapers, catalogs, historical treatises, correspondence, abstracts, patents, pamphlets, offprints, and index cards. Although this is largely a story about the consolidation of a particular format for communicating science, the drive toward standardization has always remained in tension with an equally strong drive toward heterogeneity — the scientific journal has never referred to any one thing.

This book locates the ascendancy of both the scientific journal and scientific authorship in their increasingly central role in judgment, representation, and credit. The circulation of scientific information — the presumed raison d'être of the journal — is not really my focus. I do not attempt to trace the networks of individuals, publications, and societies through which knowledge was communicated. And while no book on this topic can ignore the economic aspects of publishing, I do not follow this thread in a systematic way. Instead, the book aims to explore how scientific authority was reconfigured around periodical authorship and reading. In France, the eighteenth-century apparatus of collective judgments associated with the Royal Academy of Sciences gave way to the notion that periodical authorship was the essential activity that delineated a scientific identity, and that no small body of elites could be entrusted with assessing claims to scientific discovery. In Britain, not only did authorship come increasingly to define a career in science, but a new personage modeled on anonymous readers — the referee — emerged as the epitome of legitimate scientific assessment. In both nations, personal relationships and patronage never ceased to loom large in actual personal advancement, but this gradually went behind the scenes. Claims to authoritative consensus became more likely to be based on the imagined assent of a community of expert peers figured as expert readers than on a hierarchy of authorized judges.

If I want to make the case that the early modern origin of the scientific journal is something of a mythology, then this is where the book itself must begin. Chapter 1 opens by asking what formats, genres, and regimes of judgment were developed by early modern academies and societies, and how those were related to the emergence of scholarly journals. While elite academies became tribunals for discoveries and inventions whose authority derived from their political associations and privileges, journals became associated with public acts of criticism as expressions of the universal Republic of Letters. By the early nineteenth century, the reading public came to be figured as a literary market to which writers could appeal to prove their reputation as authors, rather than through their association with aristocratic patrons. Following the French Revolution, journals focused on science were increasingly put forward as providing alternative conceptions of the legitimate public for scientific knowledge, putting pressure on academies and societies. Chapter 2 focuses on one crucial form that such pressure took: the publicity that journalists gave to the meetings of scientific societies and academies. Eventually academies and societies decided to compete by launching journals of their own modeled on these commercial upstarts. The rest of the book is largely about the consequences of this development.

In Britain, the subject of Chapter 3, reformist agitation migrated from radical outsiders to elite men of science who zeroed in on authorship as a key criterion by which to distinguish active researchers from enthusiasts and aristocratic pretenders. In response, the Royal Society pursued several reforms that invested scientific reading with new significance, including empowering a new kind of reader — the referee — who was to be an agent for conferring rewards and publicity on deserving authors. In France, meanwhile, the architect of the Academy of Sciences' weekly journal, François Arago, worked to bind the life of the Academy ever more intimately to print publicity. Chapter 4 traces his and others' efforts to harness both historical and legal precedents in intellectual property to put forward rival visions for the future of science. Chapter 5 explores early attempts to place bounds on the emerging category of the scientific journal even as other forces pushed the landscape of scientific publishing toward greater heterogeneity.

After the Franco-Prussian War ended in 1871, savants in Western Europe embarked on public campaigns arguing for the centrality of scientific activity to national welfare. These claims rested in part on the role of science as providing a model of openness and solidarity among its far-flung participants. The book's final chapter takes place during this period, which is also when the "scientific literature" became a common point of reference. This powerful imagined entity was simultaneously a virtual storehouse of discoveries and a system of efficient and public information. But the clearest indication of the arrival of this concept was not simply that the scientific literature existed, but that it was perceived as being in need of fixing. Indeed, it was as much an object of vexation as of veneration. "To multiply Journals is to multiply sorrow," declared the British chemist William Ramsay in 1894, voicing what was then widespread frustration with a scientific institution that seemed to be careening out of control.[39]

Entrepreneurs on the periphery of the scientific elite seized on new technologies for information distribution as a means of democratizing science, while self-appointed defenders of science attempted to wrest control of scientific publishing from the periodical marketplace by standardizing the array of practices by which scientific papers were accepted, distributed, and located. The success of such efforts in bringing scientific publishing under control was slight at best, but they helped foster the idea that the scientific literature was the glue that bound together knowledge communities scattered across a nation, discipline, and even the globe. It is from this moment that histories of science began habitually to celebrate the heroic invention of the scientific journal.

The turn of the twentieth century saw a flowering of movements focused on extending access to scientific and scholarly knowledge. Diverse as these were, they were underpinned by broadly similar ideas about what it meant for science to be open and public. Through much of the nineteenth century, the dominant vision of extending access was one in which knowledge would be bestowed on a broad public through lectures, periodicals, or books. One offshoot of this was the rise of scientific popularization, which later became a profession in its own right. But late in the century a distinct access fantasy arose in which scientific knowledge was instead imagined as embodied in the medium of scientific papers, archived in libraries and catalogs, to be made available to scientific workers. The morselization of knowledge implied by this idea was a prerequisite not only for the subsequent emergence of the modern discourse of information, but for subsequent efforts to consolidate and commodify scientific publishing.

Such visions constitute one more thread that I follow throughout this book and which leads directly to our present predicament. We are currently witnessing the proliferation of new scientific access fantasies. Many of them proclaim that knowledge will be set free by the mediation of some information technology that will solve some problem of objectivity, equity, or trust. But terms such as "open," "free," and "transparent" are meaningless if we do not ask for whom, of what, and to which end? Open data initiatives, for example, take the tensions inherent in the archival access fantasy of the fin de siècle to a new level, as increasing quantities of information are being made accessible but which are of use to fewer and fewer people. Let us be wary not only of alarmist claims that the end of the journal will bring with it a dissolution of scientific norms, but also of triumphant claims that breakthroughs in communications technologies will bring about a new scientific revolution.

Our new experiments in knowledge expression will also generate new narratives of the past.[40] I hope that these new histories do not foreclose questions about the diverse kinds of publics that we have imagined in the pursuit of knowledge or of the legitimate bases of our trust in expertise. What follows is a modest contribution to how such a history might look.

1

The Press and Academic Judgment

I was doing research at the French National Library in 2008 when I stumbled on a prospectus for a new journal that caught my eye. It was dated 1802 and signed by the professors of the Muséum d'histoire naturelle in Paris. The occasion of launching a new periodical prompted them to reflect on the origins of scientific progress, which they traced back to the moment when savants began to investigate nature as a collaborative activity. "This fortunate development is mainly due to two institutions first imagined in the seventeenth century." The first of these were academies of science, "those bodies to which members come each day to submit to the examination of their colleagues new facts and relations they believe themselves to have discovered." The second was the invention of scholarly journals [*journaux savans*] that diffused discoveries to wider audiences and which proliferated during the eighteenth century. Unfortunately, however, academies — being "attached like all others to their original habits" — had not made much use of journals. Though they often published memoirs, they did so "slowly and in large volumes," accessible only to "a small number of rich amateurs." The two institutions had thus remained apart, and the Muséum would at last bring them together.[1]

These professors' idea that academies and journals had evolved in parallel but separately puzzled me. Raised on late modern narratives of the origins of modern science, I believed that the swift adoption of journals by academies and societies is what made the emergence of open, collective scientific progress possible. It was tempting to chalk up the Muséum's reading of scientific history to French parochialism. The genuine origins of the scientific journal were surely to be found in England, where the *Philosophical Transactions* was founded under the aegis of the Royal Society of London. But historical reflections across the Channel seemed rather to corroborate the professors' account. An 1813 prospectus for a new London journal noted that the moment the Royal Society took control of the *Transactions* in the mid-eighteenth century it transformed it into an expensive, infrequently published collection of memoirs resembling those of the French Academy. After that, "Britain no longer possessed a periodical philosophical journal" at all.[2] It turned out that all those stories about continuous and exponential growth of science, pegged to the spread of journals such as the *Philosophical Transactions*, might be based on a series of media-historical misunderstandings.[3]

What happens if we take seriously the idea that the professors of the Muséum really were doing something quite new, and even potentially controversial, by publishing a journal? These earlier accounts may have seen history differently because they were alive to distinctions of genre and format that many accounts of the origins of modern science, burdened with the knowledge of what scientific journals are supposed to be, have blurred together. If their forerunners—the Royal Academy of Sciences in Paris and the Royal Society of London—had largely eschewed journals, then we need to investigate why they did so, and what formats, genres, and record-keeping tools they used instead. Moreover, we should explore the relationship between the Academy of Sciences and the Royal Society to those mythical publications, the *Journal des savants* and the *Philosophical Transactions*. How did the conceptions of judgment embodied in these and subsequent learned publications of the eighteenth century compare to those favored by academies and societies? As we explore these questions, we should be careful to avoid separating out a dozen or so "scientific periodicals" from the broader landscape of Enlightenment journals. Why? Simply because no one in the eighteenth century recognized any special category constituted by "scientific periodicals."

When the professors at the Muséum announced their new journal in 1802, they acknowledged that journals dedicated to science had their problems. Because of "the ease of skimming a brochure of a few pages in comparison with the labor required to study a folio," the proliferation of journals might encourage the pretensions of *"demi-savants"* to participate in natural philosophy.[4] We should therefore ask why institutions such as the Muséum d'histoire naturelle decided that it was worth the effort to consider publishing a journal of their own. Conversely, we can also ask why savants decided to contribute original content to the new journals, and even to agree to edit them. This means paying especially close attention to the changing meaning and status of authorship, at a time when the enlightened reading public was coming to be figured as a literary market to which writers might appeal to establish their reputations, rather than by associating with aristocratic patrons. It is easy to imagine that writing for periodicals meant to early nineteenth-century savants what it does now: publicizing a claim among peers, establishing priority, and indicating that one's claims had passed some threshold of credibility. While it meant some of these things some of the time, the considerations involved could be a great deal more diverse, and could even be fraught with risks.

The Rise and Fall of Oldenburg's Vision

When Denis de Sallo, an obsessive compiler of extracts from books, founded the *Journal des sçavans* in Paris in January 1665, his aim was less to provide a venue for the generation of new knowledge than to provide a solution to there being too much of it. He promised to give his readers a digest of "all that is new in the Republic of Letters,"[5] and he filled each number with book reviews, extracts, translations, and bibliographical lists. There was also space for some other kinds of content: obituaries of scholars, reports of new discoveries and inventions, and decisions of ecclesiastical courts. The format gradually caught on in continental Europe, and it played a crucial role in the eighteenth-century Republic of Letters.[6] In 1710, an observer defined journals as "those sequential works which provide information from time to time about the various books which have appeared and what is contained in them . . ."[7] This was not only an intellectual service but a commercial one, for readers were also consumers. Scholarly journals fit into a long history of technologies for ordering and managing the landscape of scholarly news that seemed always already to be spinning out of control.[8]

Just weeks after acquiring a copy of the first issue of the *Journal des sçavans*,[9] Henry Oldenburg, the secretary of the Royal Society of London, presented to the Society his own take on the new format, which he had titled the *Philosophical Transactions*. It contained fewer reviews but more news of experiments, found objects, and other happenings in natural philosophy. As the secretary of the Society, Oldenburg felt perpetually overworked, constantly writing letters on behalf of the Society, taking "much pains in satisfying forran demands about philosophicall matters" and distributing "farr and near store of directions and enquiries for the society's purpose."[10] A printed bulletin like the *Transactions* was a means of accomplishing this more efficiently.

Although they were controlled by individuals, both these publications had some connection with learned societies. In different ways, they were a part of the strategies through which the Royal Academy of Sciences in Paris and the Royal Society in London eked out spaces of stability and trust in the face of the treacherous world of European printing and publishing. But they did so in remarkably different ways, and these differences were bound up with the degrees of assent they were able to command as central registers and arbiters of knowledge.[11]

The main documentary technology of the early Royal Society was not the *Transactions* but its register books. When observations, experiments, and discovery claims were presented to the Society, they were recorded in these volumes by hand, along with their dates. The register was a reliable means of recording and establishing a discovery claim because it existed in a single copy, it remained in the charge of a trusted individual, and it could be used to record successful demonstrations and experiments at meetings virtually in real time in front of other trustworthy witnesses.[12] These manuscript lists were particularly appropriate to the epistemological orientation of the Society's leading members, who valued above all else the accumulation of particulars and curiosities. The Society's first historian and publicist, Thomas Sprat, explained that "their purpose was, to heap up a mixt Mass of Experiments, without digesting them into any perfect model." Lorraine Daston and Katharine Park have called them *strange facts*, esteemed precisely because they resisted incorporation into hasty theoretical systems.[13]

In comparison, the circulation of printed texts left much to be desired. Books were subject to unauthorized and faulty replication, responsibility for their form and content was distributed between a number of different parties (including some motivated by profit), and dates of printing stood in no predictable relation to the date of the discoveries and observations recorded. The situation with periodicals was even more uncertain. The most proximate precedent in Britain were the newsbooks that appeared in the 1640s when press controls lapsed during the English Civil Wars. These publications came and went on a regular basis, their periodicity was based on calculated obsolescence, and many were associated with political dissension and disorder.[14]

But as long as the Society wished to publicize its work more broadly, the print shop was hard to avoid. The possibility of being mentioned or printed in Oldenburg's newsletter gave potential adherents to the experimental philosophy added incentive to engage in correspondence with him. And the miscellaneous nature of its content shared with the register a resistance to system-building. In its Charter, the state had given the Society the power to license books for publication, and it used this to allow Oldenburg to publish the *Philosophical Transactions*. Still, the Society's association with a periodical publication remained a fraught proposition. Issues of the *Transactions* were subject to unauthorized reprinting, translation, and appropriation across Europe. Oldenburg was often on the defensive about his publication. He was careful to strike a balance between playing up its utility and minimizing the significance of its contents, and he was careful to take primary responsibility for its contents.[15] After Oldenburg passed away in 1677, the publication went through decades of turmoil, and it even perished for certain periods.[16]

Across the Channel, the Royal Academy of Sciences in Paris experienced similar tensions, but it had a different hierarchical organization, more intimate ties to the monarchy, and a stronger set of privileges attaching to publications and inventions.[17] While the philosophical gentlemen in London maintained a strategically ambiguous link to the *Philosophical Transactions* well into the eighteenth century, the Parisian Academy generally kept its ties to learned journals out of view entirely.

After its founding in 1666, the Academy had experimented with various forms of collective organization and publishing, but nothing quite seemed to stick. The major works with which it was first identified were multivolume projects on the natural history of animals and of plants in which individual authorship was subsumed by the whole. Yet academicians turned out to be reluctant to have their discovery claims obscured in this way,[18] and many continued to publish elsewhere. Letters to the *Journal des sçavans* were a convenient means of staking individual claims, although the Academy insisted on screening these letters first, for the sake of protecting both its reputation and its collective property. The Academy tried out other arrangements as well. For a short period beginning in 1692 they attempted to run their own monthly publication made up of short memoirs by its own members. (They explained that academicians were sometimes distracted from their principal work by un-related chance discoveries.) This too did not last.[19]

In 1699 the Academy was reorganized and given a new hierarchical struc-ture that in some ways mirrored the orders of French society itself. The publi-cation format that emerged from this "*Renouvellement*" was called the *Histoire de l'Académie royale des sciences avec les mémoires de mathématique et physique*[20] (*Histoire et mémoires*, for short). It became the pattern not only of its own publication strategy but of many other learned academies and societies for the next century. Its lavish quarto volumes, with large margins and expensive illustrated plates, covered a year's worth of the Academy's work and consisted of two parts.[21] The *Histoire* was an overview of the works of the Academy as a whole, consisting largely of a curated summary by the secretaries of notable work that had been presented at its meetings. The *Mémoires*, which made up the bulk of most volumes, were collections of polished versions of papers writ-ten by its own members. It was thus a compromise between collective report and individual authorship. No one confused these volumes with learned jour-nals. They were issued far less frequently, and the histories and tracts they con-tained often commemorated experiments and papers that had been presented several years earlier.[22] Many other academies made no pretense to regularity of publication at all, issuing volumes of memoirs whenever they had collected sufficient matter to put out a volume.

It is tempting to regard the *Histoire et mémoires* as a rather deficient prod-uct. The infrequent publishing schedule, the long delays between presenta-tion and publication, and the fact that it tended only to publish their mem-bers' work have been cited as signs that such royal academies were unable to meet the pressing needs of science. In this reading, the Paris Academy pub-lished a journal, but despite the considerable resources of the state, it did a remarkably bad job of it.[23]

But consider the Academy's publishing strategy in the context of its role within the absolutist French state, and it takes on a different appearance. Arguably the most important genre attaching to the Academy's collective identity was not the *mémoire* but rather the *rapport*. The 1699 *Renouvellement* provided the Academy a set of privileges that gave it the right and responsi-bility to oversee developments in science and technology for the King.[24] The Academy acted as an examination board for new inventions, deciding whether inventors (or importers of inventions from abroad) ought to be granted com-mercial protections. Second, like the Royal Society, the Academy came to possess the powers of a censor with respect to natural philosophical writings. Decisions about the value of new inventions, as well as about the suitability of scientific and technical memoirs, came to involve the writing of reports by commissions consisting of two or three academicians. These would be read out at meetings and recorded in the minutes, and some would even find their way into print. The collective authority of the Academy was increasingly concentrated in these judgments. To exercise the powers of an academician was to be a *rapporteur*.[25]

Although the initial raison d'être of the *rapports* may have been legal, they came to possess other kinds of significance. Antoine Lavoisier later observed that the Academy had gradually become "a voluntary tribunal to which individuals appeal directly for judgment."[26] To present one's scientific manuscript or invention to the Academy was to present it for approval to the only public of science that was supposed to matter. Inventors who received a favorable report might advertise their product by printing an excerpt from the report, or at least the phrase "Approuvé par l'Académie Royale des Sciences." Authors sometimes printed a favorable academic report they had received as a preface to a book and pamphlet. (See for example Figure 1.1.) More generally, submitting one's research to the Academy was a crucial means of finding favor among the elites, and for a lucky few it might eventually lead to election to the Academy itself. (The 1699 statutes promised that preference in elections would be given to "those savants that have been most exact" in their correspondence with the Academy.)[27] The submission of a manuscript to the Academy was effectively a gift to a potential patron.

Until mid-century, a favorable judgment from the Academy was normally the only direct reward a non-academician could expect for the submission of a discovery claim. Except in the case of prize competitions, the Academy did not publish outsiders' work. But in 1750, the Academy began publishing memoirs of non-academicians in a regular series of volumes known as the *Mémoires des savants étrangers*. The new publication mirrored the *Histoire et mémoires* in its expensive appearance, but it outdid it in the slow, deliberate pace with which it was published (about one volume every five years).[28] From then on, commissions appointed to evaluate manuscripts also influenced decisions about publication, although in practice such decisions were made by a separate Publications Committee set up for this purpose.[29]

Along with the importance of *rapports*, earlier epistemological commitments to accumulated particulars gave way to circumspection. While learned

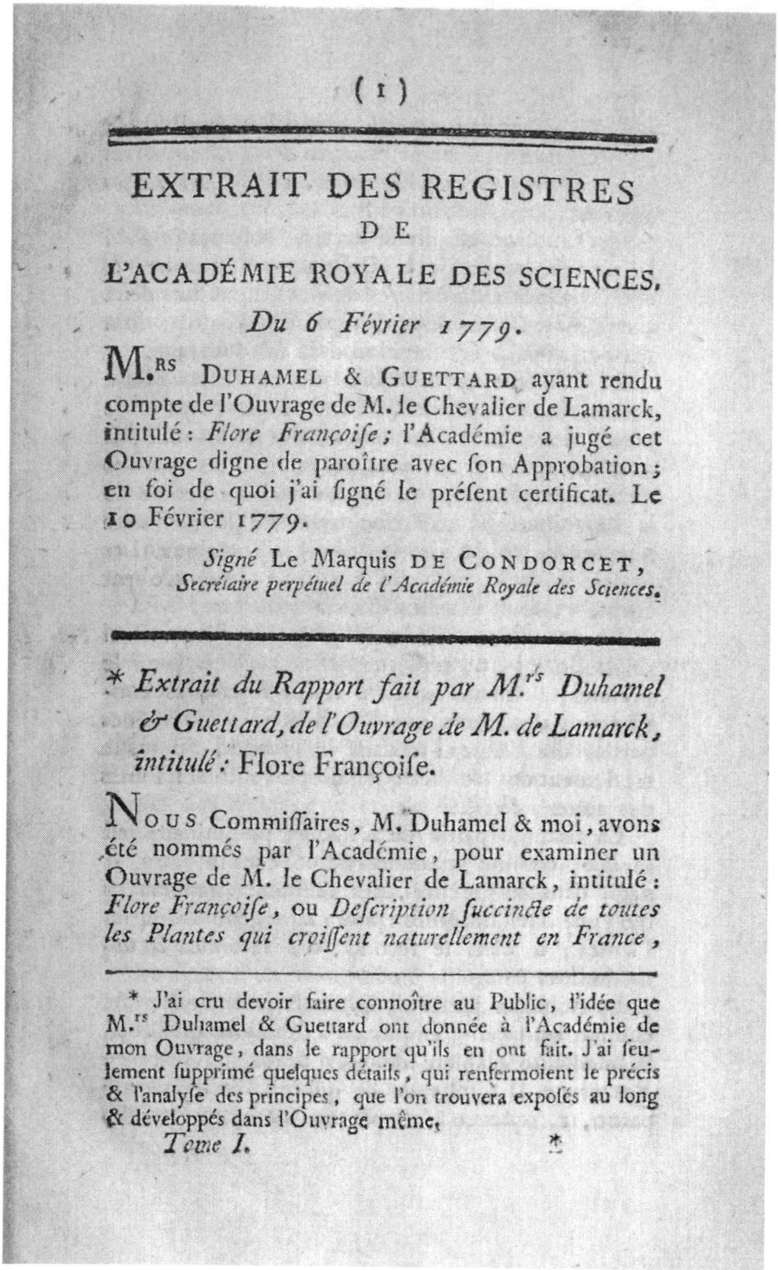

EXTRAIT DES REGISTRES
DE
L'ACADÉMIE ROYALE DES SCIENCES.

Du 6 Février 1779.

M.ʀꜱ DUHAMEL & GUETTARD ayant rendu compte de l'Ouvrage de M. le Chevalier de Lamarck, intitulé : *Flore Françoise ;* l'Académie a jugé cet Ouvrage digne de paroître avec son Approbation ; en foi de quoi j'ai signé le préfent certificat. Le 10 Février 1779.

Signé Le Marquis DE CONDORCET,
Secrétaire perpétuel de l'Académie Royale des Sciences.

* *Extrait du Rapport fait par M.ʳˢ Duhamel & Guettard, de l'Ouvrage de M. de Lamarck, intitulé :* Flore Françoise.

NOUS Commiffaires, M. Duhamel & moi, avons été nommés par l'Académie, pour examiner un Ouvrage de M. le Chevalier de Lamarck, intitulé : *Flore Françoife, ou Defcription fuccincte de toutes les Plantes qui croiffent naturellement en France,*

---

* J'ai cru devoir faire connoître au Public, l'idée que M.ʳˢ Duhamel & Guettard ont donnée à l'Académie de mon Ouvrage, dans le rapport qu'ils en ont fait. J'ai feulement fupprimé quelques détails, qui renfermoient le précis & l'analyfe des principes, que l'on trouvera expofés au long & développés dans l'Ouvrage même.

Tome I. ⁎

Figure 1.1 Approbation of the Royal Academy of Sciences followed by the Academy's report, printed by Jean-Baptiste Lamarck as a preface to his *Flore françoise, ou description succincte de toutes les plantes qui croissent naturellement en France*, vol. 1 (Paris: Imprimerie Royale, 1778). Courtesy of the BnF.

journals such as the *Philosophical Transactions* might meet with all manner of strange facts, academic science came to eschew unchecked curiosities. Where openness to wonders had earlier been a valued trait in savants, disinterested persons were now urged "to give themselves the trouble to perform contested experiments, to mark doubtful facts and suspect observations, so that one grounds nothing upon them."[30] In a sense, academic circumspection was a characteristic of French cultural life more generally. On the one hand, the system of state censorship employed scores of readers to write reports on books and decide whether they merited publication: these reports went beyond criteria of religious and philosophical orthodoxy to include questions of accuracy, style, and taste. They were not unlike book reviews themselves, and indeed the ranks of censors, journalists, and academicians overlapped to a remarkable extent.[31] But unlike the assumption of equality that defined the Republic of Letters, the Academy made it clear that only a very limited group constituted a legitimate judge in matters of natural science.

The Academy's proximity to the King also helps explain this epistemological restraint, for its public image could be seen as reflecting on the monarchy as a whole.[32] Thus, while individual academicians sometimes did express frustration over the leisurely pace of the Academy's publishing schedule, a more hurried approach brought risks. The *Histoire et mémoires* was an important prop in the various rituals that tied the Academy to the representative publicness of the monarchy, including ceremonies in which each volume was presented by the Academy to the King.[33] The academic *mémoire* was generally shorter than an individually published book, but this was not because it was a compressed summary or was produced quickly to establish priority. When the Marseille Observatory launched a *Mémoires* series in 1755, its editor reflected on the recent popularity of the *mémoire* genre, noting that it was "easier for an author to treat an object in depth by taking it in parts, than by embracing it in its totality."[34] Academic memoirs came to be associated with a full, prolix style, in which literary merit was valued alongside scientific acumen. When the academician Pierre Louis Maupertuis sent a mathematical memoir destined for the Academy to a Swiss correspondent, he apologized in advance: "you will certainly find it too long, but it is the style of our Academy."[35] Authors were given several opportunities to get things just right: "In the French Academy, after the first proofs were corrected at the royal printing house, each author corrected his own several times, up to two or three proofs."[36]

While some academicians were connected to the *Journal des sçavans*, the

Academy had no official connection to the journal. In 1771 the statesman and honorary academician Malesherbes shot down the idea of the Academy publishing its own journal, arguing that "it would not be honorable for the Academy of Sciences to produce in its own name a work that many individuals can produce just as well . . . it would only be one more journal." Later in 1787 when the idea resurfaced among chemists in the Academy, Antoine Lavoisier was informed not only that the Academy's publishing privilege did not extend to publishing a journal but that the Academy would not even receive approval from state censors to do so if it applied through regular channels.[37]

Conversely, it was this same proximity to the King and the several powers that came with it that allowed the Academy to count on savants and informants to supply it with useful facts. The Academy's identity as an authoritative tribunal in itself overshadowed its identity as a publisher: "There are many in the Academy for whom one can pass as an able man as long as one reads," noted Maupertuis, referring to the practice of presenting memoirs at meetings.[38] Plenty of short, undigested facts came in, but only polished contributions to knowledge came out.[39] The potential benefits of academic patronage, a positive report, and the hope of one day being elected were enticing enough that aspiring savants were willing to present their work directly to this corporation, independent of the possibility of publishing. So powerful was the Academy that it was able to exercise a certain amount of control over those journals that took an interest in science and the Academy's business.[40]

The *Histoire et mémoires* was a model much imitated, not only in France but also abroad.[41] In some cases this happened by the exportation of French academicians themselves. When Frederick the Great hired Pierre Louis Maupertuis to preside over a reformed Prussian Academy in 1746, the French mathematician spared nothing in designing a publication worthy of his new royal patron in Berlin. He supervised all aspects of the new *Histoire de l'Académie royale des sciences et belles lettres*. Appealing to the legacy of Fontenelle, he insisted on "giving our book the best possible appearance,"[42] with the finest paper, the best printing, and the most carefully edited and proofread texts. A system of reports like the one in Paris was also adopted. Prussian academicians found occasion to invoke the same status distinction between the Academy's publications and the press when a mathematician—Johann Samuel König— used a journal to launch brash attacks on the principle of least action (a principle closely associated with Maupertuis). Maupertuis and Leonard Euler marshalled the prestige of the academic genres of the *rapport* and the *Histoire* to fight back. Euler condemned the "chicaneurs publiques" running the literary gazettes who dared question the Academy's judgment. The latter was the only body qualified to decide on matters that "demand a profound knowledge of the Sciences to which they are related."[43]

As learned journals became central to the Republic of Letters' self-conception as a universal and rational public, most eighteenth-century scientific academies followed the lead of Paris by publishing their members' writings in lavish volumes of memoirs on an annual or occasional basis. Maintaining some distance from the wider world of periodical publishing made it easier for academies to claim special authority as judges of natural philosophical claims, even if in practice a wide variety of periodical formats continued to play important roles in natural philosophical interchange. In an important sense, it was the very power of the Academy that allowed it to distance itself from associating with a journal, while the Royal Society of London chose to maintain a strategically ambiguous connection with the *Philosophical Transactions*.

But as the *Mémoires* format spread during the first half of the eighteenth century, the relationship between the Royal Society and the *Philosophical Transactions* was increasingly a subject of confusion. Many assumed it was a Royal Society publication, but it continued to resemble other learned journals.[44] Its issues were published relatively frequently (every one to three months), and they were made up of short fragments (the median length hovered around three to four pages) along with the occasional more substantial memoir. And while its editors may have been somewhat more careful about what they published, the journal continued to include short reports of odd occurrences, epistolary curiosities, and other sorts of philosophical news about which readers were invited to decide for themselves. Monstrous births and outlandish creatures continued to appear, if more rarely.[45]

But in mid-century, things changed. John Hill, a botanist who felt slighted by the Society, launched a concerted campaign to discredit it. Citing the old motto "Nullius in verba," he argued that the Society had a duty to eschew "taking things on trust." It was improper, in his view, that doubtful claims should appear "among the transactions of a society whose name should give a sanction to the truth of what is ushered into the world under it."[46] To drive the point home, Hill published an elaborate hoax paper titled *Lucina Sine Concubitu, A Letter Humbly address'd to the Royal Society*, in which a doctor described "a wonderful cylindrical, catoptrical, rotundo-concavo-convex Machine" that he had invented, able to collect animalcules from the air that could be used to make a woman "conceive without any Commerce with Man." Hill was by no means the first to lampoon the Society, but this particular satire caught on (much to the Society's dismay), and it was widely reprinted and translated into French (twice, in several editions) as well as German.[47]

Hill's subsequent *Dissertation on Royal Societies* developed a comparison with the Parisian Academy. There, "every thing was Wisdom, Regularity, and Order" while the Royal Society was all "Irregularity and Confusion." Worst were the scandalous contents of the *Philosophical Transactions*, teeming with "Instructions for *catching Fish with Thorns*, and *making Weavers with burnt Packthread*: With Stories of *Stittlebacks* and *Cockchafers*."[48] Why, Hill asked, should the editing of the *Transactions* be in the charge of a single person, the secretary? "Tis not the office of one man, but of several . . ." Besides, the secretary was in a conflict of interest because his job security depended on the Society's officers. He must, therefore, "unless he be false to his own interest, print nonsense, if it come from a man, who has interest and weight among the body."[49]

It was high time, argued Hill, that the Royal Society owned up to its role in the *Transactions*. The Society's disavowals of the *Transactions* were "about of a Piece with that of the Fellow, who after Condemnation for a Murder, persisted in affirming his Innocence . . . upon the excellent Distinction that it was not *he*, but *his Hand* that cut the Throat." The Society should publish only what it deemed worthwhile instead of scrambling to fill up pages on a monthly basis. He suggested a *"Committee of Inspection of Papers"* be set up to decide what should first be read, and then "Let the Sense of the whole Society be heard upon them, and let them pass a second Examination before they are ordered to be printed." The pressure would then be off the poor officers, so that "the Society might, some time or other, invent an *Automaton* of Wood and Wires, which should fill the presidential Chair with proper Gravity."[50]

Hill was promptly barred from showing his face at any Royal Society meeting. But it wasn't long before the Council took into consideration precisely those suggestions he had put forward (excepting, however, the automaton-president). On the one hand, the Council denied that it ever had any role in publishing decisions at all. As one exasperated Council member put it, this misinformation had been "very detrimental to the Public Opinion of the Society's Judgment & Sagacity":

In Fact, the <u>Society</u>, have never, <u>As a Body</u>, declared <u>any Opinion at all</u> concerning the Merit of such Communications & Exhibitions . . . vast Numbers of such Projects, Inventions, & imagined Curiosities have appeared to the <u>particular Members</u>, to be <u>trifling and ridiculous</u>; Others, to be <u>useless and impracticable</u> (however ingenious;) and too many of them, to be <u>absurd, wild, & extravagant</u>.[51]

On the other hand, the Council had to admit that it was not really sure what past roles the Society had played in running the *Transactions*. The Council thus commissioned a history of the periodical to learn the truth. They found to their surprise that the Society had been more involved in its history than they had thought. The Council decided that there was nothing to be done but to take it over. They would now appoint a Committee to assemble together on occasion and select which papers should be printed. To print or not would be "always decided by ballot, and never by voices," and five would be a quorum. Taking charge also meant taking financial responsibility for the publication; the decision thus radically altered the financial situation of the Society.[52]

From this point forward, the character of the *Transactions* transformed in several ways.[53] During the 1750s and 1760s, the median length of articles jumped to around six or seven pages and continued to rise. By the end of the century it reached about fifteen pages, and sometimes topped twenty.[54] At the same time, the frequency with which parts were issued was significantly reduced, fluctuating between one or two parts per year, stabilizing at two per year in 1773.[55] Finally, the number of contributions published per year dropped precipitously.[56]

The Society later made aesthetic changes to match. In 1789 its printing committee complained of the rather shoddy appearance of the *Transactions* — "the ink foul and of a bad colour" — noting that "foreigners particularly took notice that the Philosophical Transactions were worse printed than the publications of most of the other learned societies of Europe."[57] Joseph Banks, the president of the Royal Society since 1778, went in search of a new printer. He found his man in William Bulmer, who had recently been hired by the King's printers to run the Shakspeare Printing Company, specifically tasked with producing a lavish new folio edition of the works of the English playwright. Bulmer was said to have brought to England "what is technichally [*sic*] called *Fine Printing*": it used the best types, ink, and presses, and "a sufficient

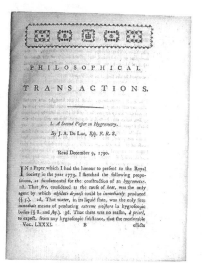
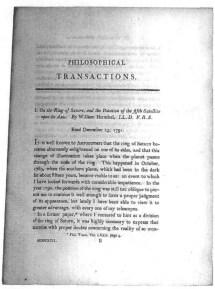

Figure 1.2 The first page of the *Philosophical Transactions* for 1791 (*left*) and for 1792 (*right*). In an effort to improve the beauty of the *Transactions*, Joseph Banks hired William Bulmer's new Shakspeare Printing Company and changed the type, layout, and paper size. Special Collections, Kenneth Spencer Research Library, University of Kansas Libraries.

time allowed to the Pressmen for extraordinary attention, and last, not least, an inclination in the Employer to pay a considerably advanced price."[58] Paper size was also increased on most copies (Figure 1.2). Banks boasted to friends that Bulmer's "printing is acknowledgd as the best in the Country," and would surely "Augment the Beauty" of the *Transactions*.[59]

With these changes, the *Philosophical Transactions* no longer had much in common with its origins; in many ways, Oldenburg's fabled creation was no more. The Royal Academy's *Histoire et mémoires*, rather than learned journals, had become the favored format not only among academies but among the scientific societies more common in Britain. The transformation of the *Philosophical Transactions* took place just as an array of specialized and regional learned societies came into being. The Royal Society of Edinburgh (1783), the Royal Irish Academy (1787), and the Linnean Society (1791) all inaugurated their own *Transactions* in the new style: they were large volumes, published every few years, consisting almost entirely of substantive, signed articles.[60] But these were less the legacy of Henry Oldenburg than of Bernard Fontenelle. A wide gulf seemed now to have opened between the sorts of genres that societies and academies were willing to print—durable contributions to knowledge, directed as much to posterity as to contemporaries—and learned journals, which dealt in news. But late in the century, more journals began to appear that complicated this divide.

Synopsis and Correspondence

The periodical press had long been central to the Republic of Letters' self-conception as the vanguard of an enlightened public, a key site for acts of public criticism.[61] But this public was also increasingly being figured as a literary marketplace of reader-consumers. Editors and writers who helped produce periodicals took themselves to be providing an important service to their fellow scholars, but most also expected to gain something thereby. Running a periodical might be a means of enlarging one's circle of contacts, of building a reputation, and in some cases of attracting wealthy and powerful patrons. But for many it was increasingly a direct means of generating an income. In the second half of the century, many young men looked to make a career through their participation in literary culture, and periodical authorship was a tempting revenue source.

Market strategy mattered. Enterprising editors were testing out more specialized niches, looking to generate a reliable base of subscribers. Where the state permitted it, as it did in England, daily journals focused on political news flourished. Periodicals focused on particular professional topics, such as agriculture, theology, or medicine, also became common, especially after mid-century. Journals aimed at audiences of physicians or pharmacists, for example, gathered together reports of notable cases, symptoms, and novel treatments. Such medical journals often contained news of discoveries as well. But journals focused exclusively on news of scientific discoveries were less common. Nearly all attempts to set up such journals foundered within a few years; unlike physicians or theologians, there was no professional group defined specifically by an interest in natural philosophy. In some German cities, burgeoning research universities provided a viable base from which to maintain such journals, as in the case of the anatomist Albrecht von Haller's *Göttingische Anzeigen von gelehrten Sachen*. Though largely a review journal, Haller focused especially on topics that meshed with his scientific interests.[62]

But as the literary market diversified, more writers were willing to give the business of publishing the natural sciences a shot. One of these was a cleric and author of agricultural books from Lyon named François Rozier. In 1771, in the wake of a string of familial and professional misfortunes, Rozier settled in Paris. Having obtained some success as a writer, Rozier looked to turn this into gainful employment. Through an acquaintance he came into possession of the publishing privilege attached to a periodical that had gone out of business over a decade ago.[63] Rozier revived the title and launched himself on a career as a journalist. But he tried something different, and narrowed his focus to news of European scientific academies. Rozier had heard that the idea of starting just such a journal had been batted around by Paris academicians, but when they did not follow through he decided to do it himself.[64]

The *Observations sur la physique, sur l'histoire naturelle et sur les arts* was issued monthly, and it began with a focus on translating and reprinting works that had appeared in foreign academies, but its issues gradually came to include many original memoirs too. One reviewer of Rozier's new product noted that its focus on publishing whole memoirs rather than excerpts or reviews made it less like a journal and more like an academic collection [*recueil*].[65] Indeed, Rozier increasingly chose to emphasize the resemblance between the *Observations* and academic collections. In 1773, he added "*et mémoires*" to the title and changed from the tiny duodecimo format with which he had begun to a much larger quarto that rivaled the Paris Academy's *Mémoires* (Figure 1.3). A new preface announced that his publication bore "no relation to the periodical works distributed in France or in foreign countries," and he openly disdained amateur readers looking for entertainment and amateur authors under "the pleasing illusion of being initiates in sciences about which they know nothing."[66] He wanted readers to treat the *Observations* as a "continuation or supplement to the Volumes of the Academies," and his instructions to contributors warned that those looking to reprint works already published elsewhere need not apply.[67] Rozier's journal became relatively successful, with print runs reaching as high as 1,500, and it received support and subscription requests from various European academies.[68]

Given that Rozier took the unusual step of mimicking the *Mémoires* of the Paris Academy, we might expect the Academy to have objected to such competition, although it hardly seems that it did. To understand why, consider the changing political circumstances of the Academy in the mid-1770s. Rozier had come to Paris just as the French Chancellor was cracking down on the Parlements of Paris in what was to be the last major attempt to reassert the absolut-

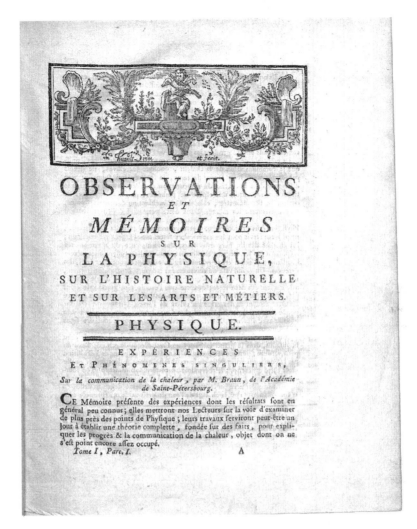

Figure 1.3 The first page of the first issue (January 1773) of François Rozier's new quarto version of the *Observations et mémoires sur la physique, sur l'histoire naturelle et sur les arts et métiers*, which mimicked the style of the Royal Academy of Sciences' *Histoire et mémoires*. Courtesy of the BnF.

ist prerogative of the French monarchy. By 1774, with the coming to power of Anne-Robert-Jacques Turgot as controller-general under Louis XVI, attempts to bolster the monarchy's power were matched by efforts at reform. Some of these were directed at making the state more accountable to public opinion, a concept that was just then transitioning from its association with unruly crowds to a representative and singular consensus. This gave potential critics of the Academy a vocabulary with which to attack its corporate privileges. In 1773, the Marquis de Condorcet, a reform-minded ally of Turgot, became acting secretary of the Academy. He devised a scheme to bring Parisian academicians into collaboration with those in the provinces. The project failed—some of those academies were suspicious that Condorcet's goal was to bring them under Parisian control—but the attempt suggests the Academy was looking gradually to open itself up to others.[69] Corporate monopolies on knowledge were unraveling in other domains as well. In 1777, the laws governing the book trade were overhauled, abolishing the Paris Book Guild's monopoly on publishing the most lucrative literary works and also allowing authors to become proprietors of their own works.[70]

Rozier's journal became a model that others later emulated, within and beyond France.[71] Another was founded by Lorenz Crell, a German professor of *materia medica* at Helmstedt, in 1778. Crell, who was forever in pursuit of scholarly fame, decided to try publishing a chemical magazine. He was inspired by several medical journals[72] that had been launched to try out a publication that combined original descriptions of discoveries, translations (especially excerpts of memoirs published by the Royal Society or the Berlin Academy), book reviews, and other correspondence. It began as an irregular newsletter, but by 1784 it was appearing monthly and was renamed the *Chemische Annalen*.[73] Crell's journal was one of the first among a boom of publications launched by professors in German-language university towns, including the *Journal der Physik* (Halle, 1790), the *Journal der Erfindungen, Theorien, und Widersprüche in der Natur- und Arzneywissenschaft* (Erfurt, 1792), and the *Archiv für die Physiologie* (Halle, 1795).

Crell figured his reader as a *Mitarbeiter* (coworker), and he saw his publication as a means of binding them together in a broad-ranging correspondence. He depended on these correspondents to provide him with content, and he was forever imploring recipients of his letters for scientific intelligence. One of those was Joseph Banks, the president of the Royal Society of London. He proposed that Banks "make it to a Kind of duty to one of the Secretaries of the R.S., to Keep up with me a regular correspondence," promising that he would return the favor with a steady stream of news in return.[74] Indeed, Crell habitually concluded letters with offerings of intelligence, like this one to Charles Hatchett: "Prompted by the hopes, that my proposition won't be discarded by you, I shall subjoin a small specimen of what you may have to expect from my letters."[75] (Banks, as we will see, was not convinced by Crell's idea that the Royal Society should treat a journal editor as an equal partner in such exchanges.)

Crell's idea that a journal is the focal point of a corresponding community of enlightened readers and writers was not new. Communities of mutual obligation, bound together through personal communication, had characterized the Republic of Letters for nearly a century. Maintaining a large *commerce de lettres*—a "regular, arranged correspondence" sometimes between strangers—involved a range of practices and forms of life that accomplished far more than the diffusion of knowledge.[76] It was a means of building relationships, making a literary or scientific reputation, and appealing to patrons. Letters could often become public, whether through reading at a salon, further circulation, or even publication. Epistolary genres continued to be crucial to scientific periodicals, both as a source of content and as a generic conceit. The article as letter to editor implied that author and reader were on the same level, participants in the literary exchange that bound them together. This vision was particularly salient for the community Crell imagined, with German scholars scattered across several states and cities, and no one geographical location or institution able to dominate a field.

In Paris, too, several new journals focused on science soon appeared. In 1785, editorship of Rozier's journal had passed to Jean-Claude Delamétherie, who used the platform to support the chemical theory that combustion was caused by the release of a particular substance called phlogiston. This view was diametrically opposed to the one supported by the chemists in the Academy led by Antoine Lavoisier. Lavoisier and his anti-phlogiston allies decided that pitting theory against theory required pitting journal against journal. In founding their journal, the *Annales de chimie,* they claimed the same synoptic ambition as Rozier and Crell: they would provide readers "a much-needed gathering of works now dispersed across a large number of volumes."[77] But the *Annales* was another interesting hybrid. It contained many original essays, was issued quarterly in octavo (Figure 1.4), and instead of a single powerful editor, it was run by an editorial team whose meetings resembled those of a scientific society.[78]

It took several years to secure permission to publish the *Annales,* and its first issue appeared in spring 1789. Within months, however, the Revolution came, and the system of book licensing and monopoly privileges that had regulated French publishing for much of that century was swiftly dismantled.[79]

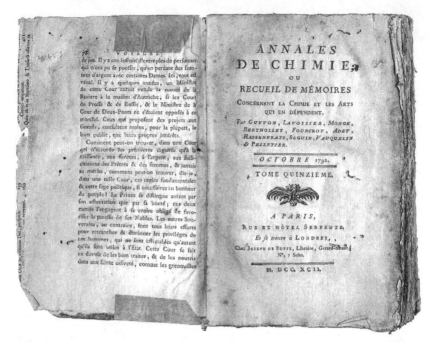

Figure 1.4 The title page of the fifteenth cahier (October 1792) of the *Annales de chimie*. This copy remains in its original cover made out of printer's waste paper from another job. Normally, the buyer would have bound it separately or with another cahier. Author's collection.

The number of print shops in Paris quadrupled, though the financial uncertainty that came with deregulation actually decimated the book industry. It was periodical publishing—cheaper, less risky, and more in tune with the swift march of events—that exploded. Journals dedicated to political news led the way, but they were not alone.[80] Several new scientific journals, including titles such as *Journal des sciences utiles* (1791) and *Journal des sciences, arts et métiers* (1792), began to appear.

A rich network of voluntary learned societies that were designed to contrast with the hierarchical Parisian Academy also materialized. Many of these made the press central to their activities. The Société philomathique, set up just before the Revolution, became a kind of mutual assistance society of scientific news for its members. Its *Bulletin des sciences* began as a newssheet of discoveries: "We ask our correspondents . . . to send as often and as fast as possible all the facts they believe they owe to our friendship and which should be inserted in the bulletin of the society."[81] The Société d'histoire naturelle, a revival of the short-lived Société Linné, was established in 1790. Many of its leading members, including Delamétherie and Aubin-Louis Millin, made their living in the publishing business. The Société assisted in the launch of various journals, including its own *Actes* and the *Journal d'histoire naturelle* (though neither survived the chaotic publishing situation of the early 1790s). The latter was printed by the Cercle sociale, a society whose veneration of the power of the press led them to install their own print shop.[82]

The renewed chaos brought by the Terror spelled the end for many of the new periodicals, but in its aftermath the Committee of Public Instruction began a program of financial encouragements and rewards to publishers of works in the arts and sciences. Scientific periodicals were singled out for their importance. Direct financial support was given to publications such as the *Journal de physique*, the *Journal des mines*, and the *Bulletin de l'Ecole de santé*. The head of the Commission in 1795, Pierre-Louis Ginguené, argued that supporting such publications was a better means of encouraging the democratic spread of knowledge than privileging a corporation of elite savants. "Despotism," he explained, "which had every reason to fear scholars and men of letters, thought that it could incorporate them, or at least . . . could guarantee their silence by placing them under their watch in academies and giving them pensions."[83] Periodicals such as the *Décade philosophique* and the *Magasin encyclopédique* provided an alternative model for intellectual life that resisted specialization in favor of an encyclopedic vision designed in part to transform society through the scientific study of man.[84] Ultimately the Academy was reestablished as part of the Institut de France, and Napoleon later doubled down on political and intellectual centralization while censoring the press, but the field of expectations for what a journal might be had nevertheless shifted.

In Britain, journals focused on natural philosophy came somewhat later. Periodicals calling themselves "magazines" had been founded in the 1780s claiming to be a storehouse of learning, and these included translated excerpts from journals such as the *Annales de chimie* and Crell's *Annalen*.[85] But a journal devoted to natural philosophical news only appeared in 1797, when the professional author and natural philosopher William Nicholson launched the *Journal of Natural Philosophy, Chemistry and the Arts*. He promised it would include "whatever the activity of men of science or of art may bring forward, of invention or improvement, in any country or nation, within the possibility of being procured." Almost immediately he had a competitor in the newspaper editor Alexander Tilloch's *Philosophical Magazine*. Tilloch pledged that his team would "consult every source of information that may be likely to furnish them with materials for their undertaking." A Latin epigraph highlighted Tilloch's synoptic aim: "The way the spiders weave, you see, is none the better because they produce the threads from their own body, nor is ours the worse because like bees we cull from the work of others."[86] Although both publications incorporated features of British magazines, both were also recognized by contemporaries as emulating the success of the new continental publications. Like Crell, Nicholson was especially proud of "the great increase of valuable correspondence" that his journal had brought him. It was the breadth of his correspondence—rather than authors who had contributed original manuscripts—that he chose to highlight in a celebratory letter to readers in 1799.[87]

The ambition to present a synoptic overview of natural philosophy set the pattern for these and other British journals of science that emerged in the next decade. However, most of them made a special point of claiming to reach out to broader audiences too. Tilloch singled out "Transactions of Academies" as works "too dear to be purchased by readers in general." Too often, these remained "unknown to a very large class of men of science," Nicholson warned, on account of "their price, their number, their extent, distance of publication, difference of language, labour of perusal." Tilloch promised to privilege "practical articles, which may be serviceable to Artists, Manufacturers, &c" and thus cater to the "the common ranks" as much as philosophers.[88] The new monthlies were usually octavos consisting of three to seven sheets (48–112 pages), while the transactions of societies were larger quartos that were issued as a rule much less frequently, from twice per year to once every few years on an irregular basis (Figure 1.5).

In France, with its academies that published volumes relatively sparingly, editors were more likely to emphasize their role as a supplement to—or, after the Revolution, a replacement for—academic collections. In central Europe, broken into many states and centers of learning, editors were more likely to dwell on the role of periodicals in binding together communities based on some particular specialization.[89] In Britain, where the pursuit of natural philosophy remained especially identified with propertied gentlemen who could afford to belong to voluntary societies and purchase expensive volumes, editors were more likely to emphasize the democratic potential of the new jour-

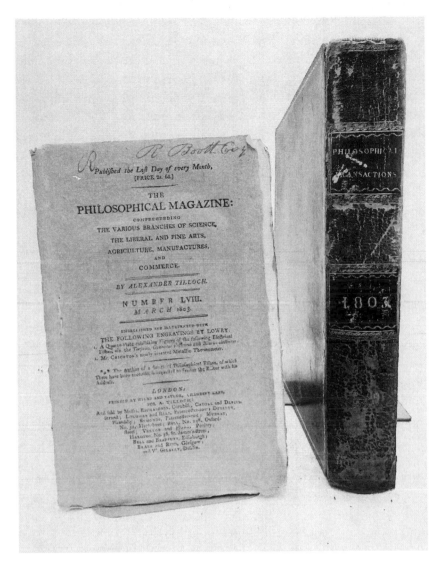

Figure 1.5 Quarto transactions and octavo journals. The volume of the *Philosophical Transactions* for 1803 alongside the March 1803 issue of Alexander Tilloch's *Philosophical Magazine*. The latter's wrapper gave pride of place to the engravings it contained rather than to authors or titles of works. *Philosophical Transactions* courtesy of the Ernst Mayr Library, Museum of Comparative Zoology, Harvard University. *Philosophical Magazine* from author's collection.

nals.[90] Despite these regional variations and their connections with earlier journals, the rising tide of specialized scientific journals was recognized by contemporaries as a Europe-wide phenomenon. They had a common synoptic mission to provide scientific information to readers, and to that end editors were part of a network reprinting content not only from other scientific journals but from other periodical sources and correspondence. Nicholson promised contributors that works published in his journal would also find their way into "the best Foreign Journals."[91] A single text or paragraph announcing a discovery might thus get translated and excerpted across Europe. It was precisely through such mixed media networks, as Iain Watts has shown, that the burgeoning science of galvanism moved across Europe in the opening years of the nineteenth century.[92]

Hewing to long-standing conventions in the Republic of Letters, for many editors and authors the boundary between manuscript letter and print publicity remained porous. Just what kinds of borrowing and reprinting were considered acceptable varied. Lorenz Crell could only conclude that "Englishmen do not like to see their letters printed" after raising the ire of Banks and other English correspondents. He could not see the harm in it, however, explaining that he printed "not private opinions about men & things; but only, matters of fact." But etiquette varied just as much within England. Thomas Young, physicist and polymath, warned his correspondent David Brewster not to "send me any information you are not prepared to have mentioned again, for I am always scribbling something anonymous, and I am very capable of introducing your experiments, where perhaps you would not wish them to appear . . ."[93] Of course, editors also depended on authors sending them articles or notes with the express purpose of having them published, and authors were obliging them in greater numbers. The next step is to explore the reasons that they did so.

Authorship and the Poor Lunar Maniac

In 1803, the directors of the *Bulletin des sciences* apologized to any who might object to their habit of "disseminating literary treasures" in short morsels. They freely admitted that their actions might "endanger the accumulated glory that an individual may obtain by the simultaneous publication of a large number of discoveries that he has amassed in the silence of his cabinet, kept from the public for a long time in order to dazzle it with greater *éclat*."[94] If the apology rang a bit hollow, the link it suggested between glory and certain kinds of authorship was quite real. Authoring a book was an activity replete with diverse meanings and uses in the world of letters, one which could enhance a literary reputation and even attract a patron. Journals such as the *Bulletin* potentially disrupted these conventions of literary and scholarly life. While learned journals had long circulated news of discovery claims and sometimes reported on memoirs presented at academic meetings, scientific journals had turned such activities into their raison d'être, sometimes without consulting the wishes of discoverers themselves.[95] But even the *Bulletin des sciences* gradually became, like Rozier's journal and the *Annales de chimie*, a venue to which savants contributed original texts.

Publicity and priority were certainly important motives for many authors. But there were other considerations too. First, many writers who contributed to commercial periodicals received payment for doing so, whether they were writing about their own or others' discoveries. Even when this was not the case, the strong association between contributions to journals and professional writing was a significant factor in deciding whether to venture into this world. Second, although contributing to commercial journals was not a particularly prestigious activity, it could be turned into a means of publishing independent pamphlets or even books. Such independent publications could be put into noncommercial circulation as a means of generating patronage and a more direct and personal kind of publicity.

Scientific journals began to spread just as the literary identity that we now recognize as the modern author was coming into focus. In the early modern Republic of Letters, cultural capital tended to be contingent on one's perceived independence from the economic considerations of the book market. Making it depended instead on the cultivation of social elites as patrons, on dedications and gifts rather than the act of publishing itself. It was not uncommon for prominent members of high society to avoid authorship of printed books altogether. But an alternative conception of literary merit emerged as some authors began to emphasize their independence in the face of the constraints of elite society. These authors were more likely to appeal to the literary market as rightful judge. As professional authorship became perceived as a viable (if ambivalent) career for *gens de lettres*, it also became possible to imagine it as the basis of a career for those engaged specifically in natural philosophy. While living by the profits of authoring books remained difficult, opportunities to edit and contribute to periodicals (for pay) made such a career choice somewhat more viable.

In practice, however, living exclusively by one's pen remained a difficult thing to pull off, and professional author remained a risky identity to take on.[96] Writing for periodicals was particularly problematic, since journalism was often singled out as a refuge for Grub Street hacks. It is no accident that the crystallization of academic *Mémoires* and *Transactions* as a format that stood apart happened around the same time that authorship took on these increasingly commercial associations. Societies and academies could offer a means of shielding philosophical authors from being beholden to the vagaries of the marketplace. Even as attachment to social elites diminished as a mark of authority, the world of academies and learned societies was a potential arena in which the logic of patronage remained legitimate, since the relevant considerations could be passed off as meritocratic.

In England especially, many forms of periodical authorship remained dangerous social territory. Such was their disrepute that until 1810, having once worked as a journalist was enough to debar one from entering the legal profession.[97] As late as the 1830s, an observer could note that "a newspaper editor is not deemed a gentleman" and that it was wise to keep such employments a secret, "as they would avoid alluding, in the presence of his brother, to a man who had been hanged."[98] Such strictures on journalism did not necessarily hold for contributors to higher-status literary reviews, but just where the new journals of science fit into the hierarchy was not entirely clear. One early review of Nicholson's *Journal* warned that periodical publications "occasionally furnish matter for reprehension." Some men of science, such as the physician Thomas Young, published anonymously as a matter of course to protect their professional reputation. Both William Nicholson and Alexander Tilloch were excluded from election to the Royal Society in part because of their association with periodical publications.[99]

On the continent, the status of periodical authorship was more complex. Lorenz Crell perceived his founding a chemical magazine to be a positive step to build up his fame, and evidence suggests that editing such publications was indeed counted a good thing among administrators of German research universities.[100] Crell was quite baffled by British men of science who desired to remain anonymous in his journal, but he promised to avoid naming them in the future.[101] Even William Nicholson, taking his cue from editors on the continent, was perplexed that most "Periodical Publications of this country have been brought forward without the name either of Author or Editor." Though he preferred to print names of contributors, he promised to do so only if he had express permission.[102]

In France, the ideological importance attributed to political journalism following the Revolution gave journalism a more positive image. British observers noted that in France "men of the highest political and literary station openly connect themselves with the daily press," and they marveled that journalism was there "regarded as a liberal profession, open to young men of honour."[103] But in reality, even if journals in France were often idealized as guardians of liberty, it was a common observation that the necessity of making a career led many young journalists to compromise their principles (Figure 1.6). Many wrote for multiple papers, switched their political allegiances on a daily basis, and thus were compelled to write against their own convictions. The lived contradiction between the duties of citizenship and succeeding in a competitive profession thus made journalism an ambivalent vocation.[104]

But in both Britain and France, there were few means by which scientific activity might be made the basis of a reliable income, and thus the emergence of scientific journals represented an intriguing possibility. Many editors of early scientific journals were either hired by a publisher or hoped to turn a profit directly from the work. This was a motive for Rozier, Tilloch, and Nicholson, and it was also crucial to journals such as the *Annales de chimie*, whose managing secretary, Pierre-Auguste Adet, counted on the income from the journal.[105] The same could be true of authors of original papers. Many early British journals, including Brande's *Journal of Science and the Arts*, the *Edinburgh Philosophical Journal*, the *Magazine of Zoology and Botany*, and the *Edinburgh Journal of Science*, paid authors for content—sometimes quite well.[106] Even in France, where Rozier's journal and the *Annales de chimie* imitated the Academy's volumes, contributors might be paid for content.[107] But remuneration was not the only reason to contribute to the new scientific journals. While the new journals had begun as complements to the judgment-apparatus of academies and their publications, they took on other features that overlapped with the latter. Writers submitted not simply news and extracts of their works to journal editors, but sometimes bypassed the academies entirely by submitting original memoirs to editors.

What it meant for aspiring natural philosophers to publish in these journals could be truly confusing. Consider the situation of two young English

Figure 1.6 "The Balance of the Journalist." Professional writers for journals were often depicted as caught in the tension between a noble profession that sought to speak truth and one that compromised such principles to make a career. *La Charge* 1, no. 4 (2 December 1832). Courtesy of the BnF.

men of science looking to establish themselves in the 1810s. Charles Babbage and John Herschel would become celebrated natural philosophers, but in 1815 they were twenty-three-year-old university graduates looking to make a career by following their passion—for mathematics. Although he received support from his wealthy father, Babbage wanted financial independence.[108] "Happiness may I am convinced be obtained by an Analyst," he wrote to Herschel, "but how he is to obtain that sine qua non of this world (money to wit), I have not yet discovered."[109] Herschel too was looking for a steady income, but with an astronomer for a father, he could not expect much help from his family. At Cambridge, Babbage and Herschel had set about bringing a revolution to English mathematics by introducing continental analysis. Their *Memoirs of the Analytical Society* (1813) was a book of essays that aped the volumes of continental academics. Full of abstruse mathematical essays for which there were exceedingly few competent readers in England, it was never likely to show much of a profit. Although Babbage imagined several schemes to bring in money by science (the construction of a calculating engine being perhaps the most outlandish), he clearly did not consider periodical authorship a worthy career choice. He was often offered opportunities to write about science for money, in both encyclopedias and journals, and while he took advantage of them often enough, he had a low view of career journalists. On one occasion, Babbage showed his disdain for the "poor Lunar maniac who spits forth his monthly venom for his daily bread" by proposing to Herschel a hoax. He suggested it might be fun to submit to a journal an anonymous attack on their own *Memoirs*: "We might each attack our own castle and being the Architects we should most probably know the weakest parts," he explained.[110]

Herschel, for his part, was more interested in exploring the possibility of living by his pen. He chose to interpret Babbage's hoax idea as a ploy for money, replying that he was up for "any scheme of iniquity that may ultimately tend to the <u>sale</u> of the Analytical Mem."[111] While at Cambridge, Herschel had begun publishing with a few anonymous notes in Nicholson's *Journal* (which could not afford to pay contributors), and he had even sent a series of memoirs to the Royal Society. But he was on the lookout for paying gigs. In 1816 he wrote to Babbage about the new *Journal of Science and the Arts*: "Who are the Editors of it? Will they pay one for writing? or do you know anybody that will? I should like to get to write articles for some of these Encyclopaedias if I could . . . In a word I want to get money <u>and reputation at the same time</u>, and to give up cramming pupils, which is a bore & does one no credit but very much the contrary."[112]

Herschel's inquiry highlights the ambiguities of periodical authorship for men of science. The *Journal of Science and the Arts* was advertised as a periodical that aimed to exist between commercial journals that published news for profit and the more prestigious transactions of societies. But Babbage reported that its editor, William T. Brande, "seemed darkly and distantly to intimate that cash was to be had."[113]

Herschel himself, though eager to publish in periodicals, was under no illusions about their inferior status. In congratulating Edward Daniel Clarke for a paper he published in Brande's *Journal* on the chemistry of volcanic eruptions, he ventured to "express something like regret that [the Royal Society's] *Transactions* had not to announce to the Scientific world such wonderful results." Though a speedier medium, he thought that "the periodical work of an individual (however excellent) seems to me a vehicle unworthy [of the] magnitude of the discovery."[114] When in 1820 Herschel and Babbage helped found the Astronomical Society of London, they decided that it ought to publish a memoir series. But what then to do with all of the miscellaneous correspondence regarding astronomical and meteorological events that they were receiving? Herschel didn't think that the Society was "bound, or authorized to print it." The distinction between "private letters" and those that had "the air of a formal communication to the Society" seemed to Herschel razor-thin: "If we print it in the Writer's own words, we must frequently make garbled extracts of his letters. — If merely as intelligence, our Memoirs become an Astronomical journal."[115] That outcome — producing a journal — was clearly one that Herschel wanted the Society to avoid.

While Herschel and Babbage managed their associations with periodicals with some care, others immersed themselves in commercial periodical publishing. Their Scottish acquaintance David Brewster supported himself largely by contributing to periodicals as well as editing them, including the *Edinburgh Philosophical Journal* (from 1819) and the *Edinburgh Journal of Science* (which he founded in 1824). But no one complained more bitterly than Brewster that it was the difficulty of making a career in science that drove them to engage in "professional authorship." He held particular scorn for "the composition of treatises for periodical works and popular compilations," for these took time away from "the fascination of original research."[116] Privately, Brewster warned a younger James David Forbes away from such a career path:

I do not object to you making money by your writing, but I am sure that it wd be injurious to your happiness to rely on such a source for a <u>permanent portion</u> of your income. — The moment you do that you become a professional author — who is the worst of all authors — following the worst of all professions.[117]

Brewster insisted that to become a writer for money was to "renounce your independence as a man of science," for not only was it an insecure way to make a living, but the temptation it brought to publish quickly and frequently might permanently harm one's reputation as a discoverer.[118]

In the 1820s, whether contributors to scientific journals ought to be paid remained a matter of controversy. The *Chemist*, a journal that aimed to open science to the working classes, criticized those scientific journals based on "the principle of gratuitous contribution," arguing that it brought down the literary quality of such publications, and made them dependent on "the chance discoveries of the month or quarter in which they are concocted."[119] The *Magazine of Natural History* insisted in contrast that only "voluntary contributions of their readers" could be "the legitimate object of a Journal of Science." Those based on paid contributions were like exotic plants, "continually nursed in a hot-house," while the former arose naturally "from the wants of the times." Literary quality was beside the point, for scientific journals were designed for a different kind of reading: "Those who peruse a scientific magazine, as they would glance over a merely literary periodical, are spending their time to very little purpose."[120]

Many editors of commercial scientific journals took a view similar to the *Magazine of Natural History* and did not pay contributors (often they simply could not afford to do so). But authors might choose to publish in a journal because it meant getting into print more quickly and with less fuss. One might also reach a wider readership than the audience connected with elite or specialized societies. Others might have felt compelled to go to journals after failing to win the approval or attention of a society or academy. Also, since many journals across Europe allowed anonymous publication, one might publicize a series of claims without taking public responsibility for them until reader reaction could be gauged. John Herschel first published under the name "A Lover of the Modern Analysis" in Nicholson's *Journal*. James David Forbes chose the pseudonym Δ for his first papers published in Brewster's *Journal*.[121] Even signed articles in journals could be seen to carry less risk to their author than books or memoirs. As Babbage reasoned in the case of a set of essays of mathematical philosophy, by publishing in the *Edinburgh Philosophical Journal* he could "secure a great circulation without appearing to overrate their value."[122] Even better, since the latter journal did pay its authors, he could use the "remuneration they would produce" to help finance an eventual book publication on the topic.[123] The general idea that publishing in a periodical was a step toward publishing a fuller account in a longer single-authored book was not new. But, through the distribution of separate copies, this was becoming possible in new ways.

The separate publication of texts that appeared in journals or other compilations had probably taken place as long as such publications had existed. But the circulation of separate copies — "tirages à part," "Separatabdrücke," or "offprints" (as they would later become known in English)[124] — expanded in the last decades of the eighteenth century from an ad hoc, occasional phenomenon to become a crucial aspect of learned publishing during the nineteenth century. For learned societies and academies whose volumes were published slowly or irregularly, publishing separately allowed authors to circulate their work more quickly. It was just when the *Philosophical Transactions* took on the character of a memoir series in the second half of the eighteenth century that there was a dramatic rise in the production of separate copies of its memoirs. They become more common in the 1760s, so that by 1766 the Society's printer was setting up type so that each memoir began on a fresh page.[125] By 1790, the Royal Society had a standard policy regarding distribution, allowing authors routinely to take fifty copies for their private use, and fifty more upon special authorization.[126]

In many cases these were not so much reprints as *preprints*. Since they were printed off at the same time and using the same type as the periodical versions, it was often possible to distribute them before the full issues to which they belonged became available. Memoirs to be published in the *Philosophical Transactions* were sent to the printer as they were voted to be published, and the printer often gave them to authors prior to the publication of the volume itself. Such pre-distribution of memoirs was common among learned societies whose publications appeared infrequently.[127] This seems often to have been based on arrangements between authors and printers, but the speedy delivery of separate copies turned out to be an important way in which societies whose volumes were published at irregular and infrequent intervals were able to rival the new monthly and quarterly journals. Already in 1790 the Literary and Philosophical Society in Manchester made it official policy that any manuscript voted to be printed "shall be sent to the press without delay" and the thirty separate copies allowed to authors dispatched immediately if desired.[128]

Allowing authors to receive separate copies of their papers from printers gradually became routine for many publishers. Independent journals that printed original memoirs submitted by authors, such as Rozier's journal and the *Annales de chimie*, usually allowed authors to have separate copies produced at cost.[129] The practice was rarer for early British scientific journals, which more often consisted of reprints and intelligence. But there too separate copies picked up once men of science began to submit original memoirs more regularly.[130] Academies and societies were generally more liberal in their policies, and in the 1820s some began to allow authors a certain number (usually fifteen to twenty-five) free of charge.[131] Some authors even made a point of requesting them at the same moment that they submitted their paper, specifying an address to which they might be sent. When Richard Taylor began the *Annals of Natural History* in 1838, he weighed an offer from William Swainson to provide him a series of original essays for publication for which he would ask no payment if he were allowed 125 separate copies. Taylor declined.[132] Acquiring separates thus became a consideration and point of negotiation in the submission process itself.

Where it was legal to do so, authors could self-publish short works for private distribution, but doing so by acquiring separate copies was distinctly cheaper, especially if any engravings were involved. Generally authors paid only the cost of printing and paper ("which you know in matters of this Kind is a trifling one," explained Banks to an author), since the type was already set up for the journal.[133] One important strategy to minimize the financial risks associated with longer works was to publish them in parts, signing up subscribers and thus acquiring a better estimate of one's audience (its size and its tastes) as the parts came out. But even this was financially risky for works with limited audiences. Far cheaper, if more labor intensive, was to publish a longer work as a series of papers in periodicals; readers who received a series of separate copies from an author might reconstitute them, binding them together.

Authors wanted separate copies not only because they might be acquired quickly but because they had different uses than articles in periodicals. Separate copies often passed for independent publications, with pagination starting from "1," and their own title page. While some featured the name of the periodical, others buried it in a footnote or omitted it entirely.[134] An article that appeared as a separate also stood a better chance of receiving a dedicated notice in a review journal, both because it could be sent directly to the editor and because it was more like a book.[135] When authors made lists of their publications in the early nineteenth century, they often did so by listing the separate copy as an independent publication, sometimes omitting mention of the serial in which it had appeared.[136] Paradoxically, then, the proliferation of commercial periodicals helped foster the circulation of printed texts based on noncommercial personal exchanges. When requesting separate copies, it was common for an author to specify that it was only "to give to particular Friends."[137] This signaled that the author would not attempt to enter into competition with the publisher by selling his copies.

The extent to which periodical authorship could be manipulated for the purpose of circulating longer synthetic works is exemplified by the examples of the French authors Georges Cuvier and André-Marie Ampère. Cuvier, who built his reputation largely on grand multivolume works, used a strategy for publishing in the *Annales*, the new periodical of the Muséum d'histoire naturelle, that led directly to producing a longer volume. For several years, whenever he published a paper in the *Annales* he asked the printer to provide him hundreds of extra copies of his paper, but instead of distributing, he hoarded. Finally, in 1812, Cuvier rearranged them in a logical manner, added a title page and framing materials, and bound them together as the four-volume *Recherches sur les ossemens fossiles de quadrupèdes*. He added a preliminary chapter that set forth "the general principles that guided they research, the fundamental ideas that they support, and the consequences that appear to follow for the physical history of the globe."[138] Cuvier apologized about the haphazard pagination that resulted, but explained that this was a means of satisfying "friends of science" quickly while at the same time synthesizing "an order of facts of so much interest for the theory of the earth."[139]

Ampère went a step further. He submitted manuscripts to a variety of journals with the principal aim of obtaining independent publications at a discount rate. His *Recueil d'observations électro-dynamique* was produced out of such separate copies, but he made sure to create uniformity by instructing the printers of each article to begin the pagination at just the right page number and to stick with a specific layout.[140] With twenty-four chapters taken from several different periodicals, the *Recueil* boasted a variety of fonts and papers, but with a specially printed title page and contents, it was easily mistaken for a book published in a more conventional way. Ampère thus put out a book of abstruse mathematical physics, paying only for paper and printing, and saving the costs associated with setting the type and making expensive engravings.

Ampère was able to pull off this virtuosic feat of publishing in part because he was a sought-after contributor to whom editors were willing to grant special favors (Cuvier, for his part, had control of the *Annales*). But in general many publishers were wary of the ways in which the wide distribution of separate copies might impact their bottom lines. For societies and academies who published at a more leisurely pace, they could contribute to a more general problem: they were being beat into print by journals willing to reprint anything they could get their hands on.

Reprinting, and Other Very Improper Acts

> In the present state of society, every discoverer of new and important facts has so many methods of making them known to the public, and the nation contains so many reading, and so many enlightened men, that merit is certain of acquiring celebrity in spite of all the obstacles which those, who think themselves already seated upon the summit, are disposed to throw in its way. A man of science, therefore, need be under no manner of uneasiness, though his discoveries are refused a place in the Transactions of the Royal Society.
> —THOMAS THOMSON (1820)[141]

In the name of diffusing knowledge — but also in order to fill the pages of each issue with content — most journal editors were eager to borrow, excerpt, and translate scientific news, no matter its source, as quickly as possible. The value placed on diffusion, and the reprinting practices that enabled it, were what especially set scientific journals apart from the proprietary codes of scientific societies and academies. Both the Academy in Paris and the Royal Society of London zealously guarded what they viewed as their right to serve as the first and primary audience for new scientific claims. Both rejected submissions of manuscripts from authors, no matter their perceived merit, if they had appeared in print in any other form. While neither institution put serious restrictions on the reprinting of a memoir after it had appeared in one of their volumes, they refused to be perceived as reprinters themselves. As press coverage of science increased in both countries, what had been unwritten convention was transformed into carefully defined rules concerning publicity, including what kinds of reprinting were to be tolerated, who could attend meetings, and what kinds of reporting would be allowed. Ultimately, as we will see, their strategies diverged.[142]

In Paris, the establishment of Rozier's journal and other venues meant savants had a choice whether to submit their work to the Academy (in hopes of a favorable judgment) or to a journal like Rozier's (providing swifter print publicity). Immediately, some authors elected to do both, hoping to combine the best of both worlds. But the Academy was quick to object to "seeing its name connected with works appearing in the different journals" when it had not yet given them approval. Some commissions simply refused to deliver reports on memoirs that were already in press, while the Committee of Publication refused to print anything that had appeared in a journal.[143] In 1775, academicians became more aggressive. When a paper on poisonous mushrooms read at the Academy appeared in Rozier's journal, they fought back in the same way they had done for decades: by threatening the editor and trying to mobilize state censors. They wrote to the authorities with their recommendation: "The only means of remedying such abuses is to order the censors who are charged with examining journals to refuse any memoirs presented to the Academy or approved by it without a document showing the approbation of the Academy signed by the Secretary."[144] Meetings themselves remained shrouded in secrecy. Even when non-academicians were allowed to attend, they were made to understand that this privilege was "a confidence whose secrecy was tacitly advised."[145]

Joseph Banks maintained a similar iron grip on the privileged secrecy of Royal Society meetings. In 1783, amid dissension regarding Banks's aristocratic management style, a Council member named Joseph Planta attempted to bring more publicity to meetings by publishing "the public proceedings of the Royal Society." Banks, believing the idea "detrimental to the interests of the Royal Society," moved to block it. Planta accused Banks of running the Society's meetings as a "Secret Court of Inquisition." He thought "the very Idea of concealing their Contents [is] impracticable," pointing out that both houses of parliament had come to this realization a decade before.[146] But Banks rejected the implicit analogy between scientific and governmental bodies: there was no reason to believe that authority in science ought to be generated in a similar manner to that of the state.

As if to confirm Planta's assertion, an alarmed Charles Blagden reported to Banks a few months later that while perusing a newspaper—the *St. James Chronicle*—he noticed an excerpt of a paper by William Hamilton on earthquakes, due to appear in the *Philosophical Transactions*: "it must have been obtained in some fraudulent manner, or put in by some of those to whom Sir Wm. presented his separate copies," Blagden suggested. "In either case it is a very improper act with regard to the R. Society." Banks agreed that it was alarming to find a Royal Society memoir printed "in the news" and observed that "other improprieties have this year happened." He suggested a new rule that "no author shall receive the Separate Copies of his work . . . till the time of Publication by the Society."[147]

Banks held off on implementing the idea, but the precirculation of separate copies continued to be a problem. Lorenz Crell, evidently doomed forever to misstep in his dealings with the Royal Society, repeatedly aggravated Banks by printing memoirs due to appear in the *Transactions*. Crell did not see how publishing news of experiments "in a foreign country, in an other language, & in a different mailer, could be thought disrespectfull . . ."[148] While the Society did waver in its prohibition on preprinting abroad (unless authors themselves initiated the translation),[149] within Britain the rule was firm. After William Nicholson launched his journal in 1797, he publicly requested readers to send him separate copies of memoirs printed by scientific societies as soon as they were received.[150] This led to a skirmish between Nicholson and Banks in 1802, with Banks threatening Nicholson that he would "enter into hostility with your Journal" if he continued to print excerpts of memoirs prior to their appearance in the *Transactions*.[151] Nicholson reasoned that if the Royal Society truly cared about the fast and free spread of knowledge to the public, then it ought to support the journal's policy.[152] Banks was unimpressed by this argument. The Royal Society allowed reprinting after publication of the *Transactions* itself, and he warned Nicholson that if "British Journalists are not Satisfied with the Liberality of the Society in giving up [to them] their Property the moment it is Realy publishd I confess I Consider them as ungratefull for a Favor which other Societies do not as far as I know indulge the Journalists of their respective Countries."[153] Banks then drafted a warning to authors to keep their preprints out of the hands of journalists: "Gentlemen who are indulged with separate Copies of their Communications, are requested to use their endeavour to prevent them from being reprinted, till one month after the publication of that part of the Philosophical Transactions in which they are inserted." This warning, which appeared on the flyleaf of the Society's separate copies for decades, was imitated by other British societies.[154] Authors did, however, continue to circulate the pamphlets themselves prior to their publication in the *Transactions*, and some continued to make it into print.[155] The prohibition on previously printed communications was so strict that when one author submitted a paper to the Society that he had had printed in an edition of just two copies ("to secure greater correctness"), the Royal Society refused it, explaining that they "never reprint Printed Papers in their Transactions."[156]

These strictures were consistent with the belief of Banks and others that certain institutions had special property rights in certain categories of natural knowledge. His well-known objections to the formation of the Geological Society of London, for example, only arose when its Council decided it wanted to publish its own *Transactions*.[157] An observer reported that when the Royal Institution, dedicated to the diffusion of science, launched a journal, Banks insisted that it not be printed in quarto, "for this cogent reason, that the Philosophical Transactions are published in quarto; and to print those journals in the same sized page might excite an unfavourable comparison!"[158] Similarly, when the Asiatic Society of London was founded in 1823, the Linnean Society attempted to convince the attorney general to block the granting of a Royal Charter on the grounds that the new Society threatened to publish memoirs on topics in natural history.[159]

The tense relationship between societies and journals remained after Banks passed away in 1820. In 1828 John Robison suggested to the Royal Society of Edinburgh that they could save money by getting rid of their *Transactions* entirely and simply send memoirs to David Brewster's *Journal*, "every thing interesting which they contain being already old and having previously appeared in some scientific journal." Rather than welcoming the extra business, however, Brewster found the idea alarming, detecting an implicit accusation that proprietors of journals were in the business of stealing scientific content from the societies. Brewster had heard the old argument that "the best papers are copied into Journals before the Transactions appear" before from "an enemy of Journals," but he claimed it to be rubbish.[160]

By the 1820s, the transactions format was coming under fire. N. A. Vigors, looking to found a society for zoologists in 1822, urged that the Linnean Society's *Transactions* was not something that new society should emulate: "they come out at intervals too distant for the constant diffusion of knowledge that is necessary; they are too costly for general circulation; and are devoted to subjects too important to take in that subordinate but still valuable mass of information that is fitted only for the pages of a periodical journal."[161] Across the Atlantic, the indomitable publisher and writer Mathew Carey hounded the American Philosophical Society for "aping the quarto volumes of the Royal Society" and thus hoarding knowledge.[162] He submitted several proposals to replace their *Transactions* with an octavo journal to be published more quickly and cheaply. The Society balked at the idea, however, pointing out that other societies might refuse to exchange their expensive quartos for an octavo journal. Carey wrung his hands that the "mania of copying European examples, led to the adoption of the quarto form originally, so very ill-suited to the circumstances of the country."[163]

In Paris, by contrast, the relationship between the Academy and the press had already changed by the 1820s. In 1789, the Revolution had turned the ensemble of the Academy's powers upside down. With the proclamation of freedom of the press and the abolition of privileges in the early days of the Revolution, it lost most legal means of maintaining any control over scientific publishing. Democratic activists viewed the academies as lingering holdouts to the elitism of the ancien régime. Many attacks on the Academy came from savants who claimed public legitimacy derived from the public press. Jean-Claude Delamétherie turned Rozier's journal (which had been renamed the *Journal de physique*) into a platform to attack the Academy as an aristocratic cabal bent on monopolizing scientific opinion. Although the Academy navigated the first waves of these attacks — as well as the Le Chapelier Law that banned corporations in 1791 — it was shut down in August 1793.[164]

When the Academy was revived by the Directory as the First Class of the new Institut de France, it was unclear just what kind of institution it was to be. On the one hand, it was built to embody a new vision of an intellectual encyclopedism integrated with society. Its various sections met together on a regular basis, shared a common administrative core, and were compelled to coordinate their publishing activities. But there was reason to believe that the Institut was still in many ways a revival of the institutional culture of the ancien régime. Many of the scientific section's members had been part of the defunct Academy, and they swiftly revived many of its customs and genres, albeit in the absence of the legal powers that that body once commanded. Ordinary meetings were declared to be strictly "private and interior" affairs. Precisely because they were dedicated to "objects of great public interest," it was important that they not be "distracted from the silence and medita-tion" required by their work.[165] The Academy again embarked on publishing its *Mémoires*, and it recommended calling for reports on papers, inventions, and technical inquiries from the state. Although these reports were no longer written as judgments on behalf of the state, and the Academy still only rarely printed memoirs of non-academicians, many savants and inventors remained interested in winning the Academy's approval.[166] Certain tendencies of its royal predecessor became engrained in new rules: reports were only to be written on the work of nonmembers, never of fellow academicians.[167] And they would never be written on papers already printed, on the argument that the author had chosen to give their work to the public by another means.[168]

Similar ambiguities attended the publications of other elite institutions established after the Revolution. The École polytechnique, founded in 1794, proclaimed that the new *Journal de l'École polytechnique* would help justify to the public its existence by publicizing useful knowledge, giving its students a public platform, and offering a model to guide other establishments dedicated to education. And while early cahiers included a range of things, including course materials, it soon became a venue dedicated almost solely to printing the research memoirs of its distinguished professors.[169] With cahiers appear-ing only every year or two in quarto form, it became indistinguishable from an academic collection. Likewise, when the professors of the Muséum d'histoire naturelle launched its *Annales* in 1802, they did justice to the revolutionary origins of their institution by announcing that this would be a true journal, cheaper and speedier than had been the custom of elite scientific bodies. But while the *Annales* was indeed published with greater frequency, it followed the form and spirit of the Institut's publications in nearly every other way, becoming a privileged venue for its members to proclaim the state of knowl-edge in natural history. (Indeed, it was later renamed *Mémoires*.) A second prospectus two years after its launch eschewed the inclusive language of the first and instead celebrated the Muséum as a central repository of authorita-tive natural historical knowledge and objects, one that had recently "grown tremendously by bringing together the various cabinets of Holland and Italy, as well as the acquisitions the FIRST CONSUL has made to augment it."[170]

Napoleon's rise to power transformed the political circumstances of French savants in several ways. The First Consul-turned-Emperor reversed the late Revolutionary period's more inclusive models of intellectual life and took measures to revive centralized institutions. At the same time, he bound savants especially strongly to his own benevolence, even making a place for elite savants in the ranks of the new class nobility that he had constructed.[171] Proclaiming loudly his patronage of the sciences, he dialed back their social and political mandate,[172] shutting down the section of the Institut dedicated to moral and political sciences in 1803. It was this era that shaped the views of powerful savants such as Georges Cuvier, who defined scientific decorum as a studied avoidance of politics and public expression: "In this country, the first rule of conduct for a savant is never to speak of what has been delivered to the grand public and to the journals—in a word, of objects of popular curiosity."[173] Cuvier's argument gained its power in part from the recent association between broader public participation in science and revolutionary excess. The fact that in reality such appeals to public approval, especially public lectures and writing for broader audiences, remained quite central to building a scientific reputation put savants such as Cuvier into a double bind that took great skill to navigate.[174]

Napoleon himself encouraged this paradoxical orientation to public opinion. On the one hand, he heavily censored the press, forcing many periodicals out of business. On the other, he made a show of supporting a daily paper, the *Moniteur*, in which he claimed to engage public opinion. (Paradoxically, he called his new censorship bureau the "Division of the freedom of the press.")[175] Napoleon offered several state-sponsored institutions, including the Institut, the right to use the *Moniteur* as a publishing venue. In return, members of the Institut agreed to "give preference to the *Moniteur* over all other journals," including all the specialized journals such as the *Annales de chimie*.[176] A daily newspaper thus became a key venue for the publishing of elite science[177] while other journals that promoted the broader encyclopedism of the Revolution, such as the *Décade philosophique*, were harassed or forced to shut down.[178] In 1810, an overhaul of the laws regulating the book trade established stricter rules still, capping the number of printers allowed to operate in Paris, forcing small printers (any with fewer than four presses) out of business, and establishing a systematic censorship regime.[179]

Despite the culling of the press during the First Empire, periodicals retained some of their revolutionary allure as a locus of scientific sociability. Many of the Institut's members were involved with other societies where they could present and publish papers, and several were involved in editing periodicals. Other genres, such as encyclopedias and dictionaries, were emerging as key venues for publishing science and contesting reputations as well.[180] In

1808, the potential risks of the Academy's hard line against publicity began to show themselves. For two meetings in a row, they reported in their minutes that "a lack of matter forced them to adjourn just a few minutes after opening their meeting."[181] Embarrassed by the dearth of new work being submitted to them, the Academy launched an investigation. Recognizing that a century earlier the Academy had been "the unique cultivator of the sciences in the nation," they acknowledged that they were now surrounded by other people and other institutions doing the same. There were many "periodical collections where members insert their memoirs as soon as they have written them," and the Academy had no real power to keep them from doing so. In response, the Class chose to dramatically reorient its publishing strategy, suggesting that its members should now feel free to submit memoirs for publication that had already been printed elsewhere: "Make a *recueil classique ou choisi* of the best of your collection of memoirs, giving yourself time to review and correct them." In sharp contrast with British societies, which continued to insist on original memoirs, the First Class would now "have the right to reprint anything that had been read at its meetings." They also considered launching their own periodical publication, but eventually decided that rather than compete with the speedier periodical press, the Academy would become an archive of authoritative science. The Commission pointed out to its members that only the Academy could guarantee them a lasting reputation, "when the periodical works that compete with it today will have long been forgotten."[182]

Academicians were now free to publish as they pleased, and non-academicians continued to present manuscripts to be read at the Institut and assigned to commissioners in hopes of an approving report. The commissioners could choose whether to comply. If the *rapporteur* was not interested or he expected that he would write a negative report, he usually did not bother.[183] The *rapporteur* might also abstain if the author had their paper printed in the meantime.[184] But the moment a report came back, particularly if it was positive, the author would likely have their work published in a journal, both so that it could be publicized and in order to acquire separate copies for personal circulation. The reports themselves often found their way into print via some journal, a practice that the Institut sometimes encouraged.[185]

Conversely, the Academy's own *Mémoires* came to consist largely of what were effectively reprints (often revised and extended) of papers that had been previously published in journals.[186] With the increasingly archival status of its *Mémoires*, the Academy had more or less gotten out of the business of publi-

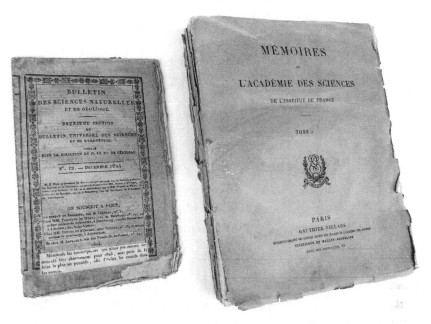

Figure 1.7 Academic collections versus the "frivolous attractions of novelty" contained in scientific journals. The quarto *Mémoires de l'Académie des sciences* 5 (1826), alongside the issue of the *Bulletin des sciences naturelles et de géologie* for December 1824, part of a suite of periodicals under the direction of the Baron Férussac, whose aim was to report news across several fields of science and technology. Note the wealth of information—including a note to subscribers glued on—on the front of the *Bulletin*. (The wrapper on the *Mémoires* was added by the publisher at a later date.) Author's collection.

cizing science. The arrangement was defended by Jean-Baptiste Biot, a physicist who would continue to take a deep interest in the relationship between the Academy and the press.

> Academic collections have nothing to fear from the competition of scientific journals that ordinarily publish discoveries first; most of these writings, since they can only give a quick and fleeting survey of results, will never keep scrupulous minds from seeking out the sources that can only really satisfy them: those that provide a more profound knowledge of the results and the methods that allowed them to be discovered.[187]

Biot argued that precise timing was irrelevant to genuine contributions to knowledge, "which had nothing to gain from the frivolous attractions of novelty."[188] Biot's determination to distinguish academic collections from most scientific journals was sustained by their larger size and graver appearance (Figure 1.7).

If Biot's defense of the Academy's *Mémoires* sounded somewhat defensive, it was. The new arrangements never quite stabilized. Others within the Academy were less convinced that they should forego competing with the journals. In the 1820s, Biot's onetime collaborator, François Arago, emerged as a key voice for an opposing vision for the Academy. He attempted to push through reforms to modernize the *Mémoires*, making it more efficient by forcing authors to submit memoirs and corrections on a tighter, stricter schedule.[189]

The Academy was also coming under pressure from other sources of serialized publicity. The old Academy's weekly meetings had been carefully guarded private affairs for members and a few select guests. Although the Directory had reaffirmed this rule for the new Institut, already during the First Empire more outsiders had been finding their way into the audience at meetings despite periodic efforts to pass stricter rules about who was allowed to attend.[190] After Napoleon's fall, more liberal press laws led to a renewed expansion of journalism as well as of public discussions of politics and culture. News of science, and news of the Academy, was increasingly finding its way into the press. The Academy of Sciences, which for most of its existence had exercised mastery of the press as effectively as any learned institution, was losing its grip on the instruments of that mastery.

2

Meeting in Public

> In our examination of the questions raised and treated by the Academies,
> we will bring to them the spirit of discussion and critique that best brings
> difficulties to light, and that contributes to clarifying, if not resolving, them.
> It is thus that the political press exercises a salutary influence with regard to
> the discussions of the chambers.
> — *Gazette Médicale de Paris*, 1833[1]

Depending on whom you asked, the spread of commercial scientific journals
was either a boon to knowledge or something of a calamity. If you were running a major scientific academy or society in the early nineteenth century, it
was very likely a nuisance. As news services, journals were beating you to the
punch in announcing new discoveries. Some authors were foregoing your
meetings altogether as venues through which to offer their work to public
judgment. Then in the 1820s, things got worse. With or without permission,
the press began to find ways to publish detailed accounts of scientific meetings, insisting that societies and academies had an obligation of transparency
to the public.

First academies and societies dug in their heels, highlighting the wide gulf
separating the forms of judgment and propriety they employed, on the one
hand, and those of the scientific press, on the other. But by the 1830s they

changed course. Instead of disdaining the press, they reshaped themselves in its image, creating journals of their own, often modeled directly on their market-oriented competitors. Often labeled "proceedings" in Britain and "*comptes rendus*" in France, these new publications were printed more cheaply and frequently than transactions, and they contained a wide range of scientific matter, including abstracts of papers and reports presented, discussions that had taken place, and other matters of society business. Their sudden appearance ultimately had a profound impact on elite institutions of science and on conceptions of legitimate scientific judgment. The rest of this book is largely about the consequences of this reversal.

To understand how this happened, we need to widen our view to the political press. The rising significance of printed proceedings of governmental bodies in Britain and France after the Napoleonic Wars made print publicity a hallmark of the legitimacy of public bodies in general. In Britain, the radical press took up this ethos of publicity and used it to lobby for the reform not simply of government but of science as well. In the mid-1820s, in the context of the emergence of a new, cheaper scientific and technological press initiated by radical authors and publishers, Richard Taylor's *Philosophical Magazine* took up this ethos and applied it to elite science by publishing extensive proceedings of certain societies. Gradually, scientific societies took charge of these proceedings and a new periodical genre was formed.

In France a similar transformation took place, though it took a more remarkable course because the challenge came not from scientific journals but directly from the political press itself. During the 1820s, the extent and power of the press in Paris expanded by leaps and bounds, culminating in the July Revolution of 1830, led in the main by journalists themselves. Coverage of Parisian science, and of meetings of the Academy of Sciences in particular, was central to this media world. Inaugurated by a group of liberals as part of a cultural offensive against the increasingly reactionary Bourbon regime during the 1820s, the new scientific journalism went through two distinct stages that corresponded to two distinct styles of political opposition. The liberal opposition developed an idiosyncratic theory of representation that took publicity to be a mechanism for concentrating reason in the public deliberative bodies of government. In this context, publicizing the Academy's meetings was a politically significant act, for a body of intellectual elites engaged in public deliberations was a prototype of the liberal ideal of legitimate government. But after these journalists came to power themselves in 1830, a more radical republican opposition emerged with a wholly different politics of representation, one that pitted the *journaliste* directly against the academic *rapporteur*.[2]

The Birth of Proceedings

"But for [the press], it is hardly too much to say, that in the state of feeling and opinion to which we have now arrived, our representative legislature would almost cease to be regarded as an institution of high worth or significance." When this observation on the publication of parliamentary debates was published in 1833, the idea had already become a truism in British politics. In 1771, after several decades of struggle, the House of Commons had given up barring such reporting, and since that time the journalists in the Gallery had become central personages in British political life. The publication of proceedings, though once "a practice which seemed to the most liberal statesman of the old school full of danger to the great safeguards of public liberty," had become, Thomas Babington Macaulay noted in 1828, "a safeguard, tantamount, and more than tantamount, to all the rest together."[3]

Writers in France agreed. "The publicity of debates in the chambers subjects those in power to carry out the search for justice and reason under the eyes of all," wrote the historian François Guizot in 1820, "such that each citizen may be convinced that this search has been carried out with good faith and intelligence."[4] In 1789, after the censorship machinery of the ancien régime had been dismantled, an explosion of pamphlets and periodicals had followed. Many of these specialized in reporting the meetings of France's new legislative bodies (first the Estates-General and then its successor, the National Assembly). Some, such as the *Journal des Etats-Généraux* and the *Journal logographique*, claimed to give a verbatim and unbiased account of these meetings. Others explicitly took their cue from English examples, mixing summaries of parliamentary debates with non-parliamentary news and advertisements.[5]

When Napoleon came to power, he severely restricted the press freedoms of the Revolution, but he chose to sponsor an official state journal, *Le Moniteur*, at the same time. This move partially reinforced the idea that journals were legitimate embodiments of public opinion. After he was deposed in 1814, freedom of the press was enshrined as a central tenet in the Charter that reestablished the Bourbons, the first French attempt to establish an English-style constitutional monarchy. This new regime saw the rise for the first time of a relatively stable cadre of daily journals built around reports of the meetings

Figure 2.1 "La tribune des journalistes." The seating area for journalists at the French Chamber of Deputies, as depicted in *L'Illustration*, no. 54 (9 March 1844). Courtesy of Harvard University Libraries.

of the Chamber of Deputies. The practice became so entrenched that there was a special area of the gallery marked off for journalists (see Figure 2.1).[6]

French journalists became the envy of the British press corps, where the practice remained more a tacit privilege than a right and where reporters scrambled for limited seats in the public gallery.[7] The link between an unrestrained press and the anarchy into which revolutionary France descended had become notorious. William Windham warned against allowing newspapers to detail parliamentary proceedings, noting "how those who wrote for newspapers in general had contributed to the overthrow of the different Governments of the world."[8] Radicals and Jacobins were usually quite happy to make this connection themselves, developing something like a cult of the printing press. Thus the staunch republican Richard Carlile looked forward to the day "when every man of property shall consider a printing-press a necessary piece of furniture in his house." In 1816, William Cobbett, a tory-turned-radical journalist, transformed his *Political Register* into a twopenny weekly. (He avoided the newspaper tax by labeling it a pamphlet.) It was hugely successful, and featured selected accounts of, and commentary on, parliamentary hearings. Others followed his lead, printing cheap, widely accessible weeklies that avoided the stamp tax, including the *Black Dwarf* run by Thomas Jonathan Wooler and the *Republican* by Carlile.[9]

Responding to the rise of radical journalism, Robert Southey noted that the press, "like all other powerful engines, is mighty for mischief as well as for good." While he admitted there might be a place for parliamentary reporting, the several papers that were "sold through the manufacturing districts at a halfpenny or penny each" did so via "inflammatory harangues" and "incendiary paragraphs" taken out of context.[10] Indeed, radical papers such as Cobbett's were especially known for their parliamentary reports. Some, such as Thomas Dolby's *Parliamentary Register* (1819), another cheap weekly meant to put the power of judging parliamentary meetings in the hands of those "with limited pecuniary means," also included commentary to help their readers interpret their summaries of parliament.[11]

In parallel with parliamentary proceedings, radical journals also reported extensively on meetings and assemblies of the radical movement itself. In this they were no different than many of the clubs and associations that had proliferated in late Georgian Britain. Religious tract societies and Bible societies, whose raison d'être depended on publicity, sometimes published proceedings of their public meetings. So too did the occasional learned society, but besides occasional public meetings this was rare in the first two decades of the century. Radical groups went out of their way to mimic, and at times satirize, the political establishment, both in the structure of their meetings and in the publicity they gave to them. Thus, alongside parliamentary proceedings, the *Black Dwarf* turned "from the acts of these mighty masters" to provide its readers with the "Proceedings of the People" too (Figure 2.2).

THE BLACK DWARF.

A London Weekly Publication,

EDITED, PRINTED, AND PUBLISHED BY T. J. WOOLER, 5e, SUN STREET, BISHOPSGATE.

Communications (post paid) to No. 4, Catherine-street, Strand.

No. 8, Vol. III.]   *WEDNESDAY, FEBRUARY 24, 1819.*   [Price 4d.

Satire's my weapon; but I'm too discreet, | *I only wear it in a land of Hectors,*
To run a-muck and tilt at all I meet: | *Thieves, supercargoes, sharpers, and directors.*—POPE.

PROCEEDINGS OF THE PEOPLE.

There is a voice that tyrants hear,
And while they hear, they *tremble* too.
In vain would they disguise the *fear*
That sickens in their pallid hue.

No one can contemplate the present aspect of affairs, and not perceive that a CHANGE is inevitable. It is impossible that things should continue *as they are.* Every thing is in commotion. The discordant atoms of our system are all jostling each other for precedency—some to maintain and others to acquire authority. We somewhat resemble the sandy desert, when the wind begins to disturb the repose of the infinity of particles that compose the surface—and it is certain, that the whirlwind will follow the indication, unless some master spirit shall step forth, and allay the symptoms of the storm. But where is this master spirit? Where is the man, who alone could lull the fears, and sooth the anger that are kindling into action. In a dispute between various parties in a state, it is only the sovereign that can speak with effect; and we are unfortunately forbidden to hope, that our sovereign will speak at all. He is the captive of his real enemies, and the inveterate enemies of his people. The monarchy is rivetted in the fetters of the aristocracy, and the Boroughmongers; and it is insulted with the supply of unlimited extravagance, as a compensation for resigning all power into the hands of its keepers. The palace seems but a splendid cage, in which some singular curiosity is kept for occasional exhibition. Instead of seeing the loved sovereign of a brave and free people, in full commerce and connection with the nation, we only hear of the existence of the Regent at the banquet or the board of revelry. We are told that he distrusts the people. He is told that the people entertain no respect for him. He is taught to believe that bayonets are necessary for his protection from those who would love and honor him, if he would condescend to step out of his eastern parade, and convince them that he really deserves their esteem, and is anxious for their welfare.

We are left now entirely to choose for ourselves the road that we prefer. Our leaders have deserted us; and it is truly fortunate that they have done so. They were only clogs upon our march—dead weights to impede our progress, rather than generous stimulators to deeds of honor, and honest advocates of our undoubted rights.

The Whigs see that a change must eventually come; and abandoning all hopes of obtaining power under the present system, the most wily of them are endeavouring to be ready to seize upon its transit, and be again the heroes of another revolution. But they will be disappointed. They have not now the simpletons of 1688 to mould into devotion to their purposes: nor would the temper of the times suffer another Prince of Orange to declare that " he would not be the president of a republic—he would be king, or nothing!"

The political information so recently spread among the great bulk of the people, has produced an effect at which both Whigs and Tories are astonished. They start at finding politicians in every village, and orators in every town, that shame the boasted talents of wealth and education. They see the public meetings, which they hoped to destroy by their absence, conducted with more skill, more temper, more sense, and more effect, than ever distinguished those at which they attended. Men now reason, who were before taught to riot—and the violence of factious contests, recedes before the wisdom of the people. The tens of thousands of Manchester and its vicinity now assemble, they discuss important topics, prepare able statements, and retire quietly to their homes; while a borough election cannot be conducted without tumult, and frequent bloodshed.

It is this reason and temper of the people that alarms their oppressors. They like tumult, because it furnishes them with pretexts for violence in return. If the people, therefore, will not riot, it is the business of the loyal faction to make them riot—if they dare; but the experiment is a dangerous one, and little calculated to succeed.

A Meeting was advertised to be held at Stockport; and the Courier impudently and coolly remarks,—" it was dispersed by " the yeomanry, before the evidently seditious business was " entered upon."

In the Manchester Observer, however, (one of the few honest country newspapers) we have a different version of the affair, which we shall transcribe for the benefit of those who may have been deceived:—

" It had been contemplated by those enlightened patriots, who had undertaken the arrangement of the respective documents that were to be proposed for the adoption or rejection of the Meeting, and to inspire the noble souls of Britons with the remembrance of their ancient constitutional birthright, that the Cap of Liberty, and two handsome banners bearing the inscription " NO CORN LAWS" and " RIGHTS OF MAN" should be hoisted. It is not surprising, that these emblems were beheld by the vassals of corruption, with marks of reprobation. The profligate engines of despotism were in motion for several days previous to the Meeting—and it is confidently asserted, that a regiment of returned transports, under the command of the ALL POTENT Nadin, were in actual array against the People.

In the course of his observations, Mr. Saxton, alluding to the expected attack, " pointed most significantly to the cap of " liberty. This constitutional ensign, said he, a ruffian ban- " ditti are at this moment contemplating to wrest from your " grasp. For his part, should an illegal seizure be attempted, " his mind was made up to perish in its defence."

The Chairman, aware of the uproar that was about to be

Figure 2.2 "Proceedings of the People," an occasional column that mimicked parliamentary proceedings, in the *Black Dwarf*, 24 February 1819. Courtesy of Langson Library, University of California, Irvine.

New press laws in 1819 that targeted the radical press dramatically increased the stamp tax on political journals and made the weekly publications of Cobbett, Carlile, and others unsustainable and illegal. But a cheap press modeled on these publications, aimed at wider readerships and created by printers and publishers trained in the bureaus and printing rooms of radicals, nevertheless emerged in the following decade. The *Hive* in 1822, and soon after it the wildly successful *Mirror of Literature, Amusement and Instruction*, whose physical formats mimicked the radical press, showed the way.[12] By avoiding political discussions, these publications avoided the tax, but their very existence — some issues of the *Mirror* exceeded 80,000 copies, far more than the higher-priced quarterlies and monthlies — provided evidence for the existence of a large reading audience that could be marshalled to argue for educational opportunities for working classes and the extension of political representation. While these papers included some original material written by paid contributors, they specialized in providing readers a digest of the most useful and entertaining cultural information available only in more expensive periodicals.

At the same time, science began increasingly to be enrolled in the radical cause. Richard Carlile, who had edited the *Republican*, reached out to the scientific elite in his 1821 "Address to Men of Science calling upon them to Stand Forward and Vindicate the Truth from the Foul Grasp and Persecution of Superstition . . ." Science was fundamentally opposed to priestly and royal tyranny, he explained, and men of science could give "the death blow" to both. The problem was that they were too much disposed to work "in silence." Through their reticence they "deprived society of many of those benefits which it was their bounden duty to have conferred upon it."[13] Carlile was certain that a truly public science would be a natural ally in the fight against religious belief and monarchical power.

Some of the most successful of the new weekly miscellanies to emerge in the *Mirror*'s wake focused on bringing science, technology, and medicine to the people. The *Mechanics' Magazine* (Figure 2.3) was founded in 1823 by Joseph Clinton Robertson and Thomas Hodgskin. Robertson was a patent agent with radical leanings, and Hodgskin had worked under radical publishers and also as a parliamentary reporter for the *Morning Chronicle*. Like the *Mirror*, the *Mechanics' Magazine* was to "comprehend a digested selection from all the periodical publications of the day," but with a focus on useful science and technology.[14] In 1824 Hodgskin split with Robertson and launched

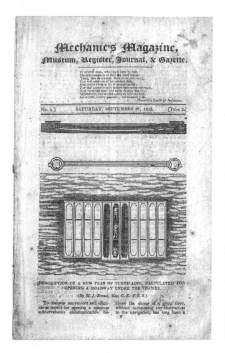 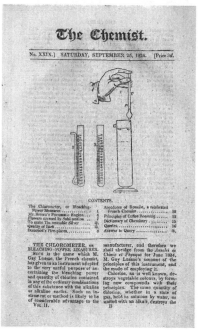

Figure 2.3 *The Mechanics' Magazine*, 27 September 1823 (*left*), and *The Chemist*, 25 September 1824 (*right*). The crowded two-column small octavo format mimicked cheap literary weeklies and brought discussion of philosophical and technical matters within reach of a wider circle of readers than journals such as Tilloch's *Philosophical Magazine*. Author's collection and Harvard University Libraries, respectively.

the *Chemist* (Figure 2.3). Both papers brought not only the cramped two-column format to science and technology, but also the radical suspicion of elite secrecy.

Hodgskin began his new scientific periodical by critiquing what he saw as the scientific aristocracy, who "keep [science] in a manner inaccessible to the profaning touch of the vulgar."

> In fact, there is some reason to believe that Royal Societies of every description partake of the opinions and apprehensions of their patrons, and, like them, are not forward to encourage that species of instruction which tends to make the great mass of mankind the accurate judges of their merits rather than submissive scholars.[15]

This "sort of royal science" set up several barriers to participation. First, societies confined their publications to expensive and difficult-to-obtain transactions, and "a vast pile of scientific rubbish" was held back "in the vaults of our Royal Society."[16] Second, admission to the Royal Society was not a matter purely of merit, for it accepted "title or fortune as a qualification for the fellowship, when knowledge happens to be wanting."[17]

A third radical weekly, Thomas Wakley's *Lancet* (founded October 1823), brought the same spirit of critique to London medicine, promising to bring transcripts of London medical lectures to a wide public via his magazine. Wakley's use of journalists to transcribe and print lectures in his weekly raised a storm of opposition from the Royal College of Surgeons, which tried every means at their disposal—from legal action to turning off the gas lamps in the lecture hall—to compel Wakley to stand down. Wakley took the case to the House of Lords, where he argued that since London surgeons had passed a bylaw giving them a monopoly on medical lecturing, they had "by their own acts constituted themselves a public functionary for a public purpose." They now had no right "to deprive the public of the means of judging whether their duties as hospital surgeon are properly discharged." Eventually, he won.[18]

These journals were part of a marked expansion in print coverage of science and technology,[19] whose existence threatened to alter the market for more established (and expensive) journals of science. Both the *Chemist* and the *Mechanics' Magazine* included a regular column called "Analysis of Scientific Journals," which their editors used to mount regular critiques of the way that "journalists of a higher class" covered science: "Such journals should be considered as the links that connect the learned with the industrious—the strainers and digesters through which the truths of philosophy must pass: to fit them for assimilating with the system of active and busy life."[20]

Hodgskin boasted that at the very least, the success of Robertson and himself in editing the *Mechanics' Magazine* and the *Chemist* had "compelled other editors to set about improvement."[21] There was truth in this. Their journals prompted a host of imitators looking to capture the larger market they had shown to exist. One of these was the *Mechanic's Oracle and Artisan's Laboratory and Workshop* (1824). It mimicked the *Mechanics' Magazine*'s two-column layout and claimed to be "devoted, principally, to the instruction and improvement of the working classes."[22] The *Mechanic's Oracle* was created by Alexander Tilloch, the proprietor of the *Philosophical Magazine*, the longest-running scientific journal in Britain (which had absorbed Nicholson's *Journal* in 1813). As it happened, Tilloch passed away within a year of its founding, but his attempt to compete on the new terrain the *Mechanics' Magazine* had opened up suggests how keenly aware the more established scientific press was of the challenge posed by the new journals targeted at a larger readership.

While the *Oracle* didn't survive the death of its founder, the *Philosophical Magazine* did, and it gradually changed the way it covered elite science. In 1822, Tilloch had brought on as coeditor and coproprietor his printer Richard Taylor, who took full charge of the publication after his death. Although associated with the upper crust of scientific publishing, Taylor's politics shared a great deal with Robertson and Hodgskin. The son of a dissenter with a penchant for composing Jacobin poetry, Taylor had been apprenticed to the Chancery Lane printer Jonas Davis in 1798. He also carried forward many of his father's political convictions. In 1824, he collaborated with the radical reformer Francis Place (who was also a mentor to Hodgskin) in pushing for the legalization of trade unions, and he also helped repeal the Test and Corporation Acts that restricted religious freedom.[23]

But Taylor was also making himself a crucial figure among London's scientific aristocracy. When he took over Davis's printing business at the beginning of the new century, he inherited the Linnean Society as a key client as well as Tilloch's *Philosophical Magazine*.[24] Over the next decades, Taylor began to corner the market in the printing of elite science in London. He joined the Linnean Society early on and even became its undersecretary in 1810. In 1822, both the Geological Society and the Astronomical Society of London hired Taylor to print their transactions. By 1828, the Royal Society and the Zoological Society were also employing him as their printer.

Though both the Astronomical and Geological Societies were among those "Royal Societies" that Hodgskin and Robertson disdained, both had differentiated themselves from Banks's Royal Society by making appeals to other constituencies and mores. The Geological Society became known for allowing discussion and debate at its meetings (a practice unheard of at the Royal Society itself), while the Astronomical Society's leaders were exactly those reform-minded natural philosophers pushing the Royal Society to abandon the aristocratic legacy of Joseph Banks.[25] Positioned between these societies and the commercial scientific press, Taylor offered them a means of broadening their public face. In 1823, the Astronomical Society resolved to allow Taylor to copy and publish the minutes of the Society in the *Philosophical Magazine*. The next year, the Society appointed the prolific author and mathematician Olinthus Gregory as its secretary, and the latter worked tirelessly to produce detailed and readable summaries of its memoirs for Taylor's *Magazine*, even correcting the proofs.[26] Not long after, the Geological Society entered into a similar arrangement with Taylor, so that by 1825 the *Philosophical Magazine* contained reliable, regular accounts of the memoirs that had been read at these two societies.

The *Philosophical Magazine* and other scientific journals had for very long included accounts of the proceedings of scientific societies, but these were generally irregular and short. The "Proceedings of Learned Societies" section of such journals (when it existed at all) tended to be a mélange of intelligence that happened to be on hand, often consisting of annual public meetings of a wide variety of societies and academies, lists of papers presented (sometimes with summary) at the Royal Society, translations of foreign accounts (in particular the brief *comptes rendus* of the Academy of Sciences of Paris that appeared in the *Annales de chimie* beginning in 1816), and other miscellaneous news. Meetings of societies do not seem to have been considered matter for regular public consumption. The Royal Society tended to be particularly circumspect about such public reports; the editor of the *Annals of Philosophy* complained that it forbade note-taking during meetings, so that he had to write his proceedings from memory (a complaint that mirrored those of parliamentary reporters). But in the mid-1820s those meetings also began to receive more extended publicity, if only because William T. Brande, one of its secretaries, was the editor of the *Journal of Science and the Arts*.[27]

Taylor's close relationship with the Astronomical and Geological Societies made the *Philosophical Magazine* something like a de facto bulletin of these societies' activities, and when Taylor began a new series in 1827 (having bought the *Annals of Philosophy*), he offered to print these proceedings as separate copies for the use of both societies.[28] This was a cheap proposition for the societies because, as with other separate copies, it only cost paper and presswork, since the standing type could be used from the *Magazine*.[29] But what began as separate copies very much resembled something like the monthly issue of a journal, and eventually both societies treated them as such, and took public responsibility for their contents. Soon they were not simply distributed to members but sent abroad to other societies and academies, and eventu-

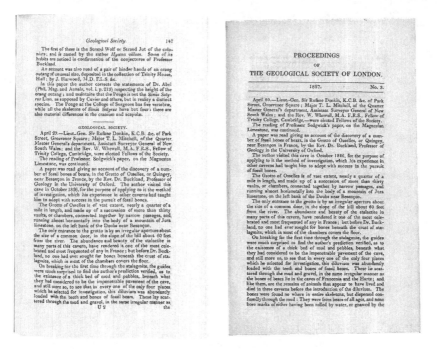

Figure 2.4 Excerpt from the proceedings of the Geological Society of London meeting of 20 April 1827 as it appeared in the *Philosophical Magazine* 2 (August 1827): 147 (*left*); and as it appeared in the third issue of the *Proceedings of the Geological Society* (*right*). Several early proceedings publications of London's learned societies were produced by reusing the type that had been set up for them in Richard Taylor's journals. Courtesy of the Biodiversity Heritage Library and Harvard University Libraries, respectively.

ally sold through booksellers. Thus, whether deliberately or not, both societies found themselves in the position of being in control of cheap monthly publications that publicized their meetings and affairs (Figure 2.4).

The societies' leaders were quick to congratulate themselves for having opened their meetings to discussion and debate outside their rooms:

> The public is hereby brought more immediately into contact with the Society—the labours of its contributors are canvassed and discussed, while the interest of the author in his subject is yet warm, and when the interchange of ideas respecting it is most beneficial, not only to the public, but to the author himself, whose views may, and probably in many instances will, be enlarged or corrected by such intercourse.[30]

Of course, this development owed as much to the growing influence of commercial journals in philosophy as it did to the benevolence of these societies. It was Richard Taylor's distinctive role as both publisher of a journal and printer to the societies that paved the way for emulating this journalistic genre. Authors, it turned out, were increasingly willing to forego whatever honor might come with publishing in the transactions of a society to publish more quickly and easily in the independent press. As the Geological Society's president noted with regret in 1830, the growing presence of independent journals had provided venues in which one could publish more efficiently, leaving the Society unable to meet "the wishes of those authors especially who have most original matter to communicate."[31]

Taylor continued to enlarge his influence. In early 1828, he successfully bid to become the Royal Society's new printer. Within weeks of the switch, the Society made arrangements to take control of the publication of abstracts of papers read at its meetings.[32] It formally prohibited the copying of its minutes and entered into an arrangement by which Taylor printed them in the *Philosophical Magazine*. When the Zoological Society of London began publishing in 1830, it also hired Taylor and began by publishing its *Proceedings* as separate copies of the proceedings that Taylor began to include in the *Philosophical Magazine*.[33]

The Politics of Representation: The Liberal Press and the Academy

In France, as in Britain, the publicity given to governmental bodies was a centerpiece of postrevolutionary political culture. In part because of the more perilous legal status of the practice of print publicity in the ancien régime, the concept of public opinion there had taken on particular significance in the decades prior to 1789. If at first an "abstract category, invoked by actors in a new kind of politics to secure the legitimacy of claims," the emergence of a robust daily press after the Revolution provided a potential embodiment of the abstraction.[34] Many new associations, some scientific, had taken this view to heart. Many of the *sociétés libres* that sprang up during this period, such as the Société philomathique and the Société d'histoire naturelle, were quick to demonstrate their democratic potential by publishing journals (the *Bulletin des sciences* and the *Actes de la Société d'histoire naturelle*, respectively).[35] Still, only rarely were these publications structured as proceedings of meetings.[36]

The Academy of Sciences had also been pressured, after its restoration as the First Class of the Institut de France in 1795, to carve out space in the quickly changing landscape of political life. As the previous chapter explored, in the absence of a clear mandate from the state, any claim to authority as a public body required appealing to new conceptions of public legitimacy. Academicians had actively debated this problem from the Napoleonic period onward, and the question became especially poignant in the decade following Napoleon's stifling patronage.[37] One result had been that the Academy gradually relaxed the stipulation that academic meetings were for members only. As one young savant reported in 1820:

> Strictly speaking, these meetings should be secret. But the Academy has seen fit to depart from the letter of the law on this point by admitting a number of young savants. It has felt that this favor, granted with due care, would be the best means of inspiring the latter toward glorious rivalry [*émulation glorieuse*].[38]

But some of these young savants did not attend meetings simply for the sake of inspiration; they also reported to others what they saw and heard. Short accounts of the proceedings of the Academy had begun to appear by the early years of the Bourbon Restoration. The *Annales de chimie* initiated such coverage in 1816, and the *Revue encyclopédique* followed up in 1820. But these accounts were extremely schematic, often consisting only of paper titles. Besides papers' contents, meetings of the French academies and societies tended to include a great deal of discussion, but tacit understandings regarding audience decorum meant that little of what was said went beyond the room. In 1824, however, things changed. The challenge came not from an upstart scientific publication at all, but instead from a new literary journal founded by a group of young Parisian radicals. It was called *Le Globe*.

After 1820, the return to power of an ultra-royalist ministry forced several liberals out of government and prompted a purge of troublesome professors in the University system (including the historian François Guizot and the philosopher Victor Cousin). Several of these liberal elites turned to journalism. At the same time, many young radicals had become involved in a secret society known as the Carbonari, though by 1823 the state had successfully exposed and broken up the group. Two of them, Paul Dubois and Pierre Leroux, founded *Le Globe* in 1824. A former student of the École normale, Dubois had taught languages and literature at several colleges before devoting himself to political agitation after being dismissed from his post in 1821. Though admitted to the École polytechnique in his youth, Leroux had been forced by family tragedy to forego his studies, and he later settled on a career as a typographer. Leroux came to believe deeply in the disruptive political power of a truly free press. Like the radical publisher Richard Carlile in England, he imagined a future in which presses would become domestic technologies, and in 1822 even imagined a prototype, the *pianotype*, that he hoped would hasten such a future.[39]

Le Globe developed into an intellectual hub of liberal opposition during the Restoration, though it was not explicitly engaged in politics at all. (The state, which regulated political journals more strictly than others, would not have allowed such a journal run by former Carbonari.)[40] While its size and two-column layout mirrored the political dailies, its editors' aim was to reform French society by starting with a reform of its intellectual life: "Paris, to judge by its journals, is nothing but one grand salon from 1778, where we fret over an actor or actress, over a bon mot or over some inanity." Dubois and Leroux looked to England in particular for inspiration, where "marvels of industry" had come about through the "constant study of all nations" and through useful instruction. They would replace the insufferable "comptes rendus of the theaters" with just this kind of instruction, whether it was "literary, industrial, or moral."[41]

Leroux in particular envisioned *Le Globe* as a cosmopolitan collection that would "keep its readers informed about all the discoveries made in science and in all branches of human activity."[42] To this end, he brought on board his close friend Alexandre Bertrand, another ex-Carbonari who had also attended the same lycée in Rennes. Bertrand had once been enrolled in the École polytechnique, but quit after the Hundred Days, later studying medicine. After his thesis in 1819, he attempted to support himself through professional authorship, writing on scientific topics—including somnambulism, animal magnetism, and ecstasy—that he thought were of public importance and yet shunned by mainstream savants.

In October 1824, Bertrand set the tone for his science journalism with an opening series on human fossils, a topic of recent public excitement after several reputed specimens had arrived in Paris. Penning a sober account that focused on the scientific background necessary to evaluate the recent claims, he distinguished his journalism from speculators and charlatans looking to exploit science to create publicity for themselves: "It is by journals alone," Bertrand insisted, "that science can become popular."[43] In the first months most of the articles on science and technology in *Le Globe* focused on topics that connected science to matters of public utility, including glowing tributes to the Scottish hero of invention, James Watt, and the French naval engineer and champion of industrial development, Charles Dupin.

It was the possibility of bringing science to politics that led the way to the new style of scientific reporting that Bertrand soon came to focus on. At the end of November, a memoir presented to the Academy of Sciences by the doctor-turned-social-statistician Louis-René Villermé caught his eye. Titled "On Mortality of the Leisure Class Compared with that of the Impoverished in France," the memoir was a pioneering use of mathematical statistics to illuminate a major question of public health. It demonstrated that mortality rates were far more closely correlated to a social indicator (class) than to other physical or geographical factors, such as proximity to rivers, climate conditions, or water purity.[44] Bertrand's excitement about the memoir prompted him to devote two separate issues to a detailed summary of the work and its conclusions. He also drew out the consequences of Villermé's conclusions for the politics of social organization: "concentrating property in the hands of the few, and augmenting the misery of the people, is not simply to deprive inferior classes of joys and pleasures; it is to attack their life; and any law, any institution that leads to this result is not simply unjust: it is cruel, it is murder."[45]

The report on Villermé's memoir was not Bertrand's first report on a meeting of the Academy of Sciences, but it was the first that broke the mold of the short, schematic accounts to be found in the *Annales de chimie* and other learned journals. And it began a trend. Bertrand continued producing more extended reports on meetings. And while he continued to emphasize topics related to his own interests (especially debates about controversial phenomena such as animal magnetism) or which he took to be of particular importance for his readership, Bertrand came to see his role as providing readers a reliable overview of academic meetings as a whole. This turned out to be a political act in itself; as Leroux put it, Bertrand was "knocking down the barriers" that kept the secretive Academy "at a great distance from the public."[46] While the procès-verbaux in the specialized journals tended to be a few hundred words, *Le Globe*'s accounts normally stretched out over a thousand words each, sometimes providing details of discussions raised by memoirs and reports.[47] The shape of the genre developed in other ways as well. Unlike in the monthly or quarterly journals, Bertrand's accounts of meetings could be relied on to appear within days (usually Thursday) of each Monday meeting. Bertrand deliberately shaped the material to make it suitable for a nonspecialized audience, and in doing so he sometimes provided commentary, including critique (the raison d'être of *Le Globe*) of the Academy's judgments.

By spring 1825, such was its perceived importance that Bertrand's report was often the lead article in Thursday issues of *Le Globe*. There was a near-perfect match between the first page of *Le Globe* and the political dailies, with the procès-verbaux of the Academy taking the place where the procès-verbaux of the Chamber of Deputies were often found. (See Figure 2.5.) The accounts grew so long that the account of a single meeting was sometimes split up over multiple issues.

At just this time the Academy decided to revisit rules of attendance at its meetings. Rather than attempting to lock down the meeting to everyone but members, by then an unrealistic goal, Georges Cuvier considered ways to restrict attendance to those who were there "for the love of the sciences" rather than those who attended in order to foment disorder and threaten the "liberty of discussion." Leroux later recalled that this had been a thinly veiled attempt to keep Bertrand—and his pen—out. Bertrand, however, had allies of his own at the Academy. Joseph Fourier, who had been elected the other permanent secretary in 1824, was a supporter of his (he even employed Bertrand as a ghostwriter), and likely opposed barring him from the Academy.[48] Though Cuvier prevailed in the vote, a technicality kept the new rules from being implemented.[49]

Indeed, academicians disagreed among themselves about what to do about these new developments in Parisian journalism. André-Marie Ampère, whose physical theories were championed as revolutionary in *Le Globe*, often praised Bertrand's reports, calling them "an excellent summary of each meeting."[50] But not everyone shared this view. At the meetings of the Royal Academy of Medicine, Bertrand had also begun writing reports and was more willing to venture direct criticisms, perhaps because the Academy so often dealt with matters directly related to public welfare. In September 1825, during a major smallpox outbreak in Paris, Bertrand published an extensive account of an acrimonious debate regarding an academic report on the safety of vaccinations.[51] (At issue was how open the reporters ought to be about the potential dangers of the vaccine.) In response, the report's author, Jacques Louis Moreau, argued vigorously—as had Cuvier—that nonmembers ought

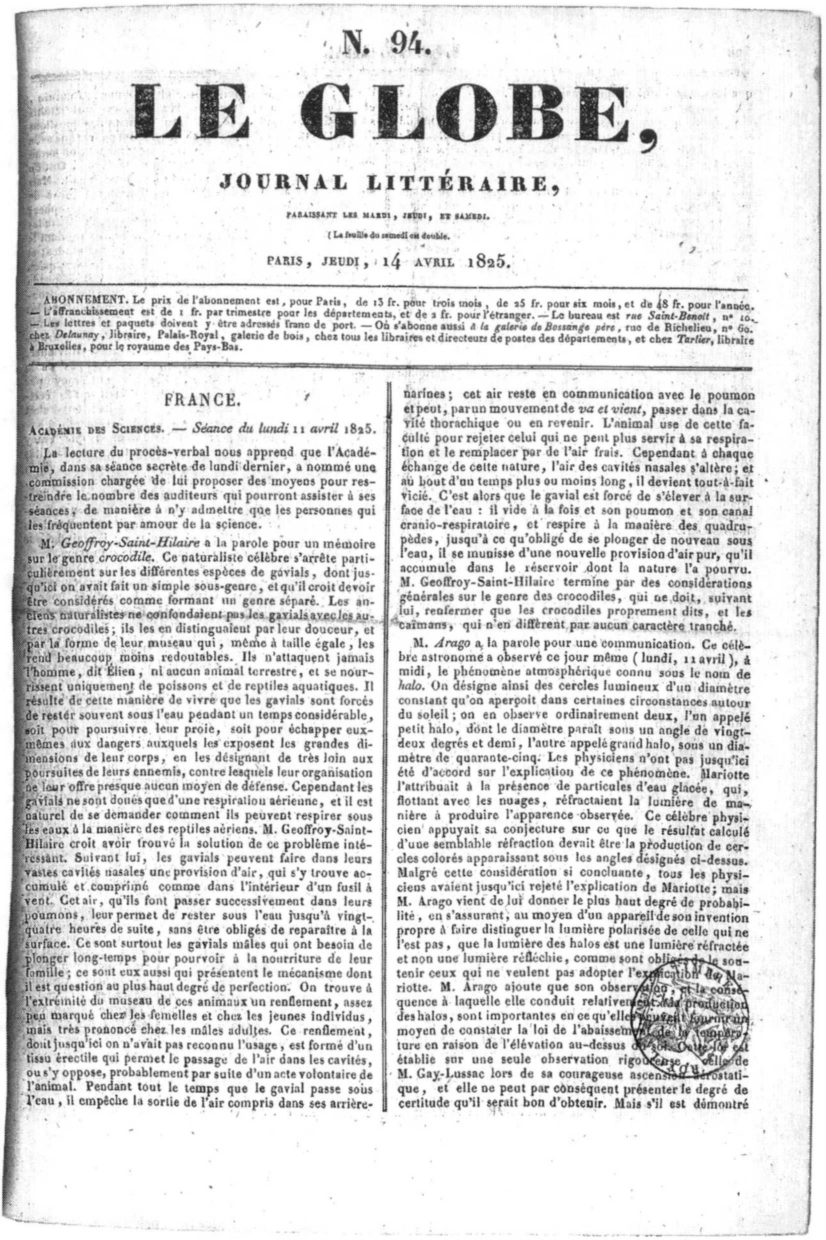

N. 94.

LE GLOBE,

JOURNAL LITTÉRAIRE,

PARAISSANT LES MARDI, JEUDI, ET SAMEDI.

(La feuille du samedi est double.)

PARIS, JEUDI, 14 AVRIL 1825.

ABONNEMENT. Le prix de l'abonnement est, pour Paris, de 13 fr. pour trois mois, de 25 fr. pour six mois, et de 48 fr. pour l'année. — L'affranchissement est de 1 fr. par trimestre pour les départemens, et de 2 fr. pour l'étranger. — Le bureau est rue Saint-Benoît, n° 10. — Les lettres et paquets doivent y être adressés franc de port. — On s'abonne aussi à la galerie de Bossange père, rue de Richelieu, n° 60, chez Delaunay, libraire, Palais-Royal, galerie de bois, chez tous les libraires et directeurs de postes des départemens, et chez Tarlier, libraire à Bruxelles, pour le royaume des Pays-Bas.

FRANCE.

ACADÉMIE DES SCIENCES. — Séance du lundi 11 avril 1825.

La lecture du procès-verbal nous apprend que l'Académie, dans sa séance secrète de lundi dernier, a nommé une commission chargée de lui proposer des moyens pour restreindre le nombre des auditeurs qui pourront assister à ses séances, de manière à n'y admettre que les personnes qui les fréquentent par amour de la science.

M. Geoffroy-Saint-Hilaire a la parole pour un mémoire sur le genre crocodile. Ce naturaliste célèbre s'arrête particulièrement sur les différentes espèces de gavials, dont jusqu'ici on n'avait fait qu'un simple sous-genre, et qu'il croit devoir être considérés comme formant un genre séparé. Les anciens naturalistes ne confondaient pas les gavials avec les autres crocodiles; ils les en distinguaient par leur douceur, et par la forme de leur museau qui, même à taille égale, les rend beaucoup moins redoutables. Ils n'attaquent jamais l'homme, dit Élien, ni aucun animal terrestre, et se nourrissent uniquement de poissons et de reptiles aquatiques. Il résulte de cette manière de vivre que les gavials sont forcés de rester souvent sous l'eau pendant un temps considérable, soit pour poursuivre leur proie, soit pour échapper eux-mêmes aux dangers auxquels les exposent la grande dimension de leur corps, en les désignant de très loin aux poursuites de leurs ennemis, contre lesquels leur organisation ne leur offre presque aucun moyen de défense. Cependant les gavials ne sont doués que d'une respiration aérienne, et il est naturel de se demander comment ils peuvent respirer sous les eaux à la manière des reptiles aériens. M. Geoffroy-Saint-Hilaire croit avoir trouvé la solution de ce problème intéressant. Suivant lui, les gavials peuvent faire dans leurs vastes cavités nasales une provision d'air, qui s'y trouve accumulé et comprimé comme dans l'intérieur d'un fusil à vent. Cet air, qu'ils font passer successivement dans leurs poumons, leur permet de rester sous l'eau jusqu'à vingt-quatre heures de suite, sans être obligés de reparaître à la surface. Ce sont surtout les gavials mâles qui ont besoin de plonger long-temps pour pourvoir à la nourriture de leur famille; ce sont eux aussi qui présentent le mécanisme dont il est question au plus haut degré de perfection. On trouve à l'extrémité du museau de ces animaux un renflement, assez peu marqué chez les femelles et chez les jeunes individus, mais très prononcé chez les mâles adultes. Ce renflement, doit jusqu'ici on n'avait pas reconnu l'usage, est formé d'un tissu érectile qui permet le passage de l'air dans les cavités, ou s'y oppose, probablement par suite d'un acte volontaire de l'animal. Pendant tout le temps que le gavial passe sous l'eau, il empêche la sortie de l'air compris dans ses arrière-

narines; cet air reste en communication avec le poumon et peut, par un mouvement de va et vient, passer dans la cavité thorachique ou en revenir. L'animal use de cette faculté pour rejeter celui qui ne peut plus servir à sa respiration et le remplacer par de l'air frais. Cependant à chaque échange de cette nature, l'air des cavités nasales s'altère; et au bout d'un temps plus ou moins long, il devient tout-à-fait vicié. C'est alors que le gavial est forcé de s'élever à la surface de l'eau : il vide à la fois et son poumon et son canal cranio-respiratoire, et respire à la manière des quadrupèdes, jusqu'à ce qu'obligé de se plonger de nouveau sous l'eau, il se munisse d'une nouvelle provision d'air pur, qu'il accumule dans le réservoir dont la nature l'a pourvu. M. Geoffroy-Saint-Hilaire termine par des considérations générales sur le genre des crocodiles, qui ne doit, suivant lui, renfermer que les crocodiles proprement dits, et les caïmans, qui n'en différent par aucun caractère tranché.

M. Arago a la parole pour une communication. Ce célèbre astronome a observé ce jour même (lundi, 11 avril), à midi, le phénomène atmosphérique connu sous le nom de halo. On désigne ainsi des cercles lumineux d'un diamètre constant qu'on aperçoit dans certaines circonstances autour du soleil; on en observe ordinairement deux, l'un appelé petit halo, dont le diamètre paraît sous un angle de vingt-deux degrés et demi, l'autre appelé grand halo, sous un diamètre de quarante-cinq. Les physiciens n'ont pas jusqu'ici été d'accord sur l'explication de ce phénomène. Mariotte l'attribuait à la présence de particules d'eau glacée, qui, flottant avec les nuages, réfractaient la lumière de manière à produire l'apparence observée. Ce célèbre physicien appuyait sa conjecture sur ce que le résultat calculé d'une semblable réfraction devait être la production de cercles colorés apparaissant sous les angles désignés ci-dessus. Malgré cette considération si concluante, tous les physiciens avaient jusqu'ici rejeté l'explication de Mariotte; mais M. Arago vient de lui donner le plus haut degré de probabilité, en s'assurant, au moyen d'un appareil de son invention propre à faire distinguer la lumière polarisée de celle qui ne l'est pas, que la lumière des halos est une lumière réfractée et non une lumière réfléchie, comme sont obligés de le soutenir ceux qui ne veulent pas adopter l'explication de Mariotte. M. Arago ajoute que son observation, et la conséquence à laquelle elle conduit relativement à la production des halos, sont importantes en ce qu'elles peuvent fournir un moyen de constater la loi de l'abaissement de la température en raison de l'élévation au-dessus du sol. Cette loi est établie sur une seule observation rigoureuse, celle de M. Gay-Lussac lors de sa courageuse ascension aérostatique, et elle ne peut par conséquent présenter le degré de certitude qu'il serait bon d'obtenir. Mais s'il est démontré

Figure 2.5 Accounts of the Academy of Sciences were sometimes the lead article in *Le Globe*, just as the proceedings of legislative bodies were often the lead article in political dailies. *Le Globe*, 14 April 1825; *Le Constitutionnel*, 26 April 1825. Courtesy of the BnF and Harvard University Libraries, respectively.

MARDI 26 AVRIL 1825. (116.)

On s'abonne rue Montmartre, nº. 121.

LE CONSTITUTIONNEL,

Prix : 18 fr. pour 3 mois, 36 fr. pour 6 mois, et 72 fr. p. l'année.

JOURNAL DU COMMERCE, POLITIQUE ET LITTÉRAIRE.

On reçoit les réclamations des personnes qui ont des griefs à exposer, et les avis qui peuvent intéresser le public. On s'abonne au bureau du journal, rue Montmartre, nº. 121, presqu'en face du MARCHÉ SAINT-JOSEPH, où l'on reçoit les insertions pour les ANNONCES GÉNÉRALES et la FEUILLE DE COMMERCE. Prix de l'abonnement : 18 fr. pour 3 mois, 36 fr. pour 6 mois, et 72 fr. pour l'année : (FEUILLE DE COMMERCE, 6 fr. pour 3 mois, 12 fr. pour 6 mois, et 24 fr. pour l'année.) — Les lettres, paquets et argent doivent être adressés, francs de port, au Directeur du CONSTITUTIONNEL, rue Montmartre, nº. 121.

CHAMBRE DES DEPUTÉS.

Séance du 25 avril.

L'ordre du jour est la discussion générale du projet de loi relatif au réglement définitif du budget de 1823.

Messieurs, dit *M. de la Bourdonnaye*, ce n'est point l'examen de la partie matérielle des comptes pour l'exercice de 1823 que je viens présenter à la chambre. Cet examen n'est pas en mon pouvoir, il n'est en celui de personne; votre commission l'a tenté vainement.

Dans l'état actuel de législation de notre comptabilité, toute recherche sur la situation vraie de nos finances est devenue impossible : la question de criminalité qui s'agite en ce moment devant la cour royale de Paris, enlèvera à cette discussion tout ce qu'une recherche de manœuvres dégoûtantes pour s'emparer des fournitures de l'armée pourrait avoir d'odieux dans cette enceinte. Strictement renfermé dans le cercle des questions de principes, je ne m'occuperai des marchés Ouvrard que sous le rapport de la responsabilité ministérielle.

Avant de puiser dans le rapport de la commission d'enquête les faits sur lesquels j'ai besoin de m'appuyer dans cette discussion, qu'il me soit permis, tout en rendant le plus juste hommage à la sagesse de son travail et à la noble indépendance qu'elle a montrée, de vous présenter quelques observations sur cette commission elle-même, et sur la nature de ses attributions, si peu en harmonie avec la dignité des deux chambres dans lesquelles les membres en ont été choisis avec une affectation remarquable; affectation qui semblerait indiquer que c'est bien moins dans la vue de découvrir des coupables que l'indignation publique n'a pas cessé de signaler, que dans l'espoir de laisser calmer cette même opinion, et surtout de prévenir la demande d'une véritable commission d'enquête; commission qui, créée dans cette chambre, nommée par elle, investie de tous ses pouvoirs, eût été assez puissante pour remonter à la source du mal, et aller saisir des coupables que l'on ne craint peut-être de mettre en cause que parce qu'ils en savent assez pour intimider leurs accusateurs. (*Vive rumeur au centre.*)

L'orateur s'appuyant sur des faits contenus dans le rapport de la commission d'enquête, et rappelant surtout ceux relatifs à la mission de M. Joinville, soutient que le conseil des ministres a commis une usurpation de pouvoir.

En ne livrant pas M. Joinville aux tribunaux militaires, dit-il, en lui continuant les marques de la confiance du gouvernement, le conseil des ministres a accepté la responsabilité des actes du commissaire extraordinaire, il a avoué sa mission, il a reconnu tacitement qu'il l'avait dignement remplie; il a fait plus, il l'a protégé contre l'indignation d'un chef qu'il avait outragé, trompé, trahi dans sa confiance; il l'a présenté à l'armée comme un modèle d'insubordination qu'il fallait imiter, comme la preuve vivante du triomphe du nouveau système d'administration qu'on veut faire prévaloir; système qui ne tend à rien moins qu'à enlever les fonctionnaires publics à l'autorité directe des ministres dans les départemens desquels la volonté royale les a placés, pour les mettre, à l'insu du monarque lui-même, sous les ordres d'un nouveau pouvoir érigé dans le sein du conseil des ministres, et privativement exercé, au moyen d'instructions verbales et secrètes, par le président de ce conseil; système qui, s'il prévalait, transformerait le gouvernement du Roi en une oligarchie ministérielle, et substituerait à nos formes constitutionnelles, où tout est public, où tout est écrit, les formes secrètes et intérieures du conseil des dix. (Mouvement dans l'assemblée.)

Sans doute, Messieurs, dit l'orateur en terminant, votre commission a rempli une partie des devoirs qui lui étaient imposés; mais si, au lieu de porter uniquement son investigation sur cette partie du mal, votre commission eût jeté plus loin ses regards, et cherché dans des fautes antérieures les causes des énormes dépenses produites par la guerre d'Espagne, elle les aurait trouvées dans le refus du conseil des ministres d'allouer au département de la guerre les supplémens de fonds qu'il réclamait pour se mettre en mesure de faire les préparatifs d'une campagne inévitable; elle les eût trouvées dans la précipitation avec laquelle il a fallu les faire dans l'intervalle des soixante-huit jours qui se sont écoulés entre le discours du trône qui annonçait la guerre, et le 5 avril, où l'armée s'ébranla pour entrer dans la péninsule; elle les aurait trouvées surtout dans l'obstination du conseil des ministres, dans son imprévoyance, dans sa résistance coupable au mouvement général qui nous entraînait à une guerre de restauration, à une guerre qu'il a faite malgré lui, sans plan, sans système politique, et sans avoir prévu qu'il serait contraint de la faire.

C'était forte de la réunion de tous ces faits, et armée de toutes les conséquences qui en découlent si nécessairement, que votre commission des comptes devait, ce me semble, se présenter devant vous, en invoquant le principe de cette responsabilité ministérielle sans laquelle il n'existe aucune garantie pour les peuples, et aucune inviolabilité morale pour les rois. C'était en vertu de cette responsabilité qu'elle devait vous demander la nomination d'une véritable commission d'enquête, d'une commission choisie dans votre sein, investie de tous vos pouvoirs, assez courageuse, assez éclairée pour remonter jusqu'aux causes premières de ces dilapidations que votre rapporteur vous a signalées, et que vous ne pouvez pas admettre dans ses comptes avant d'en avoir fait punir les auteurs. Ce que pouvait faire une commission investie de votre confiance, un député pourrait-il le tenter? (Vive sensation.) je ne le pense pas, Messieurs.

Quelqu'attendue que soit cette mesure par l'opinion du dehors, peut-être aurait-on le droit de m'accuser d'imprudence, si je la proposais avant qu'elle ait acquis dans cette enceinte un degré de maturité suffisant pour y être discutée avec calme, et surtout avec la conviction qu'il est temps de mettre un terme à cet esprit de malveillance qui se plaît à déplacer toutes les questions, toutes les responsabilités; qu'il est temps de redresser cette fausse direction donnée d'abord à l'esprit public dans l'intérêt de quelques hommes, et dont un parti toujours habile a su ensuite s'emparer pour tout pousser à l'extrême, pour tout envenimer au détriment de la France royaliste, et surtout de la monarchie.

Puisse l'époque de cette discussion n'être pas éloignée, Messieurs ! quelque divergence qu'il y ait encore dans nos votes, les esprits se rapprochent, la nécessité s'en fait sentir à tous les cœurs droits, à toutes les âmes élevées, à tous les amis éclairés de la légitimité; d'accord sur la cause du mal, nous le serons bientôt sur le remède. (Mouvement au centre.)

C'est pour ne pas mettre d'obstacle à l'accomplissement de cette mesure, qu'il convient d'ajourner l'adoption des comptes du ministère de la guerre, et d'en renvoyer l'examen à l'époque où la liquidation des dépenses relatives aux marchés Ouvrard sera terminée. Agir autrement, serait nous exposer, tout au moins, à une accusation de légèreté et de précipitation qui paraîtrait d'autant mieux fondée, que le rapport de votre commission a fait ressortir avec plus d'évidence *qu'une dissipation extraordinaire de deniers publics a eu lieu*, et qu'il est impossible, dans l'intérêt de la morale publique et de la garantie du trésor, que vous en sanctionniez les résultats avant d'en connaître la quotité, avant de savoir sur quel ministre doit peser la responsabilité des fautes qui ont donné lieu à cette dilapidation.

Je demande l'ajournement des comptes du ministère de la guerre, et, dans le cas où ils seraient admis, je vote le rejet de la loi.

M. le ministre des finances demande la parole. Messieurs, dit le ministre, j'ai recueilli, autant que j'ai pu les saisir, les observations qui vous ont été soumises par le préopinant, et j'y répondrai successivement.

Le ministre soutient d'abord qu'à aucune époque, dans aucun pays, les garanties de comptabilité n'ont existé aussi complètement que dans cette circonstance, et que les chiffres et les résultats ne sauraient être contestés. Il affirme qu'il n'existe aucun déficit, et il s'attache à prouver que, s'il existait, on ne pourrait le couvrir ni l'enlever à l'investigation de la chambre au moyen des bons royaux. Ces bons royaux ont été limités à la somme de 140 millions, et il n'y en a eu que 45 ou 46 millions mis en circulation.

Si l'orateur, auquel je réponds, continue M. de Villèle, a parlé de ce qu'il appelle l'oligarchie ministérielle (ce qu'il aurait dû nommer une monarchie ministérielle) envahissant tout, jusqu'au pouvoir royal, et s'interposant entre les ministres responsables et leurs agens, pour éluder la responsabilité ministérielle, Messieurs, la responsabilité ne peut être éludée; chacun des actes ministériels porte une signature qui constitue la responsabilité du ministre qui l'a signé. Quant à l'embarras de savoir sur quel ministre pèserait la responsabilité, il y a unanimité dans le ministère pour la réclamer et surtout pour ne pas la repousser. Ainsi je remercie l'orateur qui a bien voulu m'accuser, de faire peser plus particulièrement sur moi cette responsabilité. (On rit à gauche.)

Mais pour établir cette accusation, a-t-on prouvé qu'il était possible d'éviter les marchés Ouvrard, ou bien que le ministre ait facilité ou maintenu ces marchés? C'était là ce qu'il fallait prouver; il fallait démontrer comment le ministre des finances peut être responsable de la désobéissance d'un agent du ministre de la guerre. On dit : Un intendant militaire est envoyé à l'armée par le ministre de la guerre; il reçoit de ce ministre l'ordre de revenir, et se soustrait à cet ordre; il y a donc lieu de faire peser la responsabilité de cette désobéissance sur le ministre qui en a été la cause, car on réduit toute cette grande question à des termes si minimes!

D'abord l'orateur n'a pas accusé juste. M. Joinville n'a pas été envoyé à l'armée comme intendant militaire, mais comme commissaire extraordinaire, et sous ce rapport, je ne vois pas comment on attribuerait au ministre de la guerre plutôt qu'au président du conseil les instructions données à M. Joinville; c'est ici une simple observation que je fais; car, je le répète, nous n'écartons pas la responsabilité, nous l'appelons au contraire sur nous; mais si nous voulions éluder cette responsabilité, le système même du préopinant nous en fournirait le moyen, car le ministre de la guerre pourrait dire : Je ne réponds de rien, M. Joinville n'est pas à moi; d'un autre côté, le ministre dans le département duquel n'est pas M. Joinville, répondrait : Je ne connais pas M. Joinville; ce n'est pas mon homme. Vous voyez donc, que sous ce rapport, la question serait tout-à-fait insoluble. Dans notre système, au contraire, dans celui que nous croyons le véritable, la responsabilité pèse sur celui qu'elle doit atteindre.

L'orateur accuse le président du conseil. Qu'il produise les actes qu'il croit répréhensibles. Le président du conseil s'est-il opposé à la résiliation des marchés Ouvrard, ou a-t-il fait des efforts pour les obtenir? Le président du conseil, lorsque la résiliation des marchés fut devenue impossible, de l'aveu même de celui qui était envoyé pour les annuler, a-t-il manqué à un second devoir, qui consistait à chercher à améliorer

to be barred from meetings, at least during discussions.[52] Bertrand used the occasion to insist that his medical and scientific journalism was defined by accurate publicity. Rather than meting out praise or blame — "anyway, there is only one judge, the public" — he played an essential role in keeping the public informed about matters of public interest. The *Gazette de santé* picked up the story and defended Bertrand, insisting that "the séances of the Academy are *public*," and there could be no indiscretion in publishing what a large audience had heard in person. Several of Moreau's own academic colleagues also denied that Bertrand's journalism was harming the Academy. One of them suggested that Moreau give up on his obsession with the journals, and instead "you might, from now on, concern yourself simply with science."[53]

Several other journals had picked up the vaccination controversy, often using Bertrand's own coverage of the Academy as their source. In this case, Bertrand was happy to take credit for spreading the news, not only to readers, but to other journals and medical papers.[54] Soon, however, Bertrand complained about extensive piracy of his academic reports by other journalists. Bertrand testified that the production of these reports was painstaking work; to copy them in other publications amounted to "a form of literary felony." He singled out the *Journal des débats* and *L'Étoile* as gross offenders, noting that in contrast the directors of *L'Étoile* cried foul anytime other papers poached other kinds of news from them.[55]

It was no accident that the new style of scientific reporting developed first in *Le Globe*. Not only was Bertrand surrounded by colleagues whose literary and cultural criticism was becoming the talk of Europe, but the political principles of *Le Globe*'s most renowned contributors made the status of elite learned bodies such as the Academy — and Bertrand's new style of scientific journalism — particularly pertinent. Although Dubois and Leroux were more radical, *Le Globe* gathered its contributors from across the spectrum of the liberal opposition. One influential group were the Doctrinaires, which included the historian François Guizot, the political theorist Charles de Rémusat, and the Cousinian philosopher Théodore Jouffroy. They coalesced in the early Bourbon Restoration around a set of problems stemming from the legacy of the Revolution. Their central problem: how to establish a stable French state without endangering the civil liberties that had been introduced by the Revolution. The specific theory of representative government advocated by the Doctrinaires, based on what they called the "sovereignty of reason," was designed to accomplish this. (*Le Globe* called it "the theory of the century on that eternal question of sovereignty.")[56] According to Guizot, governing bodies such as the Chamber of Deputies were not representative in the sense that its members represented the people in a straightforward way (as republican theories of representation generally had it[57]). Rather, representation was the dynamic process by which governing bodies were put into constant communication with the broader public in such a way as to guarantee the former's legitimacy as a reasoning body.

> The totality of correct ideas and of legitimate will is dispersed among the individuals composing society, but it is not equally distributed . . . It is a matter of discovering all the elements of legitimate power dispersed throughout society, and organizing them such that they form the actual power . . . What is normally called representation is simply the means of arriving at this result. It is by no means an arithmetical machine designed to collect and enumerate individual wills. It is a natural and certain process to extract from the bosom of society public reason, which alone has the right to govern.[58]

Several aspects of Guizot's conception are notable. First, it placed the foundations of practical politics on a deeply intellectual footing and it privileged a small elite as uniquely positioned to govern. Second, the legitimacy of this elite was dependent on three principles: discussion, publicity, and freedom of the press. Discussion "obliges those in power to search for truth in common"; publicity "puts the powers occupied with this search under the eyes of citizens"; and freedom of the press "provokes citizens themselves to search for truth and to tell it to those in power."[59] This publicity was more than a check on abuses (though it was that); it was the process that *produced* public reason itself. Political reporting was thus central to the constitution of legitimate power; it was, in the words of Charles de Rémusat, "a logical consequence of our form of government."[60]

The status of elite learned bodies such as the Academy of Sciences was particularly salient to this vision of a liberal polity, for it offered a potential working model of public reason in action. Publicity, for Guizot and Rémusat, prompted a constant mutual self-revelation between government and society, and this was precisely what Bertrand aimed to do for science. Guizot later called the press "the expansion and the impulse of steam in the intellectual order."[61] It was thus the form, rather than the content, of Bertrand's reports that constituted their highest political function. Conversely, precisely because of this close parallel between political and scientific bodies, it was natural to suppose that the same conditions of publicity and openness were required to guarantee the legitimacy of the Academy's own judgments. So good was the fit that, unlike more overtly democratic political philosophies, there was little tension in these two conceptions of reasoned deliberation between the sovereignty of public opinion and of expert judgment.[62]

The liberal conception of publicity applied to science highlights what is misleading in a historiography that identifies scientific journalism with the popularization of science. First, this publicity was seen as constitutive of the rationality of the very elite groups and discussions that were its subject. Second, the public imagined by the Doctrinaires was not in itself particularly inclusive. This was not simply because of relatively restricted readership in practice (due to literacy rates and the significant expense of these periodicals[63]); its limited extent was fundamental to their conception of public deliberation. The same went for readers of academic news. Bertrand was very aware that his articles were not for everyone. He celebrated the "prompt and accurate communication" his column made possible "between the Academy and men who cultivate the sciences throughout France and abroad." This remained a relatively small—if growing—contingent, something like the *pays légal* of scientific France.[64]

Bertrand became increasingly careful to portray himself as a neutral and accurate reporter. Sensitive to his precarious position as a journalist at events whose public status remained contested, he was quick to police his interpreters and combat any suggestion that his accounts were misleading. While at times he celebrated *Le Globe*'s use as a medium for circulating information among the community of savants,[65] he also warned against taking them as verbatim accounts, since "the writing of our analyses belongs to us alone, and ought not to be attributed to anyone else."[66] At a meeting in March 1827, Charles Dupin cited a journal's account of a meeting in which Pierre-Simon Girard was reported to have referred to Dupin's plan for a dam at the mouth of the Seine as being dead in the water. Girard denied that he had said anything of the sort, implying that the journalist must have misquoted him. In response, Bertrand warned that it was the duty of "friends of publicity" to avoiding casting suspicion on journals for errors they had not committed, "particularly at a moment when the periodical press is accused of unfaithful and incorrect accounts for the most minor things."[67]

Another danger was the reader at a distance. Copies of *Le Globe* made it outside France, where some used it for news of French science. In April 1828, David Brewster published an attack on François Arago, rebutting several claims that had been attributed to him by Bertrand in *Le Globe* about the effect of the Aurora Borealis on magnetic needles. When Bertrand found this out, he became alarmed. It was just these sorts of incidents that Cuvier could cite against allowing journalists at the Academy, arguing that it allowed half-accurate accounts of academicians' work to spread across Europe. Bertrand used the occasion "to recall that however much care we take to give an accurate account of events at the Academy, it is always wise when starting a polemic to make allowance for errors we might have committed."[68]

More perilous still was when politics intervened directly. In March 1829 the *Gazette de France*, among the most conservative dailies, reported a scintillating scoop: the Academy of Sciences had just voted to ban the practice of baptism in France. "The reasoning behind this strange proposition," its story suggested, was that "a repulsive materialism" was festering in the bosom of the Academy. Worse still, when the Academy approved the report that made this recommendation—which would probably be sent to a liberal politician, "so that he might propose the abolition of baptism at the same time as of the death penalty"—not a single academician spoke up "to rebuff this tissue of impious absurdities." Bertrand moved swiftly to put out this fire, even if his own report was not the source of the problem. He laid into the *Gazette* for having completely misinterpreted, through its will-to-scandal, the proceedings. Constant Duméril, the *rapporteur*, had simply pointed out that the statistical evidence on infant mortality in the memoir under discussion suggested that the practice of bringing newborns to the prefect for registration put them at great risk of exposure. This was a matter of civil law and had nothing to do with "religious law at all." In fact, Bertrand was only partially correct in contradicting the *Gazette*: while Duméril had indeed spoken only of civil registers in his report, the original memoir, as *Le Globe* had earlier reported, had indeed discussed the problem of baptism, and called for action from religious authorities. The memoir, by Villermé and Henri Milne-Edwards, would later play a key role in French legislation concerning safe baptism practices.[69]

The political significance of Bertrand's science journalism helps explain why the genre took shape first at *Le Globe*, but it also helps account for the sudden proliferation of the genre near the end of the decade. The final years of the 1820s saw increasing instability in the Bourbon monarchy. In 1827, the authoritarian ministry headed by Jean-Baptiste de Villèle attempted to pass new, financially oppressive restrictions on journals (nicknamed the "Loi de justice et d'amour"), though these ultimately failed to be implemented.[70] Villèle was replaced in 1828 with a brief administration that attempted compromise with the moderate liberal opposition, but was quickly followed (in August 1829) by a far more aggressive ministry headed by the ultra-royalist and former émigré Jules de Polignac. The instantly unpopular Polignac prompted an immediate reaction from the press, which shifted its tone from patient critique to open hostility.

At this time, instead of science taking a backseat as the political journals pivoted toward political attack, reporting on science expanded massively. While *Le Globe* had never held a monopoly on such reports, it had dominated the field for several years.[71] But in fall 1829, both the *Journal des débats* and the *Courrier française*, two of the most widely distributed opposition papers, began publishing regular reports on the Academy of Sciences as well, while the schematic reports in the *Journal du commerce* became longer. In October the liberal publisher Jacques Coste founded *Le Temps*—a journal "of political, scientific, literary, financial, and industrial progress"—with the express message that the current regime could now only rule through oppression. *Le Temps* gave special prominence to scientific news by placing it at the bottom of the front page in what was known as the feuilleton (or the rez-de-chaussée), normally a place reserved for popular sections on the theater or book reviews (Figure 2.6).[72] In its prospectus, *Le Temps* explained that it was precisely the mutual benefit between society and science that made such reports essential. "If people of society [*gens du monde*] find in science a source of daily interest, it is perhaps also not without advantage for the sciences to be more known by the public, who will increasingly come to understand the positive influence that savants of a higher order can exercise on the improvement and welfare of society."[73] The trend continued in 1830 with the founding of the even more militant *Le National* by the historian Adolphe Thiers and the journalist Armand Carrel. *Le National* also included a feuilleton scientifique in the style of *Le Temps*, noting that "a people that ignores science and letters is not one that merits liberty."[74] Even royalist journals joined in, at least temporarily. *Le Moniteur universel*, which had become a mouthpiece for Villèle (now out of power), started regular reports on the Academy in the fall.

In the tense last months of the monarchy, as the liberal press wielded its claims to the sovereignty of reason and of public opinion, the Academy itself exploded in the most widely publicized internal dispute since its rebirth as

Figure 2.6 The front page of *Le Temps*, 13 June 1832. Its feuilleton provided an account of the meeting of the Academy of Sciences two days earlier. Author's collection.

part of the Institut de France. That spring a battle that had been brewing between Georges Cuvier and Étienne Geoffroy Saint-Hilaire over philosophical anatomy boiled over. It was largely via the weekly reports in the daily papers that the question of whether animal structure ought to be explained primarily through function or through morphological laws played out.[75] Perhaps most fascinating is that the controversy arose just after the conditions of possibility for such regular reporting had come into being. To some extent, the Cuvier-Geoffroy debate became *about* the strange new situation in which the Academy found itself. Johann Wolfgang von Goethe, who had been a great admirer of *Le Globe* from its early days, used the paper to follow the controversy from afar, and he was stunned to learn that the Academy had allowed the controversy to become a subject of debate among the reading public.[76]

In the usual telling, Geoffroy was the champion of wider publicity against Cuvier's reticence, but Geoffroy's position was more complicated. Like Cuvier, he worried that because the Academy now held its meetings "in the presence of the public," members had become more reserved in their communications. Now "those who present memoirs seem only to communicate them to the Academy in order to fix the date of their discovery." Despite the admittedly negative impacts of the Academy's open meetings, Geoffroy still believed it would be foolish to abandon the new state of affairs: "different times, different customs [*mœurs*]"; publicity "had several advantages as well." As Geoffroy suggested, and as Cuvier surely knew, not only was academic publicity an important reason that young savants still brought their work to the Academy at all, but by 1830 it had simply become impossible to imagine a legitimate body of scientific—or political—judgment that carried out its deliberations in secret.[77]

Surprisingly, this was a lesson that King Charles, or at least the Polignac ministry, proceeded to forget. In July, not long after this scientific battle ran its course, Charles X made a last, dramatic attempt to impose order with a set of ordinances suspending freedom of the press (among other things). Journalists gathered en masse at the offices of *Le National* to author a mass protest. Within a day the Revolution was on.

Journalistes vs. *Rapporteurs*

In the aftermath of July 1830, many of the leading journalists who had led the offensive against Charles X took up key positions in the constitutional monarchy negotiated with the new Orleanist monarch, Louis-Philippe. François Guizot and Adolphe Thiers in particular became two of the most influential political figures of the 1830s. There was a new normal at the Academy as well. Most of the daily journals kept up their academic reports. After a revolution that had been sparked by restrictions on the press, the Academy made no further attempts to restrict attendance at meetings. The use of the feuilleton on the first page became widespread.[78] Alexandre Bertrand himself was immobilized by an accident in early 1830 (and died the year after), but his close friend François-Désiré Roulin, another lycée-mate from Rennes, apprenticed with him at *Le Globe* and gradually took over. Soon, Roulin shifted the feuilleton to *Le Temps* after *Le Globe* became an organ of the Saint-Simonians.[79]

In the last moments of the Restoration, the astronomer François Arago had been elected to replace Joseph Fourier as the permanent secretary for mathematical sciences at the Academy. In 1832, Georges Cuvier passed away and was replaced as permanent secretary by Pierre Flourens, who had himself written reports on academic meetings as a young savant.[80] While many authors had made a habit of providing Bertrand with copies of their memoirs for his use,[81] by 1832 the Academy had established an unwritten policy that journalists could access these documents directly in the archives of the Academy.[82] Academicians came increasingly to appreciate these reports. The academician and engineer J. N. P. Hachette explained to Michael Faraday in 1833 that *Le Temps* gave summaries of the séances of the Academy of Sciences every Wednesday: "Is there a single French academician who complains of this prompt communication? No. Quite the contrary."[83] The daily political papers so dominated this scientific genre that the *Annales des sciences naturelles* simply reprinted abbreviated versions of them, and the *Annales de chimie et de physique* gave up printing proceedings altogether, reasoning that "the versions published by the daily journals seem to suffice."[84] One academician, the veterinarian and bibliophile Jean-Baptiste Huzard, made it a habit, from 1831 onward, to write out manuscript copies of these reports for his records from week to week.[85] On a request from the Royal Institution in London, *Le Temps* even started to produce its academic reports as a separate publication, with type reset as an octavo pamphlet, and sent batches of them to the academicians (Figure 2.7).[86]

Arago was quick to identify himself as the facilitator of this new publicity at the Academy, and historians have generously chronicled his role as the "public face" of the Academy.[87] He took credit—and was blamed by some others—

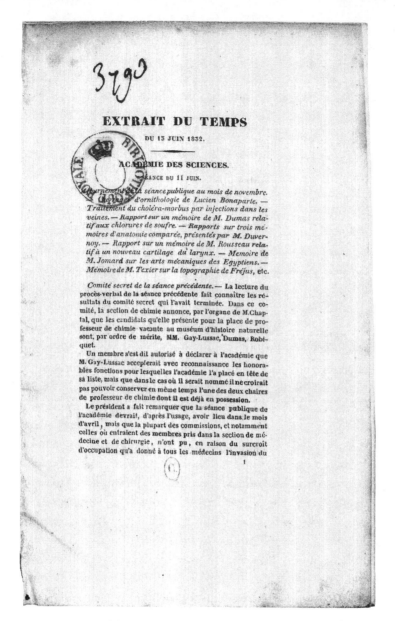

3790

EXTRAIT DU TEMPS

DU 13 JUIN 1832.

ACADÉMIE DES SCIENCES.

SÉANCE DU 11 JUIN.

Ajournement de la séance publique au mois de novembre. — Ouvrage d'ornithologie de Lucien Bonaparte. — Traitement du choléra-morbus par injections dans les veines. — Rapport sur un mémoire de M. Dumas relatif aux chlorures de soufre. — Rapports sur trois mémoires d'anatomie comparée, présentés par M. Duvernoy. — Rapport sur un mémoire de M. Rousseau relatif à un nouveau cartilage du larynx. — Mémoire de M. Jomard sur les arts mécaniques des Egyptiens. — Mémoire de M. Texier sur la topographie de Fréjus, etc.

Comité secret de la séance précédente. — La lecture du procès-verbal de la séance précédente fait connaître les résultats du comité secret qui l'avait terminée. Dans ce comité, la section de chimie annonce, par l'organe de M. Chaptal, que les candidats qu'elle présente pour la place de professeur de chimie vacante au muséum d'histoire naturelle sont, par ordre de mérite, MM. Gay-Lussac, Dumas, Robiquet.

Un membre s'est dit autorisé à déclarer à l'académie que M. Gay-Lussac accepterait avec reconnaissance les honorables fonctions pour lesquelles l'académie l'a placé en tête de sa liste, mais que dans le cas où il serait nommé il ne croirait pas pouvoir conserver en même temps l'une des deux chaires de professeur de chimie dont il est déjà en possession.

Le président a fait remarquer que la séance publique de l'académie devrait, d'après l'usage, avoir lieu dans le mois d'avril, mais que la plupart des commissions, et notamment celles où entraient des membres pris dans la section de médecine et de chirurgie, n'ont pu, en raison du surcroît d'occupation qu'a donné à tous les médecins l'invasion du

1

for the decision to open the archives of the Academy to journalists. Arago was well positioned to appreciate the significance of the Bourbon regime's failure to contend with the press. He had first come to prominence through his work to complete a measurement of the Meridian Arc in Spanish territory during the First Empire; Napoleon's decision to invade Spain in 1808 led to Arago's being arrested as a spy, though he eventually escaped, and spent the better part of a year slowly making his way back to France. When he made it back, he was feted as a scientific hero and elected to the Academy, and he became particularly adept at fashioning his public scientific image. In 1815 he had become co-director of the *Annales de chimie et de physique*,[88] and thereby held a prominent position within the scientific press as well.

Just after his election as secretary, Arago made a show of producing more extensive manuscript minutes of meetings than was customary, including summaries of memoirs and discussions. This became known as a great showdown between Cuvier and Arago over publicity, though Cuvier's actual objection was that the practice would be impossible to maintain due to limitations of both time and expertise. Cuvier's objection seems to have been well founded—Arago did not keep it up—so that while the incident served to solidify Arago's reputation as a reformer, it actually changed nothing.[89] Arago also began to build a wider public reputation after being elected to the Chamber of Deputies representing Pyrénées-Orientales in July 1831. He established his political identity as a republican critic of successive ministries, voicing support of increased workers' rights and later of extending suffrage.

The seeming entente between the scientific press and the Academy did not last long, however. Conflict soon arose that mirrored the rising opposition to the liberal monarchy as a whole. Although the Orleanists had swept to power on the coattails of journalists, many of the republicans who had allied with the liberal opposition during the July Days in 1830 were quickly disillusioned by the new administration. The "Sovereignty of Reason" of the 1820s had transformed in the 1830s to "the political preeminence of the middle classes."[90] More radical groups—militant republicans, and various flavors of socialists— took the most prominent oppositional roles that more moderate liberals had earlier filled, and the press saw the rise of more extreme voices across the political spectrum.[91]

The Doctrinaires' theory of representation had never been about calling for very broad suffrage, and indeed the liberal monarchy made scant effort to facilitate the involvement of the bulk of the population in government. For

them, participatory democracy could do little to foster public reason; it was
more likely to be a threat to public order. (In fall 1830, twin statues representing Reason and Public Order were commissioned for the Chamber of Deputies.) The property qualifications controlling who could vote and who could
sit in government were somewhat reduced, but membership in the *pays legal*
remained quite restricted. The same went for the press. Though freedom of
the press was fundamental to the liberals' theory of governance, in practice
the right to run a political journal required a hefty security deposit, and libel
laws remained on the books to allow for the prosecution of any paper deemed
a threat to order.

Radical republicans moved quickly to organize resistance, and the regime's
tolerance for dissent was put to the test often. The early 1830s saw the rise of
worker organizations, republican societies, and new publications aimed at organizing these groups. Satirical journals—*Le Charivari* and *La Caricature*—
ridiculed the hypocrisy of the bourgeois monarchy not only in words but in
images. It soon became clear that those in power were willing to make frequent and brutal use of the legal tools at their disposal for suppressing radical
opinion. Journals such as *Le National* and *La Tribune*, which had drifted republican after the revolution, were battered with fines and prison sentences
for their *gérants*. These prosecutions—themselves a frequent subject of news
and satire (Figure 2.8)—forced editors to be exceedingly cautious about what
they put into print.

A more aggressive scientific press took shape at the same time, not simply
alongside but from within the radical republican movement itself. The central
figures in this new republican science journalism were the savants François-
Vincent Raspail and Jacques-Frédéric Saigey. Writing in republican dailies,
they brought a new style of reporting to the academies, one that went beyond polite criticism of these bodies to express regular skepticism about their
legitimacy as arbiters of scientific truth. They first took control of the scientific feuilletons of *Le National* and *La Tribune*. Then, after starting a radical republican journal in fall 1834 called *Le Réformateur*, they mounted a sustained
and relentless campaign against the Academy of Sciences and Arago. For the
Doctrinaires, though the existence of publicity was crucial to the exercise of
reason in the chambers of government, the public did not actually dictate to
it.[92] Saigey and Raspail would argue instead that reasoned judgment was not
simply contingent on publicity, but was located in a public external to elite deliberative bodies.

Figure 2.8 "What a treacherous path! (luckily there are judges in France)." An allegory of the treacherous halls of justice for the independent press during the July Monarchy. *Le Charivari*, 15 October 1833. Courtesy of Ghent University Library, BIB.ACC.022156.

During the late Restoration, Raspail and Saigey had both been young savants trying to make it in Paris. They came from modest provincial backgrounds: Raspail was from Carpentras and Saigey came from Montbéliard, a manufacturing village near Switzerland. Both came to Paris just before 1820. Although each attended some lectures at the Faculté des Sciences, and Saigey was later admitted to the École normale supérieure, they were largely self-taught. Each developed research programs that placed them in the tradition of mechanical romanticism that John Tresch has recently explored (and which also included Leroux, Bertrand, and André-Marie Ampère).[93] Both celebrated the liberating potential of inexpensive laboratory instruments, and at the same time insisted on a syncretic scientific practice that did not recognize bounds of specialization. ("Nature is neither chemist, nor botanist, nor zoologist, nor mineralogist, nor physiologist," wrote Raspail.)[94] Saigey's great ambition was to produce a theory of physical bodies based on a universal force of repulsion. This anti-Newtonian view involved a reworked theory of a fluid ether that explained not only electrical and magnetic phenomena, but light, heat, and several chemical phenomena.[95] Saigey used his experience with precision tools gained while working for a manufacturer near Montbéliard (the center of watch production in France at the time) to build simple experimental systems consisting of suspended magnetic needles and discs (Figure 2.9).[96]

Raspail, on the other hand, brought together bits of chemistry, botany, zoology, physiology, and optics in a research program that made extensive use of a simple microscope of his own design.[97] The spirit of his work is encapsulated by the subtitle of his 1830 book *Essai de chimie microscopique*: "the art of transporting the laboratory to the slide in the study of organized bodies." This had a double meaning. First, Raspail brought laboratory techniques of chemistry to organic specimens observed under a microscope. Most famously, he took various starch grains and attempted to discover their chemical composition by using several reagents and dyes. Second, he developed a proto-cellular theory in which he imagined the cell as a kind of "laboratory where new quantities of gas come to be organized, successively combining together and condensing."[98] Raspail often noted with pride that the microscope he helped design was sold by a Paris instrument maker at a cost far below that of comparable instruments, making his research program accessible to a broad spectrum of contributors (Figure 2.9).

Both men came to support themselves largely through journalism, writing for periodicals and other jobs. Beginning in 1824, they took up key positions in the ambitious publishing enterprise of the Baron Férussac known as the *Bulletin universel des sciences et de l'industrie*, actually a system of eight periodicals aimed at distributing information about scientific discoveries across several fields. Saigey directed the portions dealing with physics, mathematics, and astronomy, and Raspail directed botany.[99] But the pay for such work was not great.[100] True advancement in a scientific career, Raspail believed, would likely come only by patronage from the Academy of Sciences. While Saigey shunned the Academy from the beginning (his theory of universal repulsion was not calculated to impress either the Laplacians or the anti-Laplacians), Raspail perceived academicians to be *the* audience for his research. This meant submitting a memoir and hoping for a favorable report from the Academy — and in particular from the *rapporteur* who was assigned to write it. Raspail's retellings of this period of his life were structured by these acts of submission, invariably narratives of one calamity after another.

In Raspail's vivid account of his first academic encounter in 1824, he re-

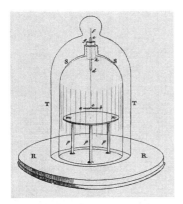

Figure 2.9 Scientific instruments accessible to all. *Left*, a "very simple apparatus" used by Saigey for his experiments on magnetism and the law of repulsion. *Annales des sciences d'observation* 2 (1829), plate 5. *Right*, the simple and cheap microscope associated with Raspail. From *Catalogue commercial ou prix courant général des drogues simples, produits pharmaceutiques et chimiques* (Paris: Typographie de Henri Plon, 1860), 556. Courtesy of Biodiversity Heritage Library and the BnF, respectively.

called how he "trembled upon entering the courtyard of the Institut for the first time," and finally mustered the courage to approach the botanist René Desfontaines with his manuscript. Upon learning that the subject was botany, the academician inquired what new species he had discovered. Raspail responded that its subject was not new species but rather new organs and analogies. "At these words Desfontaines turned his back to me, as if I had given him an insult to which he could not stoop to respond."[101] Raspail did eventually have his manuscript — on "The Formation of the Embryo in Grasses" — read at the Academy. Eventually — after Raspail had given up hope — a commission presented a report that was more or less favorable. And then nothing happened. Because Raspail had no particular patron at the Academy, his positive report won him no further notice. Raspail continued to pepper the Academy with memoirs over the next several years, most of which the Academy simply seemed to ignore. This was disappointing. But there was something worse than being ignored. While Raspail became a subject of ridicule among academicians for his many submissions, he noticed that their protégés and family members were beginning to re-present his ideas as their own.[102]

Working together at the *Bulletin*, Raspail and Saigey discovered they were kindred spirits. Although already politically radical, they were soon radicalized with respect to science as well. In 1829 they both resigned from the *Bulletin* in protest when Férussac accepted an offer from the Duc d'Angoulême, the eldest son of Charles X, to buy the journal. Backed by the publisher Alexandre Baudouin, they joined together to start their own independent scientific journal.[103] Whereas Férussac's *Bulletin* was essentially an ambitious take on the well-established genre of scientific journal as newsletter, Raspail and Saigey's *Annales des sciences d'observation* was to be an expression of their own scientific viewpoint, in much the same way that the political dailies encapsulated some particular political program. It became the mouthpiece not only of a particular scientific doctrine, but of a political economy of science as well. At the heart of their critique was the corrupt system of academic reports. "Not long ago," they wrote, "a report by a member of the Academy of Sciences carried the force of a judgment rendered, so that the vanquished was careful not to appeal, and the victor instantly became an important personage." This had become a naïve view, however. "Now that our political education has taught us to submit all power to a reasoned examination, science has . . . smashed its idols."[104]

According to Raspail, the Academy's system of reports was problematic from three points of view. First, reports were often erroneous, because those chosen to write reports were often not competent, biased, or too lazy to evaluate the work presented to them. Many academicians had simply fallen behind the research front after they had been elected. They were biased, because report writers were likely to be competitors or patrons of a competitor.[105] Laziness led many to avoid subjecting a paper to a careful analysis, a failing "just as odious and harmful as judgments obtained by corruption."[106] Critics had long questioned particular judgments of the Academy even while affirming its legitimacy as judge: "[Academicians] administer science just as the judge administers justice in a criminal court; the academician is a magistrate in equal measure," wrote one observer. Alexandre Bertrand had warned that the academies' authority as the audience for new claims was contingent on the "belief that the Academy is made up of men capable of competently judging them."[107] But for Raspail and Saigey it was obvious that the Academy had long ago disqualified itself: "It is time that public opinion be permitted to see and to judge for itself, and that from now on no one shall attempt to influence it!"[108]

Second, even a positive report was not nearly as important as concrete favors from academicians, and these depended on personal relationships. A young author had to decide between "conserving his independence with his isolation, or attaching himself to the wagon of the intriguer who is in favor."[109] Raspail observed that although the Academy claimed to represent the scientific elite, it was more accurate to say that it represented a group of social elites. To make it in the Academy, one had to know the right people, one had to gain access to "scientific coteries." Actual power was wielded behind the scenes, not in print, not even in manuscript, but in conversations and even gestures. "At first they don't attack you to your face, but they ridicule you in secret," Raspail warned. "They mock your discoveries at soirées, at banquets, at concerts."[110] We can restate the talk of coteries as a point about the importance of personal patronage relationships in Parisian science at the time. The coteries themselves were so concentrated in Paris that those on the inside might move relatively freely among them — "coteries are combined in the same city walls, at the doors of the same power, on the benches of the same academies"[111] — though it was very difficult to enter from the outside (unless one were willing to give up one's independence). The academic report system was a façade that concealed the means by which power was actually exercised.[112]

Third, the system of reports seemed designed to give academicians early access to new discoveries, allowing them to pilfer ideas from the younger savants who submitted them. Raspail wrote that for academicians such as Henri de Blainville, "every time he reads or hears a new opinion . . . his first reaction is to believe he has already professed it in his courses." The Academy advanced knowledge largely through acts of piracy. "The poor can never plagiarize with impunity; this is a privilege given only to the rich."[113] Worse still, one way they got away with it was by impugning the accuracy of the cheaper instruments used by poorer savants. Thus Raspail's own work could be dismissed because of "the inferiority of the latter's microscope in comparison with the *riche* microscopes" of academicians and their protégés.[114]

The growing importance of the daily journals' scientific feuilletons had done less than it might seem, in Raspail's view, to break the Academy's grip on science. In 1831, he reflected on the recent spread of the genre: "I don't really know whether or not this new measure is good for science . . . but what is certain is that intriguers have spared nothing in exploiting it to their own advantage." Raspail explained that most journalists were not independent witnesses. Editors-in-chief looking to start up a *feuilleton scientifique* would go to an academician they knew and ask them to recommend a student whom they could rely on to write up the meetings. As young savants, such journalists had an interest in gaining the approval of the Academy, and thus were in the pocket of an academician from the start. Raspail cited the Cuvier-Geoffroy affair, noting that Alfred Donné, who wrote for the *Journal des débats*, was at the beck and call of Cuvier. Journalism had simply given birth to a "new branch of charlatanism" for academicians to exploit.[115]

The tumult of the July Revolution spooked Saigey and Raspail's publishers and ultimately put an end to the *Annales des sciences d'observation*. (Raspail blamed Cuvier and Arago.)[116] Saigey got a job directing the scientific portion of another publication, *Le lycée*, while Raspail immersed himself fully in the Republican cause. Though offered the légion d'honneur for his part in the Revolution, Raspail refused it. He became president of the Société des amis du peuple in 1831, a major node in the republican resistance, and was involved in several other republican groups, including the Association républicaine pour la liberté de la presse. These activities quickly landed him in jail and on trial. This was a pattern that would repeat throughout most of the 1830s but which gave him extended opportunities to write political tracts as well as long and popular textbooks on organic chemistry and vegetable physiology.

In late 1833 Saigey was hired to write the scientific feuilleton for *Le National*. He brought to it the more aggressive style that he and Raspail had developed. Not only was Saigey keen to report on and criticize academicians' reports and discussions, but he also often conveyed an impression of academic meetings as something less than the solemn affairs they were meant to be, emphasizing academicians' disagreements and petty squabbles.[117] He relished the new role. "I have just left the Academy where I heard the most unbelievable things," he wrote to Raspail in December 1833. "I'll feast on them tomorrow in the feuilleton."[118] Some other journals changed their coverage likewise. The *Gazette médicale de Paris* noted in January 1833 that in the past it had "limited itself to reproducing historically" the meetings of the academies of sciences and medicine. Explicitly comparing the *comptes rendus* of the academies with those of the chambers of government, it promised now to bring to its accounts "the spirit of discussion and critique that brings difficulties into view and contributes to clarifying them."[119]

Raspail also believed deeply in the power of a free press, but did not trust that the directors of journals such as *La Tribune* or *Le National* would reliably speak truth to power, whether against the government or the Academy of Sciences. Even *La Tribune* refused, Raspail wrote, to give publicity to radical republican groups: "We needed a journal for ourselves, one in which we were the editors and the shareholders."[120] Though he was offered jobs by *Le National* and the new *Revue des deux mondes* to cover science for them, he declined them. Saigey attempted to convince Raspail to collaborate: "Fume all you want against the editor of *Le National* . . . but anger directed against a person who can do nothing is unreasonable."[121] In fact, the director of *Le National*, Armand Carrel, did begin to receive threats from Arago about its scientific coverage; Saigey reported to Raspail in spring 1834 that "Arago has already schemed to have me removed from *Le National* or to prohibit my entry to the séances and to the Bureau."[122]

In fall 1834 Raspail received the opportunity to direct a truly independent journal, when an aristocratic supporter, Théophile Guillard de Kersausie, agreed to put up 100,000 francs to back a journal controlled entirely by Raspail. Just as crucially, Kersausie was willing to put up money to cover the fines that would surely come its way with Raspail at the helm. Raspail hired several friends to help with the journal, including Saigey to write the *feuilleton scientifique* (Figure 2.10).

Le Réformateur was to be a guide to politics for the working classes, and Raspail used the journal to develop a program of social reform and political economy.[123] His insistence on the unity of science meant that politics ultimately rested on principles analogous to those of natural science: "Republican principles are intimately connected to the physiology of the masses, from which they derive."[124] The basic social unit, Raspail taught, was not the state, but the *commune* (any micro-institution, or small set of associations among men). The commune was very much like the *cellule laboratoire* of Raspail's system of organic chemistry; it was a locus of organization based on frequent exchanges and discussions among men. The key political task was to "organize communal association, and organize it on a fraternal basis and according to a progressive method."[125] Monarchy — *gouvernement d'un seul* — was fundamentally incompatible with the principle of association. First, it did nothing to encourage its growth (indeed, it normally discouraged it, for fear that it would lead to instability and revolution[126]). Second, centralized decision-making could never be sufficiently rational since it would always be based on only partial knowledge of local facts, whereas "public opinion in each commune" was in the best position to make reasoned judgments based on all the relevant social and economic data.[127] Elections and collective decision-making were the best means of regulating these groups: "Representation, brought to the level of the commune: this is the Republic in embryo."[128]

The Academy of Sciences, for its part, was according to Raspail a micro-

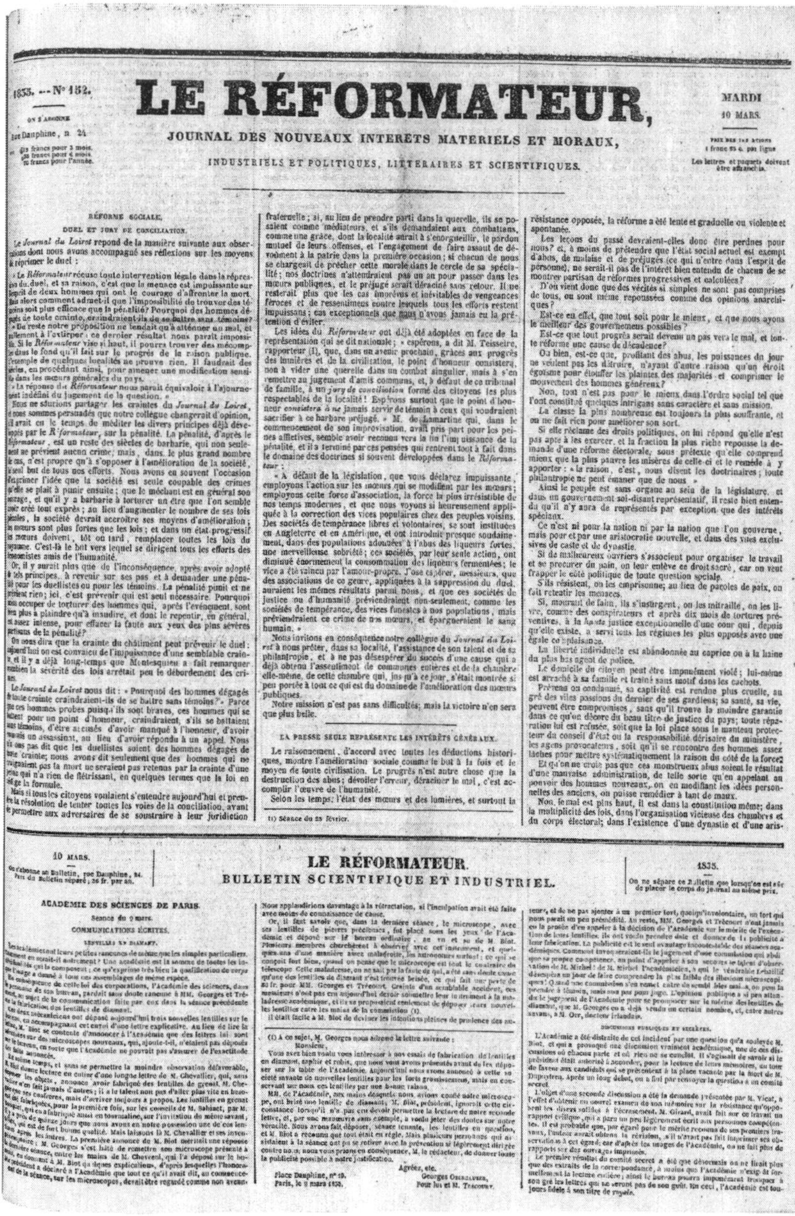

1835. — N° 152.

LE RÉFORMATEUR,

MARDI 10 MARS.

JOURNAL DES NOUVEAUX INTÉRÊTS MATÉRIELS ET MORAUX,

INDUSTRIELS ET POLITIQUES, LITTÉRAIRES ET SCIENTIFIQUES.

RÉFORME SOCIALE.

DUEL ET JURY DE CONCILIATION.

LA PRESSE SEULE REPRÉSENTE LES INTÉRÊTS GÉNÉRAUX.

10 MARS.

LE RÉFORMATEUR.

BULLETIN SCIENTIFIQUE ET INDUSTRIEL.

1835.

ACADÉMIE DES SCIENCES DE PARIS.

Séance du 9 mars.

COMMUNICATIONS ÉCRITES.

Figure 2.10 The issue of François-Vincent Raspail's *Le Réformateur*, no. 152 (10 March 1835). The main articles on the first page concern "Social Reform" and an argument that "the press alone represents general interests." The feuilleton (rez-de-chaussée) gives a severely critical account of a meeting of the Academy of Sciences involving an ongoing controversy over funding the construction of diamond-lens microscopes. Courtesy of the BnF.

cosm of the hypocrisy of the bourgeois monarchy as a whole. Although it claimed, just as the bourgeois monarchy had, to welcome public scrutiny, the taint of monarchism remained at the heart of the system: "The academies, founded under the dominion of an order of things that is no more, may have served the progress of science at a time when it was in harmony with the march of affairs," but this was no longer the case. Whereas it was in the interests of the Academy to keep knowledge fixed, "public opinion, which works quickly, and now works alone and without boundaries, has turned and left far behind these barriers to progress."[129]

If Georges Cuvier was the last bastion of the old academic order, Raspail saw Arago as the embodiment of the new hypocrisy. As an elected deputy, Arago claimed to represent the people by championing workers' rights, but at the same time he discouraged the working classes from participating in politics on their own behalf. Arago's paternalistic model of social improvement had its counterpart in the imperious manner in which he attempted to control French science by pretending to be a benevolent ruler.[130] Though he claimed to encourage scientific publicity, he actually kept a strong hold on scientific power behind the scenes through the same patronage relationships that Cuvier and Laplace had exploited. If in the era of Cuvier, patronage functioned through its public exercise, with the era of publicity represented by Arago it had simply gone behind the scenes.[131]

In October 1834, *Le Réformateur*'s scientific feuilleton joined *Le National* and *La Tribune* — the latter paper had recently restarted after heavy fines and prison sentences forced it to shut down for several months — as part of the republican press corps. Raspail opened his feuilleton with a call to renew critique, especially of "the soporific reports of *Messieurs de l'Académie.*"[132] Such critique was required now more than ever. Controversy had been swirling since the summer over the nature of the Academy's reports. The interior structure of starch had become a hot topic after several manuscripts on the subject had been submitted to the Academy; Michel Chevreul was commissioned to write a report in order to sort out the "contradictions between several competing results, and priority claims for certain discoveries." By that time Raspail's own microscope studies of starch in the 1820s had become well enough known that a previous report on the subject noted that Raspail "had opened the path to all the recent discoveries" in the area.[133] But Chevreul, while admitting that Raspail had shown that starch has an interior distinct from its exterior skin, emphasized that Raspail had been wrong about its exact nature.

Chevreul also contradicted certain academicians' own recent work on the question, including optical studies on starch recently pursued by the eminent physicist Jean-Baptiste Biot. In response, Biot, along with Gay-Lussac, raised the question of the role and responsibilities of *rapporteurs*. Had Chevreul or any of the other *members of the commission* actually performed the necessary experiments? Chevreul reasoned that the relevant experiments would have taken far too long to carry out. Biot retorted that the Academy "could not vote on facts that had not been verified," whereas Dulong insisted that as a rule "commissions never took up" experiments of this nature. Biot had only recently begun attending meetings after an extended period of self-imposed academic exile, and he claimed shock at these low standards. Back when he had sat on commissions and written reports, "he had always verified the correctness of the facts put forward by authors."[134] In *Le National*, Saigey made much of the scandalous revelation that commissions appointed to evaluate manuscripts were failing to verify and properly accredit knowledge claims. Another journal, *L'Écho du monde savant*, joined in the critique, noting that if "reports cease to be rulings [*arrêts*] to become simply services or favors, [the Academy] will lose in the eyes of the public the importance and utility that protects it against accusations of privilege and aristocracy."[135]

Raspail's first feuilleton led by recalling this incident, noting with disdain that the Academy now admitted that "reports were not meant to be verifications of the facts submitted to the examination of the Commission, who have neither the time nor the inclination to deal with such matters." The incident served as an ideal frame for Raspail's feuilleton.[136] Only the public, embodied by an independent press, could truly judge scientific merit. As he had put it in 1830, "publicity is the safeguard of science as it is of politics."[137]

Saigey and Raspail produced a narrative of contemporary Parisian science as a sordid drama in need of a major shakeup. "These miseries of science are not inherent to science itself, and with a different social organization *Isis* will once again assume her decent attire, and leave behind the sword of combat for her magic wand."[138] Academicians gave them much to work with. After a report was presented by a prize commission in December regarding several works on statistics, a debate arose that was so acrimonious that it raised the question of whether academic reports ought to be seen as representing the Academy's judgment at all. The issue was nothing less than the definition of statistics, whether it was a science dealing with the resources and goods useful to nations (said Mirbel and Ampère), or whether it was the science of things expressed numerically (said Charles Dupin, Costaz, and several others). Arago used the incident to pass a rule that reports in general should carry disclaimers that the opinions expressed in them belonged only to the commissioners themselves. Dupin lost his cool, declaring, "I protest against this decision, I protest against closure of the debate! . . . it is crucial that the Academy come to a judgment on the definition of statistics," amid general tumult and shouting.[139]

The feuilletons of Saigey and Raspail also incorporated elements of satire, the mode that characterized so much of the new journalistic opposition. In one incident, the Institut installed a heating new system that worked a little too well for many academicians. Arago polled his colleagues as to what temperature they would prefer. Raspail reported the vote, imagining a study of the "temperature at which an Academy may stagnate, and the heat necessary to cause it to flourish." The Académie des beaux-arts, he speculated, required only 10 degrees (its members possess natural heat in greatest abundance), and the Académie des sciences morales et politiques required greater than 20 degrees (since hardly anyone ever showed up to their meetings).[140] *Le Charivari* itself took notice of the persona of academician; it ran a series of caricatures documenting fifteen "Paris trades" in 1833, including an overfed "Academician" looking both pretentious and ineffectual. It happened to be followed directly by the "Plagiarist," Raspail's own preferred characterization of the Academy's research method (Figure 2.11).[141]

Academicians took notice of the new style of republican journalism and again discussed eliminating discussions during the public portions of meetings. Raspail dared them to do it, for then they would "be without the praises lavished on them by the crowd pressed in around their chairs, not to mention the notices in the journals . . ."[142] However, in late February 1835 the Academy's attention returned to microscopes. While on a trip to England and Scotland doing research for his biography of James Watt, Arago became excited by the prospect of producing simple microscopes using diamond lenses. At the time there was hope that microscopes built using precious jewels could solve the problem of chromatic aberration. He convinced his academic colleagues to sponsor — to the tune of 1,200 francs — an instrument maker named Bouquet to build two microscopes using diamond lenses and then to support the sale of equivalent instruments to the public.

For Saigey and Raspail, this represented corrupt patronage at its worst, and these luxurious microscopes were the perfect topic through which to demolish Arago. *Le National, Le Réformateur,* and *La Tribune* launched sustained at-

Figure 2.11 "Academician" and "Plagiarist," two of a series of humorous depictions of Parisian professions by the artist Edme-Jean Pigal. In *Le Charivari*, 2 and 9 October 1833. Author's collection.

tacks for weeks on end detailing Arago's abuses. First, Arago had pronounced an award to a particular favorite rather than holding a fair and open competition among the several microscope manufacturers now working in Paris. Several of the latter peppered the Academy with letters and instruments in the coming weeks, though the Academy mostly ignored them (and evidently even broke one of their instruments). Second, the Commission appointed to judge Bouquet's work (which included Arago himself) was in no position to do so, as none of its members knew much about microscope construction. Third, the Academy only seemed to become interested in microscopes once the prospect of building an extremely expensive one presented itself, whereas the scientific public had been making extensive use of the Raspail microscope—which cost only thirty francs—for nearly a decade.[143]

These attacks were the final straw for Arago. He once again attempted to have Saigey fired from *Le National*, this time with success. He clamped down on journalists' access to the archives of the Academy by asking directors of journals to register the names of their science journalists, thus keeping the likes of Saigey out. He explained to the directors that "the public character [*publicité*] of its meeting and its works is not a right, but a favor"[144]—a favor that had been bestowed by the generosity of Arago's own will. In the process, he inadvertently painted a picture of an Academy severely hampered by the radical press.

> ... the odious personalities, the calumnies, the crude insults with which certain individuals have for the last while laced their polemics, has seriously compromised the progress of science by forcing a mass of distinguished savants to abandon our meetings. I could cite several cases of physicists and naturalists of the first rank who now prefer to keep the fruits of their labors under wraps rather than descend into an arena in which adversaries await them who are determined to trample upon the most common of decencies.[145]

On further inquiry, Arago informed Raspail that access to the Academy's offices had never been a right of the press but only a personal favor granted by "an act of my will," one that he could "take back as necessary, whenever the interests of Science make it my duty to do so."[146] The directors of *Le National* and *La Tribune* backed down immediately. Raspail meanwhile used *Le Réformateur* to rail bitterly against what he called a *Coup d'état scientifique*, implicitly comparing Arago's actions to those of the discredited Bourbon monarchy at its end. Since the sole foundation of the Academy's legitimacy was the public, Arago's use of monarchical terms such as favors over rights was tantamount to treason: "[Since 1830] it is the popular *us* that constitutes the law, and not the royal or academic *me*":

> When this savant, who claims to impose his will as if it were a law, in a country in which the law is the expression of the general will, when this savant presents his candidacy as a republican, we recommend that republicans not forget these two letters, so insulting to the republican press, and contrary to all the political principles ordinarily professed by M. Arago.[147]

Saigey was out of a job, but Raspail gave him free rein in *Le Réformateur*'s pages, and in a series of articles called "La presse indépendante et la science privilégiée" Saigey gleefully gave a detailed overview of his many journalistic transgressions against the Academy.

Most other journals stayed out of the controversy, though some came to the defense of Saigey and Raspail. The *Gazette des hôpitaux* noted that Arago's action was both improper and ineffectual:

Improper, for it affects the rights of the press and book publishers at the whim of a society, or what is worse, at the will of an individual. It is ineffectual because even prohibited journals will find ways to provide *comptes rendus* of the séances. Publicity, with or without critique, is a necessity for the Academy.[148]

Arago, it turned out, was already one step ahead.

Erecting Altar against Altar

"Pervasive murmurs, great fits of anger, whispers, academic rows, ringing bells, true scientific mayhem": Raspail relished the reaction he thought he had prompted from the Academy. "The old coquette," he boasted, "does not like seeing its reflection in an overly faithful mirror."[149] But Arago hadn't actually barred journalists from attending academic meetings (even if he did cause a few to lose their jobs), and the academic feuilleton did not go away. The real impact of March 1835 was that it gave Arago the occasion to push through an ambitious plan. The very week that access to the archives was restricted, he outlined a proposal to publish the Academy's own authorized account of its meetings in the pages of a new weekly journal.[150] Unlike most proceedings journals in Britain, the new journal would contain not only summaries of memoirs read, but also of the discussions that followed them.[151] The Academy hired François-Désiré Roulin, the science journalist at *Le Temps* (and friend and successor to Alexandre Bertrand), to direct their new experimental publishing venture, to be known as the *Comptes rendus hebdomadaires* (Figure 2.12).

The *Comptes rendus* quickly took on an identity of its own. Academicians and savants whose papers were read at meetings were responsible for the summaries that they submitted for publication (although the secretaries exercised a fair amount of power as to what was included). Academicians also submitted the text of discussions. Roulin's task was thus not to write his own journalistic accounts of memoirs or reports, but to coordinate the writings of others. No journalist was to be in a position to distort events.[152]

Raspail interpreted the move as a desperate attempt to stifle critique. Arago had "erected altar against altar, publishing a journal of its own, seizing a monopoly on the weekly feuilleton."[153] He admitted that after the appearance of the *Comptes rendus*, the scientific press was forced to take a more neu-

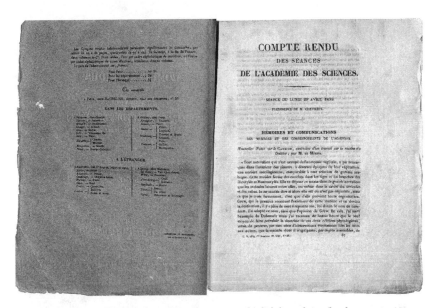

Figure 2.12 The weekly number of the *Comptes rendus hebdomadaires* for the meeting on 29 April 1839. Verso of front cover and first page. Author's collection.

tral stance, entrusting their weekly feuilletons "to docile and devoted pens."[154] While there may have been some truth to this, other factors contributed. Just as the Academy was about to launch its ambitious plan, politics again intervened. On 28 July 1835, there was a violent assassination attempt on King Louis-Philippe during the parade marking the fifth anniversary of the July Revolution. A *machine infernale* fired a hail of bullets on the royal procession as it made its way up the Boulevard du Temple. Nineteen were killed, dozens more injured, and the Prime Minister, Adolphe Thiers, was splattered with blood. The King escaped injury, but the state responded swiftly. There was a widespread crackdown on opposition publications and journalists, and Raspail was arrested as a suspected coconspirator. Restrictions on the press were stepped up in what became known as the September Laws. *Le Réformateur* limped on for a few more months, with Raspail finding ways to dispatch missives from prison (including an elaborate satire in which he put the members of the Academy on collective trial for their secrecy and incompetence), but it was finally shuttered after even more fines. Other opposition journals, including *La Tribune*, shut down as well.[155] *La Caricature* depicted the press laws as another kind of infernal machine — this one aimed at the independent press —

Figure 2.13 "The Infernal Machine of Sauzet." Paul Jean Pierre Sauzet, a member of the Chamber of Deputies, helped push through especially harsh laws against the press in the wake of the assassination attempt on the King by infernal machine in July 1835. *La Caricature*, no. 250 (20 August 1835), imagined these new measures as a form of assassination directed against the independent press. Raspail's *Le Réformateur* can be seen at the bottom right. These and many other journals were indeed put out of business. Courtesy of the BnF.

for which the assassin's weapon "served as both pretext and model" (Figure 2.13). Nearly overnight the Academy's—not to mention the government's—fiercest press critics had been silenced just as the first issue of the *Comptes rendus* was delivered. It was an auspicious beginning.

Press reporting at the Academy did not stop, although some argued that it had been diminished. At least one long-standing scientific journal shut down in the face of the *Comptes rendus*. The *Bulletin scientifique* (published in cooperation with the Société philomathique), which consisted mostly of excerpts from documents read at other academic meetings, was deemed by its editors to be superfluous.[156] The academician J.-B. Huzard, who had dutifully produced manuscript copies of the weekly feuilletons for years, ceased doing so. Even Auguste Comte, an indignant opponent of François Arago's influence on science, later admitted that the *Comptes rendus* was the only journal he read, even if it had "degenerated into a routine display of our lowest academic vanities."[157]

The structure of the French press as a whole soon underwent massive changes. The year 1836 saw a shift to a new style of daily paper that was focused more on profit than on politics. Emile de Girardin's *La Presse* and Armand Dutacq's *Le Siècle* cut by half the customary subscription price of dailies and dramatically raised the amount of advertising contained in their papers. Though both included a *feuilleton scientifique* at their inception, these were subsequently pushed from the front pages or eliminated entirely, replaced by massively successful serial novels such as Eugène Sue's *Les mystères de Paris*.[158] Bolstered by the *roman feuilleton*, the new *presse à bon marché* resulted in a far more lucrative, and less polemical, daily press.

The *Comptes rendus* brought the Academy in line with a publishing trend that was spreading across Europe. Although the specific contours of the political situation in France were certainly unique, it was more generally the case that an increasingly powerful periodical press was claiming with increasing credibility that it represented the legitimate public not only of politics but of natural knowledge. After Richard Taylor prompted the Astronomical and Geological Societies of London to begin publishing their proceedings in the late 1820s, the Royal Society of London followed suit in 1831. The Académie Royale in Brussels produced a similar publication not long after. The Paris Academy's *Comptes rendus* inspired others in turn: in early 1836 the academy in Berlin began a monthly *Bericht* and the one in Saint Petersburg founded a weekly *Bulletin*.[159] The Académie Royale de Médecine in Paris began its *Bulletin* later that fall. In Britain, proceedings publications also continued to spread, including to the Royal Society of Edinburgh and the Linnean Society in 1838.

A consequence of making authors directly responsible for writing summaries of their own manuscripts was that they tended to lose their character as abstracts and came to be treated as short research contributions themselves. Indeed, some academicians became infamous for the extreme uses to which they put the new journal. Augustin Cauchy, already alienated from his academic colleagues for his staunch Royalist views, used the journal — to the great irritation of many colleagues — to publish his research in small installments on a weekly basis, somewhat in the style of a serialized novel of mathematical analysis.[160] Thus, while the proximate impetus for the launch of the *Comptes rendus* was the weekly feuilletons, the *Comptes rendus* gave the Academy (and these other institutions) a means of competing with scientific journals that were increasingly publishing not simply excerpts but original contributions.

The sudden appearance of these new periodicals, modeled on the press but sponsored by elite scientific institutions, prompted a great deal of reflection on the nature of science and of its publics. Jean-Baptiste Biot observed the proliferation of the new genre across Europe with astonishment. In 1837 and again in 1842, he produced eloquent analyses of "the influence that these accelerated publications were exercising on the march of science."[161]

Biot admitted that the *Comptes rendus* had become inevitable once the journalists had established themselves: "given that, with or without its consent, knowledge of [the Academy's] labors and results would be instantly delivered to the world, what more wise course could there be, for the utility of science, but for itself to communicate them in all their truth?"[162] The *Comptes rendus* was an "instrument of personal defense" that the Academy could certainly not do without. Biot was also quick to admit that the Academy had actually gained a great deal thereby. By offering to savants, within and without, a medium of publicity that was both efficient and prestigious, the Academy had once again established itself as a major hub of scientific intelligence, after several years during which this had come increasingly into doubt: "When you consider that this ease of swift publication has made the Academy the focus of a scientific correspondence that arrives daily from all parts of the globe, you will see what service the creation of such a glowing center of enlightenment has been."[163]

But recouping a dominant position on scientific intelligence came at a cost. Most disturbing was that the new customs of publishing were changing both the nature of meetings and the sensibilities of savants themselves.[164] Already the Academy's deliberations, being "in the habitual presence of the public through the intermediary of the journals," had changed their shape.[165] It was not simply, as Étienne Geoffroy Saint-Hilaire had already pointed out in 1830, that increased publicity had actually forced savants to be less open, and less innovative, as the meetings of the Academy had ceased to offer a protected sanctum in which savants exchanged genuine, helpful criticisms of work in progress. The presence of a wider public also affected "if not the content of the work itself, at least the form of its exposition." This public encouraged savants "to seek after *éclat,* or the appearance of novelty, rather than severity and depth."[166]

The *Comptes rendus,* though it provided a new, written format to combat the "excessive and illegitimate power" that speech had obtained, brought other difficulties. Savants were increasingly eager to publish their work in the form of the short summaries that they had gotten accustomed to at the Academy itself. Biot worried that if scientific writers were constrained to present their work "as so many aphorisms, each detached one from the other, their value will be but imperfectly understood, and only the authority of the author's name can provide confidence in the result — always an extremely dangerous situation in science." Paradoxically, the increasing reliance on publicity in short articles forced savants to trust in the authority of a prestigious name rather than that of reason and evidence.

> But give him enough space to indicate the chemical trials he has undertaken, the elements of organization that they have enabled him to discover, and the consequences that will lead to applications: then this publication will possess all the useful characteristics wanted, both for the present and the future.[167]

As long as the press had been kept at some remove from the publishing genres employed by academic science, its influence had been kept contained. But the prospect of a blurring of genres, or worse, the progressive substitution of the short *article* for the polished *mémoire,* meant that scientific knowledge in general might become "less perfect, less durable."[168] In Biot's reading of history, the early years of organized natural philosophy corresponded to a prelapsarian age in which the broader public did not care much about science and thus savants were afforded the time and space they needed to perfect their research in carefully considered memoirs and books. But once the audience for science ceased to be limited to the "universal, but exclusive, society of savants,"[169] savants had become "subjected to nearly all the agitations of public life." The paradise was now part of a distant and irretrievable past: "To return to this state is as impossible for us as to be reborn; it is in vain to attempt to return to the sciences the calm that has been lost."[170]

There were other ways to understand Biot's discomfort with the new social reality that the Academy had helped legitimate by launching the *Comptes rendus.* Although an early collaborator of Arago, Biot had made his own scientific reputation by more traditional means as a protégé of the great mathematician Pierre-Simon Laplace. In that earlier epoch of patronage, the large master-works—Laplace's *Mécanique céleste,* Cuvier's *Règne animal*—were often in reality collaborative enterprises to which students and disciples contributed in return for protection and career advancement.[171] In this arrangement, author-ship was perhaps secondary to collaboration in establishing one's career and reputation. The potential breaking up of these large works into short, indi-vidual discovery claims could threaten to break up the immensely productive collectives that had governed French science for several decades. This risked destroying a powerful form of social cohesion.

Biot's glum reflections on the shifting publics of science stopped short of overt criticism of Arago's innovations. But others were less kind. While re-publican journalists such as Raspail saw the *Comptes rendus* as an imperious coup organized by an illegitimate sovereign against the true scientific public, some within and without the Academy saw in it a dangerous renunciation of long-standing forms of life in favor of the market-driven logic of the periodi-cal press. Arago's detractors accused him of "applying to science the absurd system of universal suffrage through which Arago has reaped so much false glory."[172] Ostensibly a gesture of openness, the *Comptes rendus* was a covert means of wielding personal power over French science:

> In the past, the Academy admitted into its collections only memoirs that it had put through a careful examination, and this favor was regarded a great honor. Today, it is simply at the pleasure of the perpetual secretary that the decision is made to insert a memoir into the *Comptes rendus,* and one can imagine that it requires great character in an author to resist the seductions of a publication which is fast, far-reaching, free, and which may be obtained easily through a few acts of deference to the reigning powers.[173]

Indeed, while academicians still did write reports on memoirs and machines, these became fewer in number over time. The role of the Académicien as *rap-porteur* gradually took a backseat to acts of authorship often in periodical works.[174] The members' adjudicatory activities were kept up with respect to prize competitions, but insofar as the *Comptes rendus* became the focus of the Academy's public life, it was stepping back from its traditional role as the rep-resentative of public scientific judgment.

As societies and academies began to compete with the scientific press on its own terms, this brought new legitimacy to the scientific publications of the commercial press itself. The fate of these bodies became tied inexorably to these new forms of publicity. While Biot worried that the Academy was in danger of becoming simply a "bureau for posting free advertisements,"[175] Raspail argued that publications like the *Comptes rendus* were an attempt to "stifle critique." It is just this tension which will be explored in the following chapters, as elite scientific institutions refashioned their purposes and publics through the new periodical format that they created in the 1830s.

The new journalism at the Academy has usually been glossed by historians in terms of the rise of science popularization in France, but this sells short its significance in three ways. First, these accounts of meetings were read widely by savants themselves both inside and outside the Academy and thus had a profound effect on how the Academy functioned. Second, this scientific genre was as much about the changing politics of science as it was about its popularity.[176] Its legitimacy as a form of scientific writing was at the heart of debates about the precise nature of the Academy as a public institution. Finally, the radical version of the genre that emerged in the early 1830s influenced the Academy's decision to found its own journal in 1835, a periodical that changed the very nature of this institution. One aim was to bury the hostile arguments of the republican scientific press. But in another, more significant, sense it was acquiescing in the radical argument. The articles and discussions in the *Comptes rendus* ultimately became the heart of the Academy's identity, as the figure of the *rapporteur* gradually faded from view over the succeeding decades.

When historians of science map out the social landscape of nineteenth-century science, we often separate popular journals from professional journals. We sometimes speak of the importance of "low scientific cultures" that have been unjustly overshadowed by the elite science of the Royal Society of London or the continental academies in Paris, Stockholm, and elsewhere. But partitioning the landscape of science in this way can lead to confusion if it cuts off inquiry into the ways in which seemingly peripheral communities and practices might have shaped elite science itself.[177]

The rise of formats and genres for communicating about the natural world was important not simply because it allowed new groups to participate in science. These formats and genres have sometimes been incorporated into elite science itself, transforming institutions in the image of those publics.[178] During the first half of the nineteenth century, the institutions through which the authors of discovery claims sought recognition were increasingly uncertain. Journalism offered a powerful—and seemingly unavoidable—medium through which to imagine the future of discovery.

3

The Author and the Referee

As the locus of scientific legitimacy began to diffuse into the marketplace of periodicals, the leaders of elite scientific societies and academies began to reevaluate the grounds of their own claims to be sites for collective knowledge making. Scientific journals were becoming venues not simply for diffusing scientific news, but for conferring scientific identities, rewards, and credibility. Yet the marketplace of periodical print remained an unwieldy instrument for these tasks. What did it mean for an author to publish in scientific journals? What was the status and credibility of the claims embodied in scientific papers? What was the relationship between oral presentations at meetings and their printed versions? Were certain publications to be counted more credible than others? Could the outer bounds of legitimate scientific publishing be identified? The next three chapters each examine struggles in Britain and France over how to answer these questions and thus to carve out a stable space in which periodical publishing could serve an expanding range of social and epistemic roles in scientific life.

The present chapter focuses on authorship and judgment. It takes place wholly in Britain, where professional authors and publicly recognized expert judges remained uncertain identities in the early decades of the nineteenth century. The attacks on elite science launched by radicals in Britain and France discussed in the previous chapter both had roots in the expanding reach and power of public opinion. But they took distinct forms on either side of the Channel, in part because they were directed at different kinds of targets. In France, François-Vincent Raspail's crusade against academic judgment concerned the legitimacy of state-sponsored scientific experts. But in British natural philosophy, the expert was an identity that had a more tenuous existence, particularly in natural philosophy. The Royal Society of London had no formal procedures by which to exercise public judgment comparable to those of continental academies. And while membership in such societies might indicate something about one's social status or general interest in philosophical topics, it was no indication of specialized competence.

In the late 1820s, working-class radicals' critiques of aristocratic science in Britain were taken up by middle-class and aristocratic reformers within elite science itself. British political institutions were going through their most fundamental transformation in over a century, inspiring visions of parallel reforms in science. Among other things, reformers wanted means of identifying true scientific practitioners in the hopes that delimiting a core of professionals would prompt the state to take more interest in natural philosophy. But it was not clear what a scientific professional should be. No course of studies qualified one to be an authority on natural philosophical topics, and scientific practitioners did not produce any obvious product or service. Faced with this dilemma, some began to promote particular forms of authorship as criteria by which to pick out true contributors to knowledge.

Twinned with the rising fortunes of scientific authorship was the rising profile of scientific reading, which took on new significance and power at the same time. In the 1830s, the Royal Society of London gradually introduced changes that turned it increasingly into a publisher, collector, and curator of printed texts. Following the examples of the Astronomical and Geological Societies, it launched a proceedings journal in 1831. And the Society's library was gradually changed from a repository of objects to a site designed for scientific research. Finally, the Society implemented a system for delegating readers to write reports on memoirs submitted to the Society. Out of this gradually emerged a new scientific personage that would have a profound effect on scientific life in general — the referee.

But when the scientific referee appeared on the scene of British science, it was not certain who he was, or what he was supposed to be for. Organizations such as the Geological and Astronomical Societies had begun to experiment with procedures that delegated special readers to appraise scientific claims. At the Royal Society of London, William Whewell suggested importing the system of reports used by Paris academicians. Whewell was not much concerned about preventing shoddy papers from being printed. Instead, he was looking to raise the public visibility of science and forge a more coherent identity for the scientific enterprise in England. But translating the report-writing practice across the Channel proved more complicated than Whewell expected. Public reports soon became shrouded in secrecy, and Whewell's public authorities came to be known as "referees," who were increasingly viewed as agents who had the power to confer rewards on authors while putting down "the obtrusive and turbulent pretensions" of others. In this personage came to be juxtaposed elements of the legal expert, the trustworthy gentleman, the state bureaucrat, and finally the anonymous book reviewer.

These changes in the Royal Society's mandate and mode of working paralleled a massive expansion of the publishing activities of the British government itself, which entered the print marketplace on a grand scale in the 1830s as a publisher of official documents. The explosion of paper documents was encouraged by reformist imaginations of the public sphere as a (mostly middle-class) reading audience, conceiving the commercial press as the model of legitimate political exchange. As the relevant public of the Royal Society's activities was similarly reimagined as a reading public,[1] the twinned rise of the scientific author and the referee offered two linked identities for the public-oriented scientific expert.

Authorship and Distinction

In the late 1820s, British scientific practitioners joined political crusaders calling for the reform of institutions of governance. Beginning with Catholic Emancipation in 1829, the political reform movement reached its symbolic apex with the 1832 Reform Acts in England, Scotland, and Ireland. Each of them was aimed at modernizing parliamentary elections by eliminating seats under the effective control of aristocratic landowners while extending the franchise to parts of the middle class. At the same time, conceptual schemas were being invented through which to articulate new political identities, including various forms of radicalism, liberalism, and socialism.[2] As in France, public opinion was now widely viewed as a crucial source for claims to political legitimacy, and its expression was coupled in the minds of many to the periodical press.[3]

Scientific reformers interrogated the Royal Society in a parallel manner, as a representative body for science whose legitimacy depended on public confidence. "So long as this learned body spent only 'their own property,' the public had no right to interfere," a letter to the *Times* declared in early 1830. "Now, however, that they have sold their independence for 'grants of public money,' their conduct is become a legitimate object of 'public investigation.'"[4] Participants consciously framed these disputes in the contemporary language of political debate.[5] "The Great Contest," wrote Davies Gilbert to Robert Peel regarding the election to the Royal Society presidency in 1827, "is the Conflict of Aristocratic and Democratic Power. I wish the Royal Society rescued from the latter." In 1830, when John Herschel learned that Gilbert (by then president) was conspiring to choose his own successor, he used the language of reformist indignation to condemn Gilbert for imagining "that he can hand it over like a rotten borough to any successor be his rank or station what they may . . ."[6] How to choose the Royal Society's president also depended on the nature of the relationship between scientific collectives and the state. Were representative bodies or appointed bodies of experts to be called on by the state to give their collective opinion on matters of public concern in return for recognition and financial support, or was the connection to governmental power best managed by means of personal patronage? How was membership in such a group to be determined, and what exactly did it mean?

Although calls for reform of the Royal Society had been mounting since 1820, when its long-standing president, Joseph Banks, passed away, matters came to a head in 1830 in a flood of printed polemics. Charles Babbage's long tract, *Reflections on the Decline of Science in England,* remains the most well-known product of this controversy, but the press war began with a controversy over the Royal Society's library.

In November 1829, a plan by the Society's Council to exchange a collection of valuable manuscripts donated to the Society long before by descendants of the 21st Earl of Arundel for duplicate books from the British Museum became public information.[7] Babbage had attempted to intervene at the Royal Society's annual public meeting, arguing that the books offered by the Museum were scientifically worthless, and that the Royal Society would be better off selling the manuscripts to the highest bidder and using the resulting windfall to renovate the library. A correspondent in the *Times* (Babbage denied writing it) brought the matter before the general public, noting how much better the situation was in Paris:

... the library of the National Institute, the pride and envy of Europe, where there is to be found almost every work that can tend to illustrate the most minute branch of science, and where the facility of access, and the attention to strangers, are alike proverbial ... to rival, or even to emulate, such an establishment as that, would indeed be worthy the labours of the Royal Society.[8]

Complaints that the Royal Society was not doing its duty as a steward of scientific print would continue to be a point of controversy. The astronomer James South, Babbage's most irascible ally in the controversy, noted among his "charges" against the Society that it kept its "*books* in Cellars, pronounced by the geologists, too damp even for the reception of *flints*." It spent money on "mace gilding, picture cleaning, and other frivolities" while its library deteriorated. "It is a fact which will be scarcely credited in other countries," he lamented, "that the library of the Royal society, does not contain a single number of the *Annales de chimie* ! ! !" Meanwhile, John Herschel had complained that the *Annales* was far superior to any periodical publication produced in Britain.[9]

But the state of its library turned out to be a microcosm of many other organizational failings. Babbage and David Brewster argued that these had made it inevitable that the government would neglect natural philosophy. In order to legitimize science as an activity, Babbage called for modernizing the Society through "full publicity, printed statements of accounts, and occasional discussions and inquiries at general meetings." They also thought the Society had to play a key role in delimiting who was a legitimate member of the corps of scientific practitioners. Both constantly invoked the example of France to demonstrate that given the right political and scientific culture, natural philosophy could thrive under governmental patronage. Its endowed academies and civil honors granted to savants — an amalgamation of prerevolutionary privilege, Napoleonic-era patronage, and meritocratic zeal — seemed to them the foundation for a more noble social identity for men of science. If British institutions could follow a similar course, science might become a profession alongside such classical professions as law and medicine.[10]

In 1827, the Royal Society's president, Humphry Davy, had permitted a Committee to be established for the purpose of exploring ways to limit the membership of the Society. A subcommittee was tasked with undertaking a historical study of the criteria used in admission. Babbage and a number of self-described reformers were a part of these committees, and while their recommendations were ignored by the Royal Society's new Council in 1828, they became the starting point for Babbage's book. The Committee began with the premise that "the utility of the Society is in direct proportion to its respectability." Respectability, in turn, derived from "the public conviction, that to belong to the Society is an honour." But at over 700 Fellows, with dozens more elected each year in a haphazard way, belonging to the Society seemed virtually meaningless.[11] The Committee recommended capping membership at 400 and making the election process far more rigorous.

In order to make their case, Babbage et al. also wanted to show that there truly had been a decline in science. The Committee decided that the way to measure this was by looking for trends in the number of individuals actively engaged in natural philosophical pursuits over time. But just who counted as an active natural philosopher? There was no educational certification necessary to engage in science, and the closest approximation to a guild or professional organization were scientific societies, but belonging to one of these did not imply much about one's active pursuit of new knowledge. Of course, they could have investigated who had been credited with making certain discoveries or inventions, who was active in teaching, who was an active collaborator in ongoing research, or even who had published more generally on topics related to natural philosophy in a variety of formats and genres. Instead, they kept things simple by concentrating on data that they had easy access to: which Fellows had published memoirs in the *Philosophical Transactions*.

Comparing periods from the early to mid-eighteenth century with those of the nineteenth century, the Committee showed that the proportion of Fellows elected in a given year who had published in the *Transactions* was "materially less than in the middle of last century." Even worse, this falling off was particularly stark among Fellows who were most active in recommending new Fellows for election, a situation likely to perpetuate mediocrity. (They did not seem to be aware that the radical changes in format and selection criteria that the *Transactions* underwent after the mid-eighteenth century rendered the comparison dubious.)[12] The Committee was certain that the dire situation they had identified required corrective action. Realizing that it would be far into the future when membership numbers were brought under control, they recommended turning publication in the *Transactions* into a kind of honor in itself. Special lists might be published of just those Fellows who had published papers in the *Philosophical Transactions*, and these Fellows ought even to be given a discount on their membership dues.

In *Decline of Science*, Babbage reiterated this call to bring about a "division of the Society into two classes," one made of active contributors to knowledge and another containing those for whom science was more spectator sport. He continued to insist that the best criterion for activity was who had published in the *Transactions*. The numbers worked out very nicely: of the 714 Fellows of the Royal Society, only 109 were authors in the *Transactions*.[13] A more select class could be formed of those who had published at least *two* papers in the *Transactions*. This gave a list of seventy-two names, which Babbage noted was about the size of the Parisian Academy of Sciences. Restrict further to three papers, and you were down to fifty-five Fellows. Giving these classes official status, Babbage thought, would help establish a scientific meritocracy, and it would give real natural philosophers a solid basis on which to claim honors and rewards, ideally from the state.

Others found inspiration in Babbage's idea. Later that year a small book appeared with the provocative title *Science Without a Head, or the Royal Society Dissected*. Although it was framed as a rebuttal of "the noisy ones" who were decrying the decline of English science, it largely extended many of Babbage's arguments about the Royal Society. It was signed "ONE OF THE 687 F.R.S. — s s s," highlighting the unwieldy bulk of the Society's membership, but its author was Augustus Bozzi Granville, an Italian physician who had lived in London for many years. His pamphlet heaped ridicule on what he saw as the faulty logic of Babbage's claims that English science was declining. Yet he agreed very much that the Royal Society itself needed fixing, and he also agreed with the premise that the "measure of the labours of the Royal Society may be said to be found in its Transactions." He launched into an ambitious study of the Royal Society, its publishing activities, and its Fellows. He wanted to size up its membership, to understand "through what claim any particular person has been admitted into the Royal Society," and to determine "the class to which he belongs." This was among the first efforts at a collective biography of a scientific group. By correlating scientific achievement with social class, Granville hoped to arrive at "a fair approximation to a real representation of the scientific public in England." In a series of tables, Granville "dissected" the List of Fellows of the Society according to their profession or social rank, placing "against each individual member his claim to the honour of having been admitted as such, based upon what he may have done in the way of 'improving natural knowledge.'" Just like Babbage, his sole criterion was authorship in the *Philosophical Transactions*. By this limited criterion, very few Fellows had ever contributed anything at all to knowledge. Nevertheless,

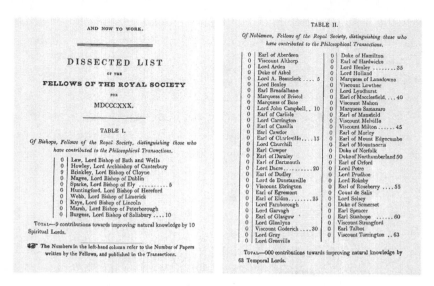

Figure 3.1 Tables 1 and 2 in Augustus Bozzi Granville's anonymous *Science Without a Head* (1830) listing the Fellows of the Royal Society of London who were bishops and nobles, respectively, alongside their "contributions towards improving natural knowledge," measured by the number of memoirs each had published in the *Philosophical Transactions*. Courtesy of Princeton University Library.

there was some interesting variation according to rank and profession. Considering only the bishops in the Royal Society, one found nine publications by ten individuals (albeit very unevenly distributed). Clergymen were rather less productive as a group: eight papers out of seventy-four Fellows. Physicians fared somewhat better. The class of noblemen was the most embarrassing: "000 contributions towards improving natural knowledge by 63 Temporal Lords" (Figure 3.1).[14]

For both Babbage and Granville, contributions to the *Transactions* were roughly to be identified with contributions to science *tout court*. Active men of science could thus be demarcated from "mere lookers on—indifferent spectators—or, at most, cultivators of what beds of flowers they found in the rich garden of natural knowledge when they first entered it."[15] Granville interpreted the exceptionally poor showing of noblemen as evidence of the especial futility of this particular constituency in advancing British science. And yet this was perhaps just as much an indication of the ambivalence that still attached to authorship for those of the highest rank in British society. The former president of the Royal Society, Joseph Banks, had often taken this view of authorship, "Quietly Contenting myself with the station I hold in Life & not like the dog with his Shoulder of mutton hazarding what I Feel to be Comfortable in search of the Reputation of an Author which in Fact I do not Consider as a gentlemanly vocation."[16] Some were quite happy to allow others to write up and publish reports on their own research. Adrian Johns has shown that, far from clamoring to be known as authors, practitioners of natural philosophy in Britain sometimes had recourse to anti-authorial practices, including cloaking their works in anonymity, putting out their works as if they were unauthorized printings, and resisting print entirely in favor of manuscript circulation.[17] Among the 000's in Granville's chart of the scientific nobility, for example, was Edward St. Maur, the Duke of Somerset. If the Duke's name was not connected directly with any great discoveries, his many engagements in London scientific life — he was vice president of the Zoological Society at the time and president of the Linnean a few years later — certainly went beyond patronage and were contributions to the collaborative advance of knowledge.[18] He later warned his son that it was best to avoid ostentatious displays of philosophical authorship, and to focus instead on practical works.[19] Those in the aristocracy who did decide to take the authorial plunge, such as the Duke's brother Webb Seymour, were aggravated to find strings attached to hazarding a work for the press. Upon deciding to publish a memoir in the Edinburgh *Transactions* on the geology of a region of Scotland, he reported to his brother his despair at being confronted by the "new and tiresome duty of correcting proof sheets," a complaint he bemoaned in his diary day after day.[20] Somerset made certain never to commit the same mistake and never published anything at all.

The ambiguities of authorship are exemplified by the bizarre case of Everard Home, whose nearly 100 papers in the *Transactions* during the period Granville analyzed made him its most prolific contributor by an order of magnitude. It was a lightly concealed fact that the source for many of these papers was the anatomical collections and manuscripts of his mentor, the eminent Scottish surgeon John Hunter, who had chosen to publish only a fraction of his original investigations in his lifetime. Home later called his papers the basis for a *catalogue raisonné* of Hunter's collections. In 1834, when a parliamentary committee investigation made it widely known that Home had decided to burn most of Hunter's manuscripts in 1823, he became known as a literary thief and plagiarist.[21]

If in Britain the meaning of authorship remained thorny, some forms of authorship had long been valued by the continental institutions to which Babbage looked for inspiration. The rules of the Royal Academy of Sciences in Paris had always included a requirement involving print publication as a possible qualification — among others — for admission to membership: all candidates were to be "known by some substantial printed work, by some celebrated course, by some machine of his invention, or by some particular discovery."[22] In eighteenth-century Prussia, publication criteria became increasingly important in deciding appointments to professorships in the emerging research university. In 1749, authorship had become a required qualification for obtaining university positions.[23] But the critical consideration was a prospective candidate's "fame." What this meant could vary, from the "applause" that "resided in the size and success of one's lectures," to the renown that came with having authored well-known works. What was most prized was fame among the broad reading public rather than a reputation "within a small group of specialists gained through original contributions from one's research."[24] Babbage was not interested in such general acclaim. Instead, by restricting his purview to a single prestigious form of publishing, Babbage was zeroing in on the collective opinion of just such specialized groups: "Few estimations will be found generally more correct," he argued in *Decline*, "than the opinion of a whole profession on the merits of any one of its body."[25]

Fixing on authorship as the measure of a life in science highlighted certain kinds of contributions while it buried others. The romantics' elevation of the literary author as creative genius — and the sidelining of others involved in the production of books — was recapitulated in this argument for conferring special status on the scientific author. While a man of science might collaborate with a wide variety of philosophical friends, and might draw on the special skills, inventions, and capacities of a wide array of individuals (artisans, inventors, assistants, students, collectors, local observers, editors, publishers, patrons), authorship was generally conceived as the act of an individual. While a team might be responsible for editing a periodical or assembling a catalog, those genres that were associated with original discoveries were generally figured as singular, comparable to the creative genius associated with literary or philosophical works.[26] Although coauthorship of scientific memoirs and books was not unheard of, it was relatively rare and did not factor into the concept of authorship coopted by Babbage and others. This erased from the record not only those like the Duke of Somerset who eschewed authorship for reasons of social propriety but the artisans who built the instruments that made advances in astronomy and optics possible.[27] Thus the name of William Herschel, one of the most eminent and prolific contributors to the *Philosophical Transactions* in the decades surrounding the turn of the century, represented an astronomical workshop employing as many as two dozen observers and artisans. Foremost among these was his sister, Caroline. His son (and her nephew) John later speculated that it was precisely because of Caroline's extremely industrious collaboration that William had ever been able to accomplish his "regular and constant communication of discoveries to the public in a succession so rapid as to outstrip all competition."[28] Caroline's biographical recollections make clear that many a paper by William was written at the same time that Caroline was carrying out the telescopic observations and calculations that would be used to write the next one. She was also a gatherer together of fragments, an amanuensis, the principal calculator, and copyist. Much later on Caroline herself was honored with a medal by the Astronomical Society, but that honor was bestowed specifically for a work credited to her alone, a summary reduction of star clusters and nebulae that she compiled after her brother's death.[29]

The massive expansion of scientific coauthorship beginning in the later twentieth century has produced a great deal of confusion in the relationship between credit, responsibility, and collaboration. This has made quite plain the fissures in the romantic concept of authorship as applied to scientific work, but the roots of these troubles were already present when the identity of the scientific author was coming into its own.[30] Turning the man of science into an author may have helped produce a coherent identity for the man of science, but it also obscured the many ways in which varied forms of activity and collaboration were central to the activities of natural philosophers and scholars. The next chapter will explore the link between the twinned romantic conceptions of discovery and authorship in more detail.

Experiments in Judgment

There were other reasons to be skeptical that scientific eminence could be measured by counting certain kinds of publications. Quality, one critic pointed out, surely mattered more than quantity: "The criterion of merit adopted in estimating the scientific pretension of the Fellows of the Royal Society seems to be the *number* of their papers printed in the Society's '*Transactions*,'" noted Nicholas Harris Nicolas, "but no allusion is made either by Mr Babbage or by the author of '*Science Without a Head*,' to a far more satisfactory criterion,— the *value* of those papers."[31] The recommendations of the 1827 Committee on admissions had also raised the question of the value of what got published, if indirectly. It recommended that the Royal Society should start appointing small committees to examine submitted papers "who should report, not only their opinion, but the grounds on which that opinion is formed, for the ultimate decision of the Council."[32]

Granville was worried about the same thing. His case for the Royal Society's collective lethargy relied on the assumption that contributions to knowledge were roughly identical to contributions of text to the *Transactions*. But he knew there was a problem with this reasoning, for he was suspicious of the means by which the Society determined just what it chose to print. Granville claimed to have witnessed incidents in which memoirs were "rejected without assigning any ground." He thus decided "to go a little more behind the scenes" and launch an investigation in the Society's archives. He went back through decades of minutes of the Committee of Papers and the judgments they had rendered, looking for clues to the reasoning behind their decisions. (See Figure 3.2 for a table that summarized his findings.) What he found was, to his eyes, utterly scandalous. On the one hand, he detected ample evidence of personal and national bias. One example was the discovery of the asteroid Pallas, about which he claimed that several earlier papers submitted by continental astronomers had been rejected without cause, "nor any thing allowed to be placed on record in the *Transactions* respecting this second interesting discovery of a planet," until an insider, William Herschel, finally got around to the subject.[33] But even more important than outright corruption was that the process for accepting or rejecting papers was a travesty, and rendered the Society an incompetent judge of discovery claims.

Since the mid-eighteenth century, and following John Hill's attacks on the Society and the *Transactions*, the Society had convened a Committee of Papers whose main job was to decide which papers read at its meetings should be published. Abstracts might be read and the manuscript itself might be passed around, and then the Committee would vote on each case. Granville found this to be absurd; routinely participants at these meetings would "have not the smallest pretension to any knowledge whatever of the subject under consideration, or indeed to science in general":

To face page 59.]

ANALYTICAL AND NUMERICAL TABLE

Of all the Papers that have been read before the ROYAL SOCIETY, from the beginning of the present Century up to the current Year (1830)—distinguishing each Branch of Science on which the Papers were written, and the Number of them in each Branch for every Biennial period—noting those that have been published in the Phil. Transactions with the letter A (accepted), and those that have not been printed, or were withdrawn, or simply deposited in the Archives of the Society, with the letter R (rejected).

YEARS of the 19th Century.	Higher Mathematics.	Algebra.	Geometry & Geodesy.	Hydrodynamics.	Astronomy.	Mechanics—including Naval Architecture.	Exp. Philosophy, including Optics, Dynamics, Barom. and Thermom. Observations, Pendulums and Magnetism.	Chemistry.	Mineralogy—Geology—Metallurgy & Fossilology	Natural History, including Physical Geography.	Anatomy.	Physiology.	Vegetable Physiology.	Medicine.	Surgery.	TOTAL of Papers read, on all Subjects collectively, in each Biennial Period.
1801 1802	2A	1A	2A 1R		5A 15R	1A 1R	8A 3R	9A 2R	5A	8A 2R	1A	2A	1A	1R	2A	72, of which 25 R.
1803 1804					6A 3R	1A	3A 3R	8A 2R	5A	2A 1R	2A	1A 1R	4A			42, of which 10 R.
1805 1806	4A 1R	1A	2R		5A 3R	2R	8A	6A	6A 2R	2A 1R	5A 3R	2A	4A			57, of which 14 R.
1807 1808	1A		2A	1A	5A 2R	2R	3A 2R	12A	4A 1R	2A	2A 2R	3A 1R	4A			42, of which 10 R.
1809 1810	2A 1R			1A 3R	2A 1R	3A 1R	5A 3R	10A 1R	4A 1R	3A 2R	3A	7A	2A	3A 1R	1A	60, of which 14 R.
1811 1812	1A 2R		2A 1R		5A	2A	2A 3R	11A	4A	3A	1A 2R	5A 5R	2A	1A 1R	1R	54, of which 15 R.
1813 1814	3A				11A	7A	9A 1R	21A 1R	4A	4A 1R	2A 1R	6A 2R	1A	1A 1R		76, of which 7 R.
1815 1816	5A	1A		3A 2R	3A	1R	10A 3R	12A 5R	3A	5A	2A 1R	6A 4R	1A	2A 2R		71, of which 18 R.
1817 1818	4A 3R	1A			10A	2A	11A 3R	6A 3R	2A 1R	9A 2R	2R	6A 1R	2A 2R	1A 1R	2R	74, of which 20 R.
1819 1820	4A 3R			1A	2A	4R	13A 3R	4A 1R	2A	5A 3R	4A 2R	4A 1R	1A 2R	1R	1A	61, of which 20 R.
1821 1822		1R	1A		7A 2R	3A	14A 6R	11A	3A 1R	9A	4A 3R	8A			1A 1R	75, of which 14 R.
1823 1824	1R			1A	15A 2R		14A 5R	9A 2R	6A 2R	3A 3R	4A	2A 1R	1R	4A		77, of which 17 R.
1825 1826	3A		1A	2A	6A 5R	3A 1R	18A 8R	6A 1R	1A 1R	6A 1R	3A 2R	5A 1R			1A	75, of which 20 R.
1827 1828	1A			1A	11A 1R	2A	17A 6R	6A 5R		7A 1R	2A 2R	1A	1R			66, of which 18 R.
1829 1830	1A 1R		2A 1R	1R	2A 1R	2A 1R	24A 4R	5A 3R		1A 1R	4A 2R	4A 3R		1R	7R	67, of which 27 R.
TOTAL of Papers in each Branch for 30 Years.	31A 12R — T. 43.	4A 1R — T. 5.	10A 5R — T. 15.	10A 6R — T. 16.	95A 35R — T. 130.	26A 13R — T. 39.	159A 55R — T. 214.	136A 26R — T. 162.	49A 9R — T. 58.	71A 18R — T. 89.	39A 22R — T. 61.	62A 20R — T. 82.	22A 6R — T. 28.	12A 9R. — T. 21.	6A 7R — T. 13.	GENERAL TOTAL. 732A 244R 976

Average Number of Papers printed yearly 24⅔, ditto of Papers rejected yearly $8\frac{1}{10}$.

The Numbers in a former Table for 31 years, including the Year 1800, are quoted at 1000, or 751 accepted and 249 rejected: in that particular year, therefore, there were 19 Papers printed and 5 rejected.

Figure 3.2 Augustus Bozzi Granville's "Analytical and Numerical Table of all the Papers that have been read before the Royal Society, from the beginning of the present Century," classified by subject and distinguishing those rejected or ordered for publication. In *Science Without a Head* (1830), following p. 58. Courtesy of Princeton University Library.

> Assuredly it cannot be expected that a sculptor for instance, a painter, a secretary to the Admiralty, an astronomer royal, and a botanist, congregated together, should come to a right decision respecting the propriety of publishing a paper on physiology or internal anatomy!

Occasionally the Committee did call on more "competent judges" to give a preliminary opinion, but Granville's research had convinced him that "much oftener is the fate of a paper committed to the chances of the mere yea-and-nay box."[34]

For late modern observers habituated to institutions that refer many decisions to professional experts, Granville's shock is not very surprising. And it is true that there had been much criticism of the Committee of Papers in the past decades. But the basis of such critiques had not normally been that it was in principle unsound, only that the Committee had become corrupt and weak, a mere puppet of the president's will.[35] Granville was instead condemning the legitimacy of this process of judgment *as it was intended to function.* This was a novel criticism. It was not obvious that a system delegating power to one or two individual specialists was a more legitimate method, and less subject to prejudice, than a ballot taken by a distinguished committee of savants. Granville himself admitted that the system he was advocating was also "vastly objectionable" in many ways.[36] Then as now, specialists in the same branch of knowledge might indeed know the subject best, but they very likely also knew the author through ties of friendship, mentorship, or rivalry. The Committee that Granville dissected in terms of its members' specialized knowledge was made up of members of the Society's Council, a twenty-one-member body elected annually by vote of the Fellows. It was "on the character and conduct" of Council members, as one Fellow put it, that the "progressive success of the Society must mainly depend." The Council made decisions on nearly all matters of the Society's business. Its dual identity as the Committee of Papers was central to how it was supposed to be selected, the thought usually being that it should "combine the highest qualifications that, in each department of science, the country affords." Coverage of various sorts of special knowledge was important to maintaining an "equal balance between its own laborious and ingenious members."[37]

Notably, while Granville questioned the justice of the selection procedure, neither he nor anyone else who critiqued the Society around 1830 focused their attack on the quality of the papers that *did* get published in the *Transactions.* In contrast, the Committee of Papers had been set up in the mid-eighteenth century precisely in response to those kinds of attacks. When John Hill satirized the papers in the *Philosophical Transactions,* he himself saw the ideal solution in a committee made up of a range of respected Fellows to inspect the papers, even suggesting that before publishing papers the "sense of the whole Society be heard upon them."[38] Later, Joseph Banks would explain to would-be contributors that the method of collective voting helped avoid public embarrassment, for when questionable papers got published, "Criticks [*sic*] in the Reviews would instantly attack."[39]

Committees who passed collective judgment in this way had become a regular feature of those societies that chose to publish "transactions" in the late eighteenth and early nineteenth centuries. New societies boasted of the care and disinterestedness with which they carried out such decisions.[40] The archetype of this form of judgment was not an individual who made up his mind through solitary reading but a group who listened, perused, and discussed. Societies that set up such committees did so not to cut corners but because this was widely viewed as the most legitimate means of ensuring high-quality contributions. Many societies did include provisions to seek the help of specialists in perusing particular papers, but this tended to be informal and was by no means required — the weight of authority remained squarely with the group. The Royal Society guarded against any suggestion that these decisions would be left to individuals by insisting that any non-Council member who was brought to a meeting to comment on a particular paper also vote on all the other papers before the Committee that day.[41]

Still, the Society — and others in Britain and Ireland who employed this system — tended to rebut suggestions that the Committee's task was "to answer for the certainty of the facts." They claimed only to evaluate the "importance or singularity of their subjects."[42] This stance contrasted markedly with many continental academies, where making judgments on the veracity and utility of discovery claims and inventions had long been at the heart of their mission. The French Academy's *rapports* were an instance of a widespread genre of technical reports commissioned by the state on a variety of subjects.[43] Those possessing the authority to write such reports were commonly known as *experts*. The French *Encyclopédie* defined them as "those who are versed in the knowledge of a science, art, or some type of merchandise or other thing, who are chosen to write reports and give their opinion on a point of fact upon which depends some contested decision and which cannot be properly understood without knowledge belonging to a certain profession."[44] Until the later eighteenth century, the prototypical French expert possessed practical, local knowledge associated with surveyors and merchants.[45] But the notion that an expert's authority might also rest on theoretical knowledge gained ground in the second half of the eighteenth century. Although the basis of expert authority thus expanded over time to include not only practical but theoretical knowledge, what remained constant was that an expert was not simply one who was considered to have specialized knowledge but that this was considered the basis on which to decide a matter of public concern.

The term *expert* in this sense was rare in Britain. Early uses of the noun in English sometimes alluded to its French roots, as in the 1791 edition of *Gilbert's Laws of Evidence*: "proof from the Attestation of Persons on their Personal Knowledge, we may properly, with the French lawyers, call proof by Experts."[46]

In the legal domain English law included no special category for expert witnesses until the nineteenth century. Those with specialized competence who might be deemed relevant to a case were most likely to be called to serve as a member of a jury (thus as part of a collective), or perhaps as an advisor to a judge, rather than to give their opinion in court. The problem was that witnesses were supposed to testify as to the facts of a particular set of events, but someone who testified about a case based on their specialized competence was simply offering a personal opinion in the eyes of the court. For men of science whose authority depended on theoretical knowledge, such legitimacy was especially hard-won, and became routine only gradually in the nineteenth century. It was only then that courts began to recognize the sorts of theoretical knowledge possessed by men of science as a legitimate basis for credible opinions.[47]

Some groups of natural philosophers had begun to explore strategies to raise the public profile of specialized scientific opinion, whether by imitating continental models or borrowing from classical professions such as the law and medicine. One of those groups was the Astronomical Society of London (founded in 1820), many of whose core members were leading reformers in 1830. Astronomical knowledge was of long-standing importance to the state for its uses in navigation; individuals acquainted with the subject had long occupied positions as governmental advisors and employees, staffing public observatories, advising the Royal Navy, and populating the Board of Longitude. The precise nature of these positions, and just how they were filled, were central concerns of the Astronomical Society's founders. The Admiralty in particular was populated by individuals, many of whom achieved their positions of influence through aristocratic connections or the patronage of Joseph Banks. The new Society had been set up in opposition to this model, as a committee of specialists that ought to be called on to advise the state as an independent body. In many cases, they set about doing the work of the Admiralty and what they saw as the indolent Board of Longitude themselves, working on the Star Catalogs and Nautical Almanac that were supposed to be the latter's official business. As the secretary of that Board, Thomas Young (a frequent target of attacks by members of the Society) complained that the Society would "never rest satisfied until the Astronomical Society . . . shall succeed in dictating to the *Admiralty* and the *British Parliament* . . ."[48]

The Astronomical Society's leaders insisted on their status as a public body from early on. With the assistance of their printer, Richard Taylor, they were among the first specialized societies to provide detailed proceedings of their meetings in the periodical press. The Society's members drew inspiration from the French state-sponsored academies whose duties included the formation of expert committees to write reports on scientific and technical questions. They also followed the academies' practice of forming committees to consider manuscripts of papers for possible publication. But while the language of committees and reports in the bylaws seemed inspired by continental academies, in practice the process worked somewhat differently. Committees did not necessarily write reports at all, and when they did, they were never made public. Indeed, their comments were communicated to the author via the secretary. It could even happen that a committee would give its approval of a manuscript at the very meeting during which it was nominated.[49] The Society by no means received a glut of memoirs in its early years, and committees accepted most of what they were given to inspect, so that the stakes in founding this system do not seem to have had much to do with gatekeeping. Rather, the Society's use of such committees to report on knowledge claims was part of its attempts to adopt practices of judgment associated with continental academies, and thus to differentiate itself from the societies of gentlemen exemplified by the Royal Society of London.

The Geological Society of London was another group looking to differentiate itself from the Royal Society. It also launched an experiment using specialist judges to evaluate manuscripts, but instead of the French academic expert this one was inspired by another kind of professional judge: the referee. Though in its first decades the Society's composition resembled the Royal Society's core membership of gentlemen amateurs, its leaders were keen to embrace a more public identity. At their founding they nearly chose to break with the Royal Society model of transactions and publish a scientific journal, "printed in as cheap a manner as is consistent with neatness — an octavo number similar to Nicholson's journal." (They did not follow through with this, however, until publishing their proceedings about two decades later.) They took the unusual risk of allowing their meetings to incorporate open debate and discussion, an experiment that observers such as Babbage thought made the meetings not only "very entertaining" but more useful. The Society also pursued projects designed to be valuable to the public interest. Under the guidance of its founding president, George Greenough, the Society set itself up as a self-appointed national repository of mineral specimens and of geological facts, and Greenough initiated a Society project to assemble a geological map of British natural resources.[50]

Few fields were as riven with controversy as geology at this time, and the Council worked to police the collective identity of the Society. Theirs was to be a science in which older Enlightenment speculative theories of the earth were to be banished. There were numerous struggles over what the Society would publish in its early years, and the Council exercised a heavy editorial hand on the memoirs it received. In 1813, the mineral surveyor John Farey was so offended by the Council's attempt to rewrite and abridge his memoir on the geology of Derbyshire that he became a public antagonist of Greenough and the Society. Farey attacked what he called the "Geognostic Society" for its unjustified prejudice against the work of practical surveyors such as himself and William Smith under the guise of distrusting speculation. He cited numerous instances of the Council's duplicitous and nontransparent dealings with authors, implying that Greenough abused his power to bury Smith's cartographic contributions while pilfering them for his own geological map of England. In 1817, just when Farey published a book-length version of his Derbyshire research with a venomous preface aimed at the Society, Greenough drafted a new, highly complex set of procedures by which the Society would exercise control over what it published.[51]

The Society had started out with a simple system of Committee vote that mirrored the Royal Society model. In practice the Council might rely on some particular member to give an opinion on a paper, or to edit it for publication, but there was no requirement to do so. Greenough's new rules left no room for such flexibility. They comprised twenty-eight articles from the act of submission to eventual publication. Greenough was a lawyer by training, and this showed not only in his approach to Society business but in his geological practice more generally. Much of his geological writing betrayed a commitment that truth could be teased out through something resembling legal criticism of geological thought.[52] The new rules required a succession of judgments, starting with the secretary who could reject submissions immediately, to the Council vote which remained the final word. But now all papers that made it through the first pass of Council would be sent to a "referee," whose job was to read the paper assigned to him and write "on a separate sheet any remarks that may occur to him & any alterations which he may think it desirable to make in the paper."[53] These remarks would then be given to the Council, along with the abstract and an estimate of illustration costs.

The Council approved Greenough's suggestions in their totality. In theory and in practice, the "referee" system contrasted with the "committee" system set up by the Astronomical Society. A referee system implied individual judgments, whereas the committees of the Astronomical Society supposed collective opinions (although surviving records suggest that committees sometimes consisted of a single individual). Unlike the astronomers, the Geological Society took seriously the written nature of referee reports, even archiving them. Both systems were in principle meant to keep reviewers' identities anonymous, although this anonymity was regularly lifted. The pool of reviewers appears to have been small, normally not extending beyond the respective Councils.[54] The reports for the Geological Society that survive from the early years are usually very short, often simply suggesting yes or no to publishing, but evidence suggests that it was common for referees to correspond directly with authors over suggestions for revision.[55]

Among those societies that sometimes sought the opinion of an individual reader, it was already common to use the verb "to refer" when designating that individual, but Greenough's statutes inaugurated what seems to be the first systematic use of "referee" to indicate a specially designated reader of scientific manuscripts.[56] "Referee" was perhaps the nearest common English word to the French term *expert*, though a nearer French translation was *rapporteur*. In legal parlance, the referee was an independent authority appointed by a court or agreed on by two parties to settle a dispute by arbitration.[57] The word was also sometimes used more informally for anyone asked to give their opinion or advice on some matter of concern. Referees sometimes appeared as characters in literary works; often they were portrayed as dubious characters, prone to overvalue their capacity for sound judgment and more interested in getting paid than in producing a good decision. A verse that made its way through legal circles starting in 1821 included a typical jab at these mercenary arbiters:

> When all expences are incurr'd,
> The parties will have it referr'd;
> And you may be the referee,
> If in Court you chance to be,
> And earn a twenty guinea fee.[58]

Despite expressions of ambivalence such as these, formal uses of the term were just then spreading beyond legal contexts to designate special assessors in matters from engineering to medicine. Physicians, for example, were finding a new service role as "medical referees," working on behalf of the life insurance industry to evaluate the health of potential subscribers. The Geological Society's use of the term to designate advisors whose job it was to decide the legitimacy of knowledge claims makes sense in this context. Like the Astronomical Society, the Geological Society's system reflected an aspiration that particular forms of philosophical knowledge might qualify its bearers with the sort of competence that courts and other state bodies ought to appeal to as independent authorities.

By the 1820s a few learned societies were thus already experimenting with systems for soliciting specialist opinion on new papers. Babbage and Granville likely had these precedents in mind when they recommended that the Royal Society adopt such a system itself. But their arguments were ignored, and the matter was revived only when William Whewell attempted to import the French academic system of public reports across the Channel. It was only then that referee systems began to attract broader public attention as a new institution of scientific judgment.

The Reading Public and the Public Reader

The institution of a referee system was one element in a broader set of changes that took hold gradually at the Royal Society over the 1830s, many of which focused on its role as a publisher and steward of printed objects. The public reform crusade appeared to have fallen apart at the end of 1830 when the reformers failed to have John Herschel, the eminent astronomer, elected the Society's president instead of Augustus Frederick, Duke of Sussex (and brother to the King). In public Babbage and his allies appeared to retreat from the business of the Society for years. It was nearly two decades before electoral reform finally happened at the Royal Society. But the Royal Society nevertheless underwent crucial changes in the years immediately following this confrontation. Alongside the establishment of a system for reporting on manuscripts, it began to publish a great deal more regularly about its activities by founding its *Proceedings* journal. The function and makeup of its library was also reconceived, giving new prominence to facilitating literary research as part of the Society's mandate.

At the same moment that Charles Knight's *Penny Magazine* was highlighting the printing press as a powerful emblem for the diffusion of knowledge (Figure 3.3), the Royal Society gradually embraced an image of its own public

𝔐onthly 𝔖upplement of
THE PENNY MAGAZINE
OF THE
Society for the Diffusion of Useful Knowledge.

112.] November 30 to December 31, 1833.

THE COMMERCIAL HISTORY OF A PENNY MAGAZINE.—No. IV.
(Conclusion).

PRINTING PRESSES AND MACHINERY.—BOOKBINDING.

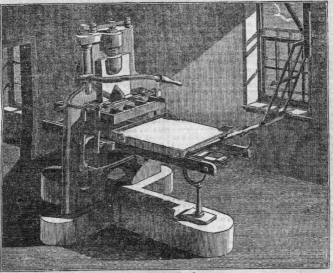

[The Stanhope Press.]

THOSE who have examined the early history of printing will scarcely have failed to see how the ordinary laws of demand and supply have regulated the progress of this art, whose productions might, at first sight, appear to form an exception to other productions required by the necessities of mankind. There can be little doubt, we think, that when several ingenious men were, at the same moment, applying their skill to the discovery or perfection of a rapid mode of multiplying copies of books, there was a demand for books which could not well be supplied by the existing process of writing. That demand had doubtless been created by the anxiety to think for themselves, which had sprung up amongst the laity of Catholic Europe. There was a very general desire amongst the wealthier classes to obtain a knowledge of the principles of their religion from the fountain-head,—the Bible. The desire could not be gratified except at an enormous cost. Printing was at last discovered; and Bibles were produced without limitation of number. The instant, therefore, that the demand for Bibles could be supplied, the supply acted upon the demand, by increasing it in every direction; and when it was found that not only Bibles but many other books of real value, such as copies of the ancient classics, could be produced with a facility equal to the wants of every purchaser, books at once became a large branch of commerce, and the presses of the first printers never lacked employment. The purchasers of books, however, in the fifteenth and sixteenth centuries, were almost wholly confined to the class of nobles, and those of the richer citizens and scholars by profession. It was a very long time before the influence of the press had produced any direct effect upon the habits of the great mass of the people. In our own country, the many hundreds of pamphlets of political and religious controversy that were issued during the times of the civil wars, were unknown to the larger portion of those who took sides in the quarrel. They were directed to the important body of landed proprietors, and the no less important leaders of the people in towns; and they were formed to influence, as they were in great part produced by, the active spirits, whether of the church, the bar, or the senate,

VOL. II. 3 T

Figure 3.3 "The Commercial History of a Penny Magazine," a series of articles in which the Society for the Diffusion of Useful Knowledge highlighted the media technologies that helped constitute its imagined public. *Monthly Supplement of the Penny Magazine,* December 1833. Courtesy of Harvard University Libraries.

in which readers were given far more prominence. This paralleled a broader shift in the politics of representation that followed the passage of the Reform Acts by the Whig parliament. Beginning with the publication of the reports of the Poor Law Commission in 1833, the British state became a prolific publisher in the 1830s, producing and selling numerous reports on parliamentary inquiries about a range of facets of British society. Public printing became a frequent subject of inquiry itself, with committees dedicated to exploring all aspects of format, genre, and distribution. The state also took steps to open up its archives, founding the Public Record Office in 1838. And while the minutes of parliamentary debates remained in the private hands of the Hansard firm, government officials relied on Hansard's experience with government reporting to establish a house style that transformed a mass of manuscript reports into printed documents whose uniform and regular appearance gave them the appearance of periodical publications.[59] Likewise, the Royal Society relied on the printer Richard Taylor, who had earlier catalyzed the proceedings journals at the Astronomical and the Geological Societies, to dictate much of the format of their own new *Proceedings* journal.

Many of the officers that were elected to run the Society under Sussex not only supported the reformers but remained on good terms with them. Peter Mark Roget, one of the two secretaries, confided to friends that without some reconciliation the Society was likely to sink lower still.[60] The new treasurer and vice president, a young banker named John William Lubbock, had maintained good relations with nearly all factions. Lubbock was in line to inherit a baronetcy and had moved to London after his student years at Cambridge to work at the family bank. At Cambridge he had begun to do work on problems in physical astronomy and counted William Whewell as a mentor. He also discussed mathematical matters with Babbage, who solicited his opinion on *Decline of Science*. Lubbock reported that he agreed with much of it, though he felt that Babbage was too much focused on the capital: "I think you might have entitled it with more truth 'on the decline of science in London,'" as science did not seem to be waning in cities like Cambridge, Edinburgh, or elsewhere.[61] Roget and Lubbock were both keen to bring Babbage, Whewell, and others back into the fold. In 1831 they helped set up a new reform committee, and invited many of the self-described reformers to participate. Most declined, protesting that they were busy with more important things. Babbage drafted a long, bitter response about the impossibility of reform as long as the current regime remained in power.[62]

But within a year Lubbock had managed to entice Babbage into collaboration. Lubbock told him that the Society had been moving forward with its plans for the library, and Babbage simply could not resist giving advice — as long as he could do so in secret. The Society had ultimately decided to follow Babbage's counsel by obtaining money for the Arundel Manuscripts instead of settling for duplicates of British Museum books. They were now taking steps to get rid of "unscientific books" as well, and would use the funds to replace them with more useful ones. For the first time, the Society began systematically to invest funds in building up a working library of key periodical works.[63] Just as important, the Society planned to publish a detailed catalog of their newly renovated, increasingly scientific, library.

Babbage was excited by this news. Even better, Lubbock and the secretaries wanted advice about what to buy. Babbage was put in touch with the bookseller Baillière, who was locating many of the books to be purchased. Babbage saw the printed book as a crucial technology for leveling of hierarchies; by this medium, even for those without access to formal education, "the greatest and wisest men of all countries and all times, may become their instructors."[64] Babbage was also an avid cataloger, just then in the midst of producing a general catalog of scientific tables, and he offered to help here too. "Do let me suggest to you an idea," he wrote to Lubbock. Wouldn't it be more useful to produce a catalog listing all of the scientific books across several existing catalogs whether or not they were in the Society's library? Printing "in octavo in the cheapest form possible" would make the volume cheaper and thus more affordable for individuals to purchase. An asterisk could be put beside those books in the Society's own library so that it could also function as its catalog. Babbage predicted that it "would be an example to other societies . . . and to the man of science it would be a treasure."[65] His imagination was fired by the possibility; he reported to Lubbock that he spent the night meditating "upon the most feasible means of executing" what he was now calling "the Grand Cat. of Science." An even longer letter set out a detailed procedure for cutting up catalogs and pasting them into folios under classified subject headings, right down to whom they could hire to do it. Babbage had in his employ "a young lad of 17 who contrives to exist on ten shillings a week by putting french polish on furniture." The boy was reading mathematical books and seemed to know "something about them in a strange way." He suggested he might be set up in the Royal Society Library to do the work of cataloguing at a bargain rate.[66]

Lubbock was somewhat alarmed by the scope of this plan, but he did welcome Babbage's enthusiasm: "Any suggestions from you on this or on any other subject connected with the Society will always find the greatest attention from me."[67] They also discussed progress in the project to publish abstracts of papers and proceedings of meetings, another project that Lubbock was spearheading and which Babbage had encouraged. Babbage wanted to ensure that the new *Proceedings* that were being sent to Fellows be made available as widely as possible. They communicated on other schemes as well, Babbage using Lubbock as a conduit for suggesting ideas to the president but insisting that his name be kept out of it: "any opinion of mine would only injure the cause with him . . ."[68]

Lubbock and the Society were moving forward on several other fronts. Eventually, Babbage's idea that membership fees should be discounted for those who had published at least one paper in the *Philosophical Transactions* was partially adopted.[69] Financial contributions to the Society were thus exchanged for contributions of papers. Lubbock had also reached out to William Whewell. Whewell had not only been a mentor of Lubbock's at Cambridge, but they had become collaborators in developing the science of tides, a significant preoccupation of both of them during the 1830s.[70] Whewell was interested in expanding the publishing activities of the Society. Like Babbage and Brewster, he saw in the Royal Academy of Sciences at Paris much to be emulated. The continental academies that had been founded on the Paris model were known especially for their reports on the state of knowledge. He explained that it was through reports by authorities such as Georges Cuvier and Joseph Fourier in Paris, or Jacob Berzelius in Sweden, that these academies were most revered abroad. If the Royal Society engaged in something similar, it "would tend much to increase both the energy of those who pursue science and make communications to the society, and the interest felt on such subjects in the world at large."[71] (Whewell later wrote similarly to Vernon Harcourt regarding arrangements for what became the British Association, where the plan of reports on the state of science was taken up.)[72] Whewell also suggested one other kind of report:

> I believe it is also the practice of the R.A. of Paris to refer most or all memoirs received by it to a committee of a few persons known to be acquainted with the subject in question and to require from these committees written reports

which are often printed in the Annales de Chimie and elsewhere, and which are often more interesting than the memoirs themselves, containing both good abstracts & judgments upon the subject by the best authorities.

Whewell suggested writing to Paris, to find out just how they did it. The advantage of such reports would be "the encouragement to writers from the certainty of being appreciated and the facility of diffusion of scientific information by abstracts and critiques."[73]

When Whewell brought the idea to Lubbock, he pointed out that these reports would be an excellent addition to the new *Proceedings*.[74] Whewell repeated this point to Harcourt in the fall: "All committees on memoirs presented or on any subject concerned with science ought to give <u>public</u> reports of their views."[75] What Whewell had in mind was nothing less than a new genre of scientific writing in England. Although the Astronomical Society had been quietly using small committees to report on manuscripts for some time, what Whewell was suggesting was a far more public role for expert readers, especially because of the prestige associated with the Royal Society.

Babbage's and Granville's vocal demands for a system of specialist judgment had gone nowhere, but Whewell's variant received more traction. Whewell and Lubbock agreed to collaborate on the first report, which became a kind of test case for the system.[76] They chose a manuscript that had been read that November by their colleague, the astronomer George Airy, called "On an Inequality of Long Period in the Motions of the Earth and Venus." Things, however, did not go well.

Airy's paper was a sequel to a paper he had submitted and published in the *Philosophical Transactions* in 1828. That paper had accounted for empirical discrepancies between recent observations at Greenwich Observatory of the sun and the predictions in the widely used solar tables computed by Joseph Delambre. In working this out, he had been led to hypothesize that there was an as-yet-unnoticed inequality in the motion of the Earth due to its interaction with Venus (which had a period of 240 years). Accounting for this based on the mathematical theory would require an arduous series expansion of the equations governing the system and a search for the correct higher-order terms. This task would take a great deal of time and effort, and he had put it off until 1831, when he at last worked through the calculations and submitted them, through John Herschel, to the Royal Society.

When Lubbock and Whewell got a chance to read through the paper in detail, they each knew exactly what they thought of it. And they completely disagreed.

Lubbock quickly concluded that the paper, admirable as it was in the intricacy of its calculations, had taken serious liberties with the mathematics of Laplace and Lagrange. He wrote a long letter to Whewell (who was in Cambridge), outlining his reservations. Lubbock did not seriously dispute that the result of the calculation was correct, but he was perplexed by a term for the epoch that Airy had included in the equation which he thought both superfluous and misleading. "The theorems of Lagrange are perhaps the most beautiful analytically considered which have ever been discovered," he reported, "but the introduction of such a term in them . . . destroys entirely their analytic beauty."[77]

Whewell did not share this concern. The vehemence of Lubbock's protests gave him pause, but he did not believe that Airy had made a serious error: "The difference between Laplace and Airy," he replied, "may be a matter of opinion."[78] Lubbock went to the author himself to see if he could convince him to make changes, but found Airy "not disposed to admit my objection about dR/dt &c."[79] Airy reported to Whewell that he was not pleased by Lubbock's visit: "I must say that I do not particularly admire being put on my trial in this manner . . . There the paper is and I am willing to let my credit rest on it." (He also noted that Lubbock "does not understand this particular branch of Physical Astronomy so well as I do.")[80]

Airy was annoyed not simply because he disagreed with Lubbock's critiques but because he disagreed that this was an appropriate way to deal with papers submitted. Through the Astronomical Society, he had been exposed to their own system of reports. In fact, in 1829 he had been asked to review a paper submitted by Lubbock himself on the calculation of the trajectory of comets. In response, Airy protested that he was "not perfectly acquainted with the rules of the Council in regard to the printing of papers" and was wary of giving an "unqualified judgment on the whole." He stuck to relatively neutral remarks.[81]

Lubbock expressed concern that requiring reporters to write joint reports would lead to constant problems such as these, since consensus would often be hard to obtain.[82] Both Whewell and Lubbock ended up drafting their own reports on Airy's paper, and they could not have been more different. Whewell's focused on the difficulty of the task Airy had set himself, the remarkable conclusions he had come to, and the place of this work in the history of physical astronomy (and British contributions to it in particular). Lubbock's went into detail about the unfortunate misconceptions Airy built into his equations, and his failure to come up with a general expression for the terms in the disturbing function he had located.[83]

It was not simply that Whewell and Lubbock disagreed about the merits of Airy's paper; more problematic was that they disagreed about what a reader's report should be. Whewell argued that their disagreements were relatively minor, not "important to our report, except perhaps for a phrase or so." He elaborated:

> For I do not think the office of reporters ought to be to criticize particular passages of a paper but to shew its place with regard to what has been done already ... I should think it by no means for the good of the R.S. that persons who sent us papers were to feel that they were put <u>upon their defence</u>.[84]

Lubbock admitted the potential difficulty, noting that he too did not wish "to throw any difficulty in the way of procuring papers," but he stood his ground. As far as he could tell, Airy had not "ever read Lagrange's theory of the variation of arbitrary constants"—an unpardonable sin for a mathematician. The problem that now arose was fundamental, he thought: "if we are to make reports upon papers I do not see how we can pass over grievous errors. This is the difficulty to be encountered in making reports & renders it impossible unless the report can be of a nature extremely favorable."[85] Lubbock felt that a reporter's duty ought to include keeping an author from falsifying the historical record: "if any one principle is worth establishing, it seems to me to be that authors should as much as possible make their work dovetail with what has been done before."[86]

Whewell and Lubbock seemed to have reached an impasse. Lubbock threatened to withdraw his name altogether: "I have read your report, which can be read to the Council as <u>yours</u>."[87] Whewell was faced with the potential failure of his new system with the very first report. But he was determined to make things work, noting that "it would be much better that we should make a joint report than separate ones." In the new year, he made some concessions, and offered to make more "by alteration of the phraseology." After a few more weeks Lubbock relented, citing the bigger picture: "As the first report which the Council have ever made, I think the Society ought to feel greatly indebted to you for setting the example & putting your shoulder to the wheel." He then assured Whewell, "[I] have signed my name under yours."[88]

The report was accepted by the Council and it, along with a report by Samuel Hunter Christie and John Bostock on one of Michael Faraday's "Experimental Researches in Electricity," was presented later that spring at the Society. As planned, it was included in the new *Proceedings* of the Society alongside abstracts of papers (Figure 3.4).

The new *Proceedings* thus contained not only abstracts of papers presented at meetings, but also occasional reports on papers. The opening up of the Society exemplified by the *Proceedings* impressed even the *Mechanics' Magazine*, not known to be kind to elite establishments. In 1834, its editor J. C. Robertson praised the Society's change in direction: "It is by no means an unpleasing spectacle to see so stiff and unbending a Society losing, either by choice or on compulsion, a good portion of its haughty spirit of exclusiveness."[89]

But there was room for cynicism about powerful institutions appropriating commercial forms that had become associated with progressive alternatives to elite culture. In France, François-Vincent Raspail had reacted in just this way to the Academy's founding of the *Comptes rendus*, calling it an attempt "to seize a monopoly" on a key genre of the scientific press. And in Britain, where the state had also entered the literary marketplace through its massive publishing efforts, publishers such as William Cobbett complained that the state's flooding of the literary marketplace with cheap, authorized accounts of inquiries into social questions was designed to control public debate, rather than to foster the free exchange of ideas. By 1838, the *Mechanics' Magazine* was still extolling the Society's embrace of more open reporting, though it began to wonder if it had gone far enough. Robertson noted that the move to publish more cheaply and regularly had "been absolutely called for by the voice of a scientific public, *rather* more extensive in its numbers, and more impatient in its demands, than the public of the unlocomotive age in which the Society began to flourish."[90] By 1838, in fact, while the *Proceedings* lived on, reports on papers had already disappeared from its pages.

From Public Authority to Anonymous Referee

The conflict that attended the inauguration of the Royal Society's system of reporting on manuscripts never became a matter of public knowledge. But it encapsulated two distinct models of what a scientific judge ought to be. By the

March 29, 1832.

GEORGE RENNIE, Esq. Vice President, in the Chair.

The following Report, drawn up by the Rev. William Whewell, M.A. F.R.S., and John William Lubbock, Esq. M.A. V.P. and Treasurer R.S., on Professor Airy's Paper, read before the Royal Society on November 24, 1831, and entitled, "On an Inequality of Long Period in the Motions of the Earth and Venus," was read.

Report.

The object of this memoir is similar to that of Laplace's celebrated investigation of the great inequality of Jupiter and Saturn, announced in the Memoirs of the Academy of Sciences for 1784, and given in the volume for the succeeding year. The occasion of that investigation was an acceleration of the mean motion of Jupiter and a retardation of that of Saturn,—which inequalities in the motions of the two planets Halley had discovered by a comparison of ancient and modern observations: and Laplace showed, in the Memoirs just referred to, that inequalities like those thus noticed would arise from the action of gravitation; that they would reach a considerable amount in consequence of twice the mean motion of Jupiter being very nearly equal to five times the mean motion of Saturn; and that their period would be nearly 900 years. The occasion of the investigation of Professor Airy was an inequality in the sun's actual motion, as compared with Delambre's Solar Tables, which appeared to result from a comparison of late observations with those of the last century,—as Professor Airy has explained in a memoir published in the Philosophical Transactions for 1828. This comparison having convinced him of the necessity of seeking for some inequality of long period in the earth's motion, it was soon perceived that such an inequality would arise from the circumstance that 8 times the mean motion of Venus is very nearly equal to 13 times the mean motion of the earth. The difference is 1,675 centesimal degrees in a year, —from which it follows, that if any such inequality exist, its period will be about 240 years.

To determine whether such an inequality arising from the action of gravitation, amounts to an appreciable magnitude, is a problem of great complexity and great labour. The coefficient of the term will be of the order 13 *minus* 8, or 5, when expressed in terms of the excentricities of the orbits of the Earth and Venus, and their mutual inclination; all which quantities are small; and the result would therefore, on this account, be very minute. But in the integrations by which the inequality is found, the small fraction expressing the difference of the mean motions of the planets enters twice as a divisor; and by the augmentation arising from this and other parts of the process, the term receives a multiplier of about 2,200,000. In the corresponding step of the investigation of the great inequality of

---

Figure 3.4 The reader's report as public genre. The published version of the report written by William Whewell and John William Lubbock on George Biddell Airy's manuscript appeared in the ninth number of the *Proceedings of the Royal Society* (1832, p. 108), later gathered together as the third volume of the *Abstracts of the Papers Printed in the Philosophical Transactions of the Royal Society of London.* Courtesy of the Biodiversity Heritage Library.

1830s, Whewell had emerged as one of the most respected figures in British natural philosophy, but he was renowned more for the breadth of his interests and his broad philosophical view than his special knowledge of any particular branch of it.[91] For many early nineteenth-century observers alarmed about the disorienting pace of the growth of knowledge, it was just such synthetic discernment that was called for in order to separate the wheat from the chaff. Lubbock was a decade younger and near the beginning of his scientific career; his own work in physical astronomy (some of which had just appeared in the *Philosophical Transactions*) was directly relevant to Airy's paper. Whewell's report writer was a generalist, familiar with many branches of knowledge and able to locate a discovery in the grander scheme. Lubbock represented the opposite tendency, a reader whose special knowledge allowed him to pick apart the details. Whewell represented authority that was above the fray, publicly declaring judgment. Lubbock's reader was more likely to be the author's equal — and a direct competitor. Whewell's was one for whom trust was the dominant principle; Lubbock's reader was more suspicious, and doubt was the ruling impulse.

It was Whewell's vision that the Society embraced at the beginning. Its president, the Duke of Sussex, waxed eloquently about the experiment at the Anniversary meeting, recalling its inspiration in the Paris Academy, made up of elites that claimed to represent authoritative scientific judgment:

> As the persons who are selected for this duty are frequently veterans in their respective sciences, who have earned by their labours an European reputation, the Reports which are thus produced prove often more valuable than the original communications upon which they are founded. The collections of them, as is well known, form a most important part of the stock of modern science.

These judges, according to the Duke, derived their authority as much from what amounted to social distinction, being "elevated by their character and reputation above the influence of personal feelings of rivalry or petty jealousy." They used that authority to "establish at once the full importance of a discovery," to "fix its relations to the existing mass of knowledge." Would-be authors could thus be assured "that their labours will be properly examined and appreciated by those who are most competent to judge of their value." Finally, a system of reports could play the role, at the level of publishing, that men such as Babbage and Granville agreed was necessary: to distinguish deserving men of science from "the obtrusive and turbulent pretensions of those who presume to claim a rank as men of science." The report writer was a publicist and a conferrer of rewards.[92]

But this initial vision transformed as the system was put into practice. What soon emerged not only looked more like Lubbock's vision, but was also more similar to the systems at the Astronomical and Geological Societies. Not long after Sussex extolled the value of public readers' reports, those same reports — and their writers — became shrouded in secrecy. Appointed readers stopped attempting to write joint reports. When an author inquired about the Society's rules in 1836, he was informed that "the usual course is to refer the paper to a Sub-Committee of two or three persons, nominated by the Council as being particularly competent to give an opinion on the subject, & who report to the Council, generally in writing, their opinion of its merits & of the expediency of publishing it in the Transactions: & the decision of the Council on this latter question is always taken by ballot."[93] Although the old system of voting was retained, referees (that term found its way to the Royal Society soon) were systematically appointed and their opinions were followed almost without exception, and the reports themselves were carefully archived.

For a time, the report system retreated into such obscurity that some who followed the workings of the Society weren't even sure that the system was even in operation. In 1837, Granville, in an expanded version of *Science Without a Head*, praised the Society for inaugurating the system but assumed it had already ceased operating altogether.[94] Authors knew better. While authors were supposed to remain ignorant not only of who had refereed their paper but of the contents of the reports, in practice the secretaries broke that rule often, and referees themselves regularly contacted authors directly to discuss their papers and offer advice or an explanation.

The system operated at a much larger scale than those of the smaller specialized societies, and it proved to be a complex and ambitious operation to maintain, especially before the establishment of the Penny Post in 1840. Since making multiple copies of manuscripts was not easy to do, referees (there were usually two) often had to read and report on papers in sequence. If a referee mislaid a paper,[95] this might very well constitute a literal loss of the author's contribution to knowledge. Moreover, while the referees' reports came to be viewed as independent of one another, in practice the second sometimes had access to the first report, since it might be passed along with the manuscript.

Other societies, including the Zoological Society of London and the Royal Society of Edinburgh, set up referee systems of their own within the next few years.[96] As more reports were commissioned, men of science became accustomed to being called upon to write them. The genre of the reader's report slowly began to take shape, even if the expectations of what a report might contain could vary radically (as it has continued to do). Some reviewers wrote a single sentence giving their general view of a paper, while others wrote longer disquisitions that resembled Whewell's original vision of public reports.

As the practice of refereeing evolved, the imagined function of writing reports also diverged from what Whewell had originally intended. Though Whewell had imagined reports as a form of publicity and encouragement to authors, he continued to defend the system even as it was gutted of what he had taken to be its key features. In 1836 his friend James David Forbes complained to him that Royal Society referees were harassing him over a paper of his about hot springs, criticizing not so much his facts but his style. Such a system, he worried, had "a dangerous tendency to degenerate into a system of minute interference."[97] Whewell's response gave this rationale for its importance: "it gives us a useful and indeed indisputable means of excluding from the Transactions papers which for matter or form, ought not to appear there."[98] The referee had become a gatekeeper, a legitimate defender of the Society's reputation.

For astronomers and mathematicians such as Lubbock and Airy, who dealt with mathematical deductions or with publicly available sets of observations, there was some hope of verifying claims simply by working with texts. But how such verification was supposed to work was less clear for the majority of papers, which concerned new observations or experiments. Referees were readers, and no one expected report writers to replicate experiments or make new observations in the field. In contrast, the committees who reported on discoveries and inventions at the Academy of Sciences did sometimes take replication to be a fundamental part of their job, and they could call on the resources of the Academy or other institutions to this end.[99] Controversy of this kind arose quite soon. In 1837, the physician Marshall Hall, upon receiving a negative verdict about a paper, demanded that the Society "*appoint a commission, to witness my experiments, to examine my plain deductions from them, to look over my Paper with care*, before you finally condemn my labours."[100] Virtual witnessing through the printed page had always depended on gentlemanly trust. By appointing judges whose job it was to take a suspicious eye to knowledge claims, the Society was throwing such gentlemanly conventions out the window. In that case, Hall implied, only total replication or true witnessing made sense. The referee as suspicious reader was an illegitimate half measure.

How did the shift to anonymous and individual reports come about at all? The archives of the Royal Society don't provide a clear answer, but it is not surprising that this should have occurred. Continental institutions could not simply be shipped across the English Channel without adapting to local circumstances. British scientific culture possessed little established role for report writers as public experts, and there was no state-sponsored scientific oligarchy. No one, not even Whewell, could truly claim to represent scientific opinion in the manner of earlier generations of French savants. (Even in France the academic report was progressively diminishing in its importance.) Nor was there much precedence for the supposed meritocratic hierarchy between *académiciens* and *étrangers* (all those who were not members of the Academy). All of these circumstances made it difficult for Whewell's model of the public reader as judge to stick. Too much had been lost in translation.

A more familiar model of public judgment in Britain was the anonymous critic epitomized by the contributions to the most prestigious quarterly and monthly periodicals. Publications such as the *Edinburgh Review*, the *Quarterly Review*, and the *Westminster Review* consisted largely of reviews of books (even if reviewers often used the platform as a pretext to hold forth freely on a given topic), and British reviewers' identities remained cloaked throughout this period.[101] The logic underlying this practice was best articulated when it came under attack. Such attacks remained rare until the 1860s, when anonymous criticism began finally to lose its dominance, although they did occasionally arise. One such moment occurred in 1832, when Edward Bulwer-Lytton took over the editorship of the *New Monthly Review*. Bulwer-Lytton decided to banish anonymity from his journal's pages. Invoking the example of France (where anonymity in the press was less extensive),[102] he argued that while concealing the identities of critics offered justifiable protection in times of political upheaval, it too often encouraged them to engage in slander and irresponsible behavior. This experiment was short-lived, however, for Bulwer-Lytton was quickly replaced. The new editor reinstituted anonymity and explained his predecessor's folly, articulating a widespread view:

> The anonymous, and the mysteriousness attached to the plural unit We, seem best adapted to the chair of criticism. The individual is merged in the court which he represents, and he speaks not in his own name, but ex cathedra. . . . The decisions are oracular. What a totally different air would they assume, and how soon would they dwindle into the insignificance of mere individual opinion, if the name of the writer of each article were appended at the end![103]

The authoritative critic, to be unencumbered by party feeling, could only act without a name. Justifications of the anonymity of the scientific referee took a similar view. Of course, the scientific referee's report was not simply anonymous but private; nevertheless, it was precisely this oracular authority that the referee came to assume. Later, in 1863, an aggrieved Royal Society author made "aspersions against the moral character of one of the referees" of his paper, threatening to publish the reports (which he had managed to obtain). The secretary, George Stokes, responded by setting him straight about the personage he was attacking: "Such ascription of unworthy private motives influencing public conduct is at variance with the usages of society and is particularly painful when made by one Fellow of the Royal Society against another."[104] The referee discharged a public service *through* privileged secrecy.

It took just a decade for the referee to become an established scientific persona, and the secrecy of these systems was soon under attack. In 1845 the *Mechanics' Magazine* turned from praise of the Society's more public-oriented character back to its more customary tone of attack. It reviewed a self-published pamphlet on factorial analysis by a schoolteacher, Thomas Tate, which the author had proudly prefaced as having been read before the Royal Society. The reviewer took pity on Tate's naïveté in revealing—to those who understood what this really meant—that his paper must thus have been rejected by the Society's referees. He took the opportunity to explain to readers how this system worked: "The Reports are secret—even the Reporter's name is not minuted—and the author is left at the mercy of this Star Chamber jurisprudence." Authors were left completely in the dark even as to criteria of acceptance, the *Magazine* reported: "We know of no conditions whatever being required to be fulfilled in the quality or subject of a paper, which the Society lays down as a criterion of its approbation or condemnation."[105]

Later that year, a short-lived monthly called *Wade's London Review* published a multipart exposé on the Royal Society. It was likely written by a Fellow of the Society or someone closely connected with it, for the author displayed intimate knowledge of its machinery of judgment. This critic also emphasized the mysterious nature of referee reports: "There is, as far as we can learn, no record kept even of the person to whom a paper is referred; much less is there any accessible means of learning what the report itself may be!"

> No remark upon the subject is recorded, no word of encouragement to the author, no hint as to the course he should pursue in such inquiries, no allusion to its particular class of defects, no suggestion as to where the same or collateral subjects have been discussed, nor to the real desiderata which should be the author's object of pursuit.

Just what kind of person was the referee? In terms of his technical competence, he was likely a close colleague or even a rival: "some person of the same class of pursuits, a rival for fame in the same line of inquiry, carrying on a similar course of investigation, meeting perhaps with obstacles which the referred paper itself may have successfully removed." In terms of moral competence, it was more difficult to say.

> The referee may be a man of integrity in general matters; he may have no personal animosity, no "green dragon" in his eye; he may even soar above all personal feelings . . . On the other hand . . . he may be full of envy, hatred, malice, and all uncharitableness . . . His enemy is in his hands, the darkness of night covers the deed, no record can exist of the part he takes in the matter, and he is overcome by the temptation![106]

The implicit accusation that a cloaked referee was a dishonest one was common. The delays occasioned by using the cumbersome referee system brought accusations that they enabled a new kind of piracy. Referees could reject papers or simply delay passing judgment on them while they went on to publish on similar topics.[107] Secrecy seemed only to invite accusation: "If the judgment be a righteous one, why not give us the clearest evidence of it? — why seek to shroud yourselves in the privacy of the Star-chamber, or the darkness of the Inquisition?"[108]

It was not just outsiders and authors who complained about the system. Referees themselves sometimes bristled at a task that was "thankless in itself" and which was likely to put them in "fear of giving offence & favour" despite the façade of anonymity. This was the opinion of the mathematician J. J.

Sylvester. In his view the Society had managed the feat of "combining the disadvantages incidental to the two systems of open judgement and secret scrutiny and achieving none of the advantages belonging to either of them." He assured the Society of his "exceeding unwillingness to be again called upon to act in the capacity of referee."[109]

Public protest of the Royal Society's secret system of reports began to mount at precisely the moment that governmental secrecy also became a hot topic in British political culture. In 1844, the scandal of the season in the London papers was the revelation that the British government had been operating a clandestine system of postal espionage. Giuseppe Mazzini was an Italian in London who was known to be connected to revolutionary elements in Italy. The Austrian ambassador, concerned about a potential insurrection in Italy, asked British officials to put Mazzini's communications under surveillance. When Mazzini realized that a letter he had received might have been intercepted and read by others, he had his correspondents set a trap by sending him a letter folded in a complicated way to see if it arrived in an altered state, which it did. Mazzini and his British supporters brought their evidence to Thomas Duncombe, a radical M.P., who brought the matter to the House of Commons. Accusations of illegal covert surveillance led to investigations launched in both houses of parliament, and calls for the resignation of the Home Secretary, Sir James Graham, who shouldered the public blame. The scandal exposed a significant secret surveillance mechanism within the British government, leading to widespread protests in the press:

> This anxiety for secrecy on the part of public officers is a growing evil. In the customs, in the stamp office, in various government departments, we hear now of common clerks sworn to secrecy, or told by their superiors that if they communicate to the public any information connected with the business of the office, they will be instantly dismissed.[110]

The image of secret reader-spies ensconced in a secret office policing English communications fired the public imagination. Both the *Illustrated London News* (Figure 3.5) and the *London Journal* published front-page illustrations of what the secret chamber of the post office might look like, with men seated at tables poring over letters, copying and breaking seals. *Punch* satirized the situation for months, lambasting Graham and his troop of "London Postmen." It even made a good business selling special "Anti-Graham wafers"

SECRET-OFFICE, AT THE GENERAL POST-OFFICE.

Figure 3.5 "Secret-Office, at the General Post-Office," *Illustrated London News* 4 (29 June 1844): 409. This was an imagined depiction of the office charged with postal espionage in which letters addressed to persons under surveillance would be opened, copied, and resealed. Courtesy of Harvard University Libraries.

with slogans such as "HANDS OFF" to be affixed to letters. Illegitimate governmental secrecy and the corruption to which it led were a major theme in the best-selling serial novel *Mysteries of London* that year. The author, George Reynolds, dedicated three chapters to what he called "The Black Chamber," the post office's espionage wing in which "the secret springs of that fearfully complicated machine [the British Empire] were all set in motion and controlled."[111]

The British state had always practiced some quantity of espionage, but the notion that publicity and transparency were hallmarks of liberal government had been gaining ground. "Secresy is an instrument of conspiracy," in the words of a Jeremy Bentham essay first published in 1839; "it *ought not, therefore, to be the system of a regular government.*"[112] Whig governments of the 1830s and 1840s had passed legislation that reduced the barriers to communication in civil society, including reducing the Newspaper Stamp Tax in 1836 and establishing the Penny Post in 1840. The Public Record Office had begun to make the archives of government available to public consultation, and parliament had flooded the marketplace with government reports. The Mazzini Affair prompted the government largely to abolish its postal espionage program. But scandals such as these also put the growth of governmental paperwork in a negative light. It helped establish a new public imaginary of the sinews of power based on the ability to manipulate the masses of documents being produced and kept by those in positions of influence. In a chapter titled "The Document," the narrator of *Mysteries of London* explained that the result was that while "brute force is now less frequently resorted to," London had become an even more perilous place. The criminal element was simply more refined, operating under a "cloak which conceals modern acts of turpitude." It was this nightmare that also animated Dickens's *Bleak House*, which portrayed the Court of Chancery as a vicious battleground focused on mastering paperwork.[113]

In a similar way, as scientific eminence came to be connected to certain forms of authorship, scientific judgment was increasingly imagined as inhering in a covert network of reports written by those with real scientific power. In 1848, after a particularly acrimonious scandal over a refereed paper that played out in the press,[114] the Royal Society system was at last codified in by-laws, with specialized committees given the job of running the system. This was the moment when the Royal Society officially incorporated a concept of specialized expertise into its structures. It also finally implemented electoral reform, restricting membership and making admission to the Royal Society an honor. The dual institution of referee reports and of restrictive membership criteria cemented the position of the anonymous referee as an iconic personality — if mysterious and mistrusted — in British science. For a new generation of scientific practitioners, some of whom were looking to make a career, the referee had become a point of obligatory passage. For a young Thomas Henry Huxley, the referee represented "the intrigues that go on in this blessed world of science":

> For instance, I know that the paper I have just sent in is very original and of some importance, and I am equally sure that if it is referred to the judgment of my "particular friend" then it will not be published . . . I have such a horror

of all this literary pettifogging. I could be so content myself, if the necessity of making a position would allow it, to work on anonymously.[115]

In making a reputation in mid-Victorian science, working anonymously had ceased to be an option. The scientific author, alongside his anonymous counterpart, the referee, had arrived.

It might have been expected that referee systems at the societies should have functioned to reinforce the eroding distinction between the *Transactions* format and other scientific periodicals. But this does not seem to have happened. In many cases, refereeing gradually spread to other society publications, including proceedings, and while it was very rare for independent journals to use formal referee systems, those that focused on original contributions did what they could to imitate the prestigious publications of societies. For example, editors of most independent journals that published original contributions came to insist that authors sign their contributions. The *Cambridge Mathematical Journal* (founded in 1837), for example, had included many anonymous contributions and cryptic acronyms in its early years, but eventually its editor decided that this was improper for a scientific journal, and insisted on publishing authors' names without exception.[116]

❦

Scientific authorship and the referee system are at the center of current debates about legitimate judgment and accountability in science. Standard histories of peer review have often given the impression not only that referee systems are a great deal more ancient than the nineteenth century, but that their imagined functions have remained relatively constant. In fact, the prominence of the author-referee configuration became truly widespread only in the second half of the twentieth century. When the referee emerged as a prominent figure in Victorian science, its status and role changed several times over. Indeed, the temptation to view referees as primarily gatekeepers obscures the varied roles that such readers and critics were imagined to play in the life of science, including that of publicist, advisor, synthesizer, and judge. By midcentury, referees were widely understood as conferrers of rewards and defenders of the reputations of scientific societies, but it was later still that they began to be viewed as gatekeepers of the scientific literature as a whole.

Practices of authorship and judgment continued to vary across political and scientific cultures as well. While the rise of referee systems in Britain was paralleled by similar systems in the United States, they remained relatively rare elsewhere. Indeed, academic reports on manuscripts in France actually diminished in importance and frequency during this same period, as the *Comptes rendus* (which involved no such system) came to dominate the publishing activities of the Academy of Sciences. The increasing visibility of scientific authorship and the rise of a system of anonymous judgment in Britain occurred as men of science reimagined the place of natural philosophy within a changing political landscape.

4

Discovery, Publication, and Property

"The rights of individuals, as to the honour due to the origination of new views, processes, or methods, are matters of constant discussion." So observed the mathematician and bibliophile Augustus De Morgan in an article titled "Invention and Discovery" for Charles Knight's *Penny Cyclopædia* in 1846. He expressed puzzlement over how it could be that such an important subject had never been the subject of any general study, and he was eager to fill this gap. But several sticky problems immediately presented themselves. Just what, for example, was the relationship between invention and discovery? Strictly speaking, these were distinct concepts, the one implying "the formation of something which would not necessarily have existed, but for the invention," and the other referring to those things which "would have existed whether the discovery had been made or not."[1] But upon inquiring into the distinction with his logician's eye, De Morgan found it to be razor-thin, for all inventions seemed to imply and rest upon some prior discovery.

De Morgan thus concentrated most of his treatise on a more specific problem: just what "constitutes a claim to discovery"? Even this more limited question hid within it several others, not the least being the identity of discovery itself ("frequently the greatest difficulty"). But he pushed forward, and hypothesized that a claim to discovery is established by "priority of publication." When two investigators hit upon the same discovery at about the same time, he claimed that "the one who first publishes is universally recognised as the discoverer." The simplicity of this response itself hid a host of new difficulties, however, and these De Morgan proceeded to explore: "publication" turned out to be a concept that was of ambiguous but wide scope, and there seemed to be many cases in which the rule of thumb failed to apply at all. In legal matters, strict rules were required to maintain "private rights and public peace," while "the object of the scientific historian is truth for its own sake."[2]

Discovery and its relationship to both history and property were hot topics during the early decades of the nineteenth century. Across Europe, savants and philosophers puzzled over how and whether discoveries could be fixed in time and space, and to what extent they could be attributed to particular individuals. Disputes about the date of, and credit due for, a discovery were very often disputes about just what that discovery was. This mattered because it was taken to be a key step in defining what marked out scientific activity from other creative forms of endeavor, and thus in defining who belonged to these communities of practitioners. For these reasons, histories of such disputes have been a rich source for understanding changing conceptions of scientific fields from the nineteenth century onward.[3]

But historians have had much less to say about the media history of priority disputes. How might transformations in the formats and genres through which discovery claims became known to others have changed the nature and significance of such disputes? What kinds of acts were considered sufficient for making a discovery claim? What was the relationship between authorship of discoveries and authorship of scientific texts? Insofar as publication mattered, what kinds of publication counted?

During the 1840s, the idea that there was a necessary connection between scientific merit, priority, and print publication emerged as a *topos* that transformed the discursive resources available to combatants in such disputes. It was not that science simply moved faster during the nineteenth century, so that savants demanded more exacting dates for establishing their priority; when sufficiently intense, disputes over priority had always come down to weeks or even days. What changed, rather, were the kinds of dates that were deemed to matter, and why. The claim emerged that publication was the essential credit-worthy act in science, and that printing was the only means of putting such claims into the public domain. This idea reflected contested conceptions of the relevant public to which scientific claims ought to be addressed, and the relative merit to be accorded manuscript and print as media for making discoveries public.

The chapter begins by examining changing conceptions of scientific discovery and its relationship to the media landscape of early nineteenth-century scientific life. Until mid-century, it was not regularly asserted that priority claims depended directly on putting those claims into print.[4] But as scientific periodicals rose in status, the role played by print in fixing acts of discovery became a focus of attention. There were two privileged resources for reflection on these matters: history and the law. François Arago marshalled a series of historical case studies as he worked to shape a general doctrine of the rights and responsibilities of discoverers. In the disputes that this gave rise to, historical method—and the relative legitimacy of manuscript and print sources—loomed large. Just as important were debates occurring at the same time over the legal protections for property in inventions. The most ardent historians of discovery—Arago in France and David Brewster in Britain—were both deeply invested in patent reform. Both had entered this domain in order to promote the role that scientific discovery played in commerce and industry. Their interventions on this score were broadly consistent with one another, but when it came to applying these ideas to property in discoveries, they diverged sharply. Their divergence is a powerful reminder that views about how to make science open and equitable have always varied according to one's geographical and institutional circumstances.

Pour Prendre Date

> Les Routes en fer ne donnent pas encore aux nouvelles Scientifiques, la vitesse de la pensée, et sous ce Rapport, elles sont encore très imparfaites.
> —J. N. P. HACHETTE, 30 August 1833[5]

A constant stream of periodical publications focused on the sciences came and went during the early decades of the nineteenth century. They were often celebrated—especially in France—for their utility in settling the dates of discovery claims. At the turn of the century, Georges Cuvier, one of the most powerful academicians in France, praised the monthly publications of the humble Société philomathique as "a very convenient means of fixing the date of discoveries," thus protecting them from others who might steal them away before ready to be published in full. The editor of the *Annales de mathématiques* (1810), the first French journal focused on mathematics, promised that it would help "guarantee everyone the priority of the results he has achieved." When the *Annales de chimie* was remodeled in 1816, with Arago and J. L. Gay-Lussac at its helm, they made much of their new articles summarizing the proceedings of academic meetings, a handy means of "fixing the true dates of discoveries, the object of so much contestation."[6]

British authors and editors at the time were more circumspect about the link between priority and periodical print. When the president of the Astronomical Society of London cited this link while celebrating the Society's new *Monthly Notices* in 1828, he added a caveat: "so far at least as priority of public communication can be regarded as evidence in questions of that nature."[7] In fact, it was a question in Britain whether scientific journals ought to focus so exclusively on novelty. There, prefaces to new journals were far more likely to emphasize their role as digests of philosophical news. "If every article in a journal of science were to be professedly original," noted William Nicholson upon founding the *Journal of Natural Philosophy, Chemistry and the Arts* in 1797, "it would be a work of comparatively much less value to Philosophers and the Public." Many commentators criticized journals that made themselves out as repositories for original speculations, often of "fourth and fifth-rate men, upon subjects of ninth and tenth-rate importance." Better "to be useful and popular, than to be original and trashy."[8] Others, however, took seriously the problem of regulating scientific controversies. When David Brewster split with his coeditor of the *Edinburgh Philosophical Journal* and founded the *Edinburgh Journal of Science* (1824), he used the opportunity to inaugurate a new feature called "Decisions on Disputed Inventions and Discoveries." Brewster promised to use his accumulated expertise in the history of science to sort through complicated cases and to render justice where it was due. "The task of deciding such causes as these, though a difficult one, is not," he assured the readers, "of an ungenerous character."[9]

The idea that periodicals might be useful for settling such disputes was by no means new. Indeed, Henry Oldenburg, the founder of the *Philosophical Transactions*, suggested this very thing. Writing about a claim staked by Christopher Wren in the *Transactions* for "A Way of *Injecting liquors into Veines*," Oldenburg reflected: "Surely all ingenuous Men will acknowledge, that the *certain* way of deciding such Controversies as these, is a publick Record, either written or printed, declaring the Time & Place of an Invention first proposed, the Contrivance of the Method to practise it, and the Instances of the Success in the Execution."[10] We might suppose that what Oldenburg intended by "declaring the Time & Place" was the date attached to the published record, but the "time" to which Oldenburg referred in his model case had nothing to do with publication at all. The report on Wren's theory he had cited read as follows: "'Tis notorious, that at least six years since (a good while before it was heard off, that any one did pretend to have so much as thought of it) the Learned and Ingenious Dr. *Christopher Wren* did propose in the *University of Oxford* . . . to that Noble Benefactor to Experimental Philosophy, Mr. *Robert Boyle*, Dr. Wilkins, *and other deserving* Persons, That he thought, he could easily contrive a Way to conveigh any liquid thing immediately into the Mass of Blood."[11] Wren's publication in the *Transactions* was simply publicizing an already publicly attested claim, along with the highly credible witnesses to which one might go for confirmation. There is little to suggest that printed records were preferred over written ones. Quite the contrary. Oldenburg went on: "For the trial of which latter, [the Royal Society] give order at their Publick Meeting of 17 May 1665, as may be seen in their *Journal*, where 'twas registred by the care of their Secretaries obliged by Oath to fidelity." The word "journal" was a reference not to any printed periodical but rather to the manuscript register kept by the Royal Society and entrusted to Oldenburg's care. Not only was the register, by virtue of its being a single, closely guarded object, more reliable, but its entries — especially when they were made, dated, and signed by witnesses on the occasion of a particular experimental performance — would have been more faithful temporal renderings of discovery than print publications.

Far from disappearing, complications such as these remained just as relevant in the nineteenth century. The ways in which some printed text might be used for the purpose of fixing discovery claims varied a great deal. For Cuvier, publishing in the Société philomathique's *Bulletin* was only a temporary measure before proper publication happened later on. The *Annales de chimie*'s summaries of Academic meetings were third-party reports of oral events and manuscripts. Brewster, like many other editors of journals, used the platform to wade through the evidence in disputes — whether oral, manuscript, or print — and to offer his recommendations to the public. If authorship of *discoveries* was coming to define the romantic man of science, this did not equate to the authorship of *texts* in any straightforward way.[12]

Nor were the dates attached to the new generation of scientific journals to be trusted. In 1834, while sorting several matters of priority involving French chemists, Justus von Liebig complained that issues of the *Annales de chimie* were habitually dated six months in advance of the time they were actually published. "In no country," he wrote, "are ridiculous priority disputes more the order of the day than in France, and we thus permit ourselves to make our friends and associates aware of a circumstance, the ignorance of which puts them at an obvious disadvantage with respect to their French neighbors."[13] When Michael Faraday gathered together the series of papers making up his epoch-making "Experimental Researches in Electricity" in book form in 1838, he lamented the great misfortune that "the date of a scientific paper"—"a matter of serious importance"—was often impossible to ascertain: individual memoirs were rarely dated, and "the journals in which they appear [have] such as are inaccurate."[14] Faraday had been traumatized by an incident in 1831 involving the first in his famous series of papers where he first reported the phenomenon known as magnetic induction. After reading his memoir at the Royal Society on 24 November, he sent a letter summarizing his results to his friend Pierre Hachette in Paris. Hachette did what he took to be the natural thing and read the letter to his Academic colleagues at their next meeting (26 December). The journalists attending the meeting[15] printed summaries of the letter in the daily papers, and the report from *Le Temps* made it to Italy, where a physicist attempted to reproduce the result. A subsequent report of this research appeared in an issue of an Italian journal in February 1832. This issue—backdated to November 1831—then traveled back to England, where Faraday received it and mentioned the result casually to various acquaintances. The news of new Italian work on induction traveled by word of mouth until it was reported by the *Literary Gazette* on 24 March, implying that Faraday—"rapidly tending to the same discovery"—had been scooped. His Scottish friend, James David Forbes, characterized this "history of the transfusion of information" as a lesson to discoverers. Faraday was not amused.[16]

The confusions raised in this episode were by no means uncommon, and they were due only in part to inaccurate dating. They also arose from differing ideas among savants about the functions of, and relationships among, the various formats and media involved. Faraday extricated himself from this controversy with his discovery claim intact, but he never forgave his old French friend for having betrayed him by making his private letter public. Hachette, for his part, was bemused by the idea that he had done anything improper at all.

You can go ahead and say "O, unfortunate Letter!" — but don't you read with pleasure the French journal, *Le Temps*, which presents the Monday séances of the Academy of Sciences every Wednesday? Is there a single French academician who complains of this prompt communication? No. Quite the contrary. The railroads may not yet give to scientific news the speed of thought, and from this point of view, they remain quite imperfect.[17]

In invoking the railroad — the first tracks in France had opened just months before — Hachette seemed to imply that the only conceivable limits to fast and open communication in science were technological ones. But cultures of secrecy and publicity varied in ways that did not map in any simple way onto the limitations of communications technologies. The relationships between correspondence and print, between oral performances at meetings and reports in papers, between journals and transactions, were uncertain and changing.

Confusion was especially likely to arise across national boundaries, where conventions separating public and private, and the media that embodied them, often diverged. Investigators outside France were wont to note that in that country "the emulation for such reputation is perhaps more vigilant and anxious than it is elsewhere."[18] The Swedish chemist Jacob Berzelius complained of the French propensity for the act of *"prendre date,"* which he defined as follows: "as soon as you believe that you are on track of something new, you announce it in the journals, promising to develop it in the future." This was, Berzelius argued, "a means of hindering scientific research" because it deterred others from pursuing open questions.[19] Even François Arago himself scolded French savants "who in our weekly meetings demand, raising a hue and cry, permission to communicate the smallest remark, trivial reflection, or little note."[20] This was no failing of character, however; it was a defining characteristic of the institutional culture of French science. Academicians understood the settling of priority disputes to be part of their job description: the Academy's commissioned reports on the merit of new discoveries and inventions were in part reports on the question of originality and the settling of rival claims. Indeed, the very fact that there existed such paid positions at all coincided with a culture in which it was considered acceptable, indeed necessary, for those aspiring to join that elite to be their own advocates as discoverers. What for Faraday was a painful act of setting the record straight was for Hachette and his French colleagues simply a part of being a savant.[21]

Indeed, the reformers who advocated adopting French savants' institutional culture in Britain embraced precisely these aspects of it. For David Brewster and Charles Babbage, the orderly settling of priority disputes was a crucial first step in recognizing and rewarding (both morally and financially) active men of science. Brewster's "Decisions on Disputed Inventions and Discoveries" was an attempt to provide for British men of science a service accomplished by the academies in France. Babbage, for his part, insisted that scientific discoveries and inventions ought to be understood as property, and he spearheaded attempts in the early years of the British Association to advocate for legislation that would secure to men of science their "scientific and literary property."[22]

Even within the same country, city, and discipline, disagreements arose about what constituted proper and improper communication, especially as new venues and formats came and went. Thus, while oral readings of papers at meetings were normally assumed to be admissible evidence in establishing priority claims, new kinds of institutions, such as the British Association, led to new problems. In the early years of the Association, a violent dispute over the status of such readings nearly tore its Council apart, leading its founding Council secretary, James Yates, to resign in disgust and to an overhaul of its procedures and publishing routines.[23] Alexander Nasmyth, the Queen's dentist, had presented a series of papers on a theory of the constitution of the ossified matter of teeth at the meeting in Birmingham in 1839. Some months later he came into conflict with Richard Owen, who published a short notice giving his own theory of the same thing in the *Comptes rendus* of the Academy of Sciences (Owen was a corresponding member) that fall. Nasmyth had since circulated a printed abstract of his speech as a separate copy from the forthcoming British Association *Report*, and Owen objected that this report was not an accurate record of what Nasmyth had presented. The dispute was acrimonious and complicated: the Association's Council spent nearly a year examining printed reports, collecting and evaluating manuscript evidence, consulting a committee of experts, and even interviewing witnesses, all to determine just what Nasmyth had *said* at a particular time and place. The Council ultimately failed to come to a determination, and instead published *in extenso* much of the manuscript evidence, and accumulated correspondence, leaving readers to judge for themselves. Following the incident, its Council attempted to clarify that oral communications made at its meetings were not the stuff of priority claims.[24]

Conflicts of this kind were normally rare because it was a strict rule of societies set up on the Royal Society model to retain manuscript copies of papers read at meetings. Authors were not permitted even to borrow their own manuscripts for the purpose of making copies, as the manuscript was considered the ultimate authority with respect not only to reading events but to the printed paper itself. The British Association did not follow these procedures, but its Council admitted that it was understandable that members might have assumed that such readings conferred the same rights as they did elsewhere. Problems of dating led to a proliferation of dates and of administrative rules for fixing them. Among the first reforms carried out by the Royal Society of London in its push for reform in the late 1820s had been complex new rules defining precisely the dates that memoirs were to be deemed as having been received and the order in which they were chosen to be read. Dates of the reception of memoirs were soon printed in the *Transactions* along with the title and date of reading.[25] Societies also began to pass rules in which authors wishing to revise manuscripts before printing them, but after reading them, were forced to mark such changes with a date.[26]

At the Academy of Sciences, careful attention to such dates, and of rules to determine the order in which papers were read, were long-standing concerns.[27] But the Academy tended to view manuscripts read at meetings as distinct documents from the printed versions, in part because there was so often such a long period between reading and printing.[28] In France, a more revealing index of the increasing fixation on dates was an explosion in the practice of submitting sealed notes — *paquets cachetés* — to the Academy. The practice of using sealed notes deposited with some authoritative body *"pour prendre date"* went back a long way, but it had never been an everyday occurrence: the eighteenth-century Academy received just a few such notes per year. But it picked up dramatically in the late 1830s, and by the 1840s the Academy was accepting about one hundred such notes per year.[29] The Royal Society also accepted sealed notes for keeping in its strong box (Figure 4.1), but this service was used exceedingly rarely — with the single exception of David Brewster — and it was abolished altogether in 1854.[30]

Despite the importance widely ascribed to fixing the date of — and the credit due for — discovery events, the difficulties inherent in actually accomplishing this were widely acknowledged. Even those most zealous about apportioning honors recognized the problems. "The most valuable inventions and discoveries are often completed by different individuals, and at dis-

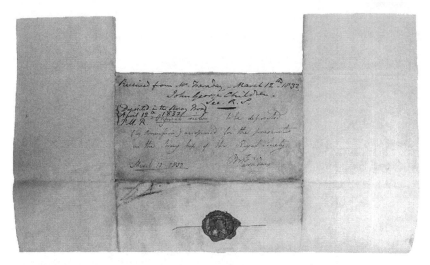

Figure 4.1 A sealed note by Michael Faraday containing "Original views," which he requested be deposited unopened "in the strong box of the Royal Society." (The note has since been opened.) Such sealed notes were exceedingly rare at the Royal Society, while the Academy of Sciences in Paris received as many as one hundred per year by mid-century, although they remained controversial. Royal Society of London, MM/10/178.

tant intervals," admitted David Brewster in 1824. "In such cases . . . it is extremely difficult to apportion to each author his due share of merit." "The most historically-informed watchmaker," noted François Arago, "would be dumbstruck if asked to name the inventor of the watch." The point was echoed by De Morgan in his later treatise, when he pointed out that in the voluminous literature on the question of who had invented the steam engine, "the principal point at issue is, what is the steam-engine?" Simultaneous discovery was also a well-known issue: "When the period arrives at which a discovery becomes possible, there are many courses which lead to it, and many ships sailing on each of these courses."[31]

Such acknowledged difficulties did not usually stop anyone from trying to sort things out, but they did lead to more sustained reflection on discovery and invention that went beyond particular, contemporary disputes. Such reflection was overwhelmingly historical in nature. Many of the most active participants in the public regulation of priority disputes in Britain and France during this period — Arago, Brewster, Jean-Baptiste Biot, and De Morgan — were also active biographers. The link between their historical works and the regulation of discovery claims was explicit: Brewster, for example, claimed to be practicing "scientific history" when he decided disputed inventions and discoveries.

Debates about the nature of discovery were thus also debates about how to write historical narratives of science. The lives of certain figures — Galileo, Newton, and Watt — became privileged vehicles through which to fashion exemplary images of both the inventor and the man of science.[32] Watt's status as both discoverer and inventor made him an ideal case for reflecting on the broader question of what constituted a discovery claim and how one might best lay claim to it. On the one hand, Watt's invention of the steam engine became iconic of the relationship between scientific advance and national prosperity. His complicated history with British patent laws provided a great deal of material for a legal history of claims to invention, especially for those interested in reforming these laws. On the other hand, Watt's purported role in discovering the composition of water became the focus of a long Franco-British debate beginning in the 1830s. The dual role of Watt in such narratives made it natural to compare property in inventive ideas to priority in scientific ideas. The first person to do this was Arago, in his *Éloge* of James Watt, first outlined in a public lecture in 1834. The trope was repeated by many engaged in this debate, as it was in Robert Muirhead's preface to his published edition of Watt's correspondence on water:

> . . . is honour to be meted out with a less liberal hand, or guarded with less jealous care, than those pecuniary rewards, which the true philosopher does not covet, and which few men would with equal ardour desire? Are learned Societies, or the individual followers and friends of Science, to be guided by less exact principles of justice, in their award of praise to a *first inventor*, than those impartial Tribunals where, in similar cases, but with other interests at stake, the great improver of the steam-engine found his rights vindicated, and his inventions sacredly protected, by the strong arm of the Law?[33]

Many agreed with Muirhead's sentiment, even if there was less consensus about just *how* the protection of pecuniary rewards for invention was related to those of honor for discovery. But the widespread discourse of patent reform during these decades, much of which focused on the detailed minutiae and mechanics of claiming priority — including the status and nature of the documents needed to secure such protection — provided a ready set of conceptual tools that could be adapted for use in discussions of priority in discovery.

Arago, in particular, came to fashion a doctrine of priority that could be readily applied to contemporary discovery claims. These ideas began to take shape during the controversy over the composition of water, but his doctrine was developed in systematic fashion in the early 1840s in a controversy at the Academy of Sciences ostensibly over Galileo's role in the discovery of sunspots. This became a trial over the new regime of print publicity that Arago had brought to the Academy during the 1830s.

How to Write the History of Science

After founding the *Comptes rendus* in 1835, the members of the Academy of Sciences found themselves in control of a different kind of institution. In some ways it was a more powerful one, for the weekly journal allowed it to compete with, and participate directly in, the independent press. From one point of view, they had helped penetrate the barrier between *la science académique* and *la science publique*. The most vocal journalists had been silenced, the most critical journals shuttered, and the *presse à bon marché* that emerged in 1836 gradually pushed science from the front pages of the most popular dailies to be replaced by serial novels.

But press reporting on the Academy by no means stopped. New critics of Arago emerged, some focused on his career in the Chamber of Deputies, some on his scientific deeds, and others on both. In 1840, a new flood of attacks coincided with Arago's well-publicized, controversial speech in the chamber advocating universal suffrage. Writers from both the right and left condemned Arago's republican principles as misplaced or disingenuous, and took the opportunity to point out that he had been wreaking havoc by "applying to science the absurd system of universal suffrage through which Arago has reaped so much false glory." Arago, it was claimed, had set up the *Comptes rendus*, an ostensible gesture of openness, to wield personal power over French savants.[34]

Several writers seem to have been behind these attacks, although their public face was Gustave de Pontécoulant, an astronomer who had fallen afoul of Arago by critiquing the observational work of the Paris Observatory that Arago also directed. Encouraging him behind the scenes, however, and setting the agenda for the assault on Arago's scientific and political program, was a Tuscan expatriate, mathematician, historian, bibliophile, and erstwhile revolutionary named Guglielmo Libri.[35]

Libri, of aristocratic stock,[36] had first come to Paris in 1824 as a precocious professor of mathematics in Pisa on a combined mission to meet the French masters of mathematics and to rescue his wayward father who had been languishing in a French cell for counterfeiting Florentine notes of credit. By the time Libri returned to Tuscany later in 1825, he had not only set his father free, but he had established excellent ties to the French scientific elite. Eventually, feeling cramped by the parochial confines of the Tuscan academic world, he visited Paris again in June 1830. On 27 July, when the revolution happened, he stepped out of the Bibliothèque Mazarine, where he had been poring over manuscripts of Leonardo da Vinci, and joined the front lines with the revolutionary rioters, achieving a name as a freedom fighter.[37] Imbued with the spirit of revolution, he returned to Tuscany and organized an insurrection to force a constitution on his erstwhile protector, the Grand Duke. This revolution failed, however, and by the spring of 1831 he was on a boat headed for Marseille in exile. Once back in Paris, equipped not only with mathematical but with revolutionary credentials, he was welcomed enthusiastically by the Parisian liberal intelligentsia. Arago became his most ardent patron and friend, and Libri enjoyed a meteoric rise through the Parisian scientific establishment: corresponding member of the Academy of Sciences and lecturer at the Collège de France in 1832, full member of the Academy in March 1833, and Professeur Suppléant at the Sorbonne in 1834.

Such was his ascent that Libri became a center of power and influence himself, establishing strong ties with François Guizot, the historian and powerful minister under Louis-Philippe, in part through their shared passion for historical manuscripts. By late 1835, Libri and Arago had broken decisively. Precisely what touched off the split is uncertain, but it was not surprising. Arago had not long before become known as an enemy of Guizot in the Chamber of Deputies, when a rumor circulated that Guizot had excluded him from the jury of the 1834 Industrial Exposition.[38]

The divergence between Arago and Libri also coincided with Arago's founding the *Comptes rendus* and the resulting reorientation of the Academy's relationship to print. In October 1835, François Double read a report he had prepared on a memoir by Jean Civiale, an early attempt to apply statistical methods to demonstrate the efficacy of a medical procedure (in this case lithotrity, a method for removing stones from the bladder). Double praised the data collection, and suggested the Academy encourage its author, but was not convinced by the applicability of statistical methods to the subject and did not recommend publication. The ambivalent report raised a firestorm of debate. Arago weighed in, insisting that at least some portion of the memoir, if only the data, ought to be printed as soon as possible. "The publication of tables . . . would be useful even supposing that these tables contained several errors, for they would necessarily lead to new information, and little by little that which is erroneous in the results would be corrected." That is, printing of new scientific information, even when imperfect, ultimately contributed to useful knowledge; Arago pointed out the analogy with astronomical data, which was regularly published by the state despite known imperfections. Libri spoke up, arguing precisely the opposite: it was precisely the task and privilege of the Academy to encourage works in progress such as these *without printing*, so that the work would be pursued further and completed. Printing indiscriminately might actually put a premature *end* to a fruitful line of research.[39]

This otherwise routine disagreement over printing was a sign of things to come. No one came to detest the new journal and the alterations it brought to the customs of elite scientific life more than Libri.[40] As he explained in a series of attacks on the Academy published in the *Revue des deux mondes*, his ideal of scientific community was that of the small learned elite, bound together by personal trust relationships and face-to-face discussion, an ideal that was fostered during his early career in Tuscany under the personal patronage of the Grand Duke. While scientific Paris of 1825 — still largely Cuvier's dominion — had maintained some of this, the Academy's opening itself to the press was destroying whatever was left of the old forms of scientific life. He agreed with Biot that the ease of publishing in the *Comptes rendus* allowed — demanded — that savants rush headlong into print rather than nurture their work in the private discussions that the Academy should have afforded them. The "incomplete notes that are printed therein show so much evidence of haste, that they can hardly contribute to the progress of science nor to the reputations of their authors." If the Academy did not proceed with more caution, Libri warned, "its effect will be to put all other forms of publication into abeyance." The Academy, which had been very careful to protect its dignified reputation by publishing only the most carefully vetted and completed work of its members and a few chosen outsiders, "would be reduced to publishing nothing but this journal, edited in haste, in which everyone might aspire to become a collaborator."[41] The *Comptes rendus* threatened to dissolve whatever authority the Academy still retained.

According to Libri, journals such as the *Comptes rendus* were purpose-built for minds such as Arago's. Prone to quick, sudden bursts of fragmentary ideas, but incapable of deeper conceptions developed over longer periods of time, Arago tended to focus on useful applications and to neglect the development of the mathematical principles that were necessary to bring discoveries to scientific fruition. By opening its doors wholesale to the norms of periodical publicity, the institutional life of the Academy of Sciences would come to resemble Arago's own fragmented intelligence.[42]

Disagreements between Libri and Arago were often over cultures of trust and of the material forms of communication and registration that were supposed to underlie them. Libri gradually established himself as the foremost expert on historical manuscripts (especially, but by no means exclusively, scientific manuscripts) in France, while Arago systematically developed a program to make the progress of science synonymous with the control of science in print. Libri, the keeper of manuscripts, developed an epistemology to suit, while Arago, orchestrator of the scientific press, did likewise.

For several years following their break in 1835, Arago and Libri used the weekly meetings of the Academy to wage battles over reputations, elections, and claims to discovery and invention. Their domains of expertise in astronomy and mathematics did not overlap sufficiently to engage in direct combat very often, but they did so anyway via proxies. Libri was a protector of Pontécoulant who battled with Arago over astronomical matters, while Arago took the up-and-coming mathematical star Joseph Liouville as a protégé. Like Arago, Liouville was also a journal editor, and directed the *Journal de mathématiques pures et appliquées* beginning in 1836. In the early 1840s, upon establishing himself as an important mathematician, Liouville used both his journal and the Academy's meetings to launch a series of direct assaults on Libri's mathematical accomplishments. He targeted exactly those findings by which Libri had made his mathematical reputation, arguing that they had been anticipated, or could easily be reduced to other known results, or contained fatal flaws.[43]

The particular way in which these disputes played out is itself of interest, for, in the first decade of the *Comptes rendus*, the relationship between the verbal disputes that took place on the Academy's floor and the confrontations that took place in the printed pages of its weekly journal was by no means settled. Because the Academy's journal had been modeled on newspaper journalism, it included records of discussions alongside memoirs and correspondence. All parties were very conscious of the publicity the new journal accorded their arguments, and because they were themselves responsible, after each meeting, for re-presenting in print their parts in the discussions, there were struggles over just where the core of debates should lie: whether in print or in speech. Combatants would often complain that their opponents had sent replies to the printers that had been supplemented with substantially new arguments, or that their printed arguments put words in the mouths of their interlocutors. When pressed, however, all parties reserved the right to develop their statements in print—whether of reports or discussions—beyond what had actually been said at meetings. In contrast to the sense of British men of science for whom fidelity to verbal events remained crucial, academicians generally agreed that the authority of printed statements could not be made to depend on their accurately representing oral performances. But the matter was delicate, and led to repeated problems.

For example, in August 1843 Libri objected to a report written by Liouville for a Commission reporting on the work of the young Charles Hermite on elliptic functions.[44] Libri saw in the report an implicit dismissal of his own work on the general theory of the classification of elliptic functions. Liouville answered by boldly dismissing Libri's claims to credit, and he promised to produce a detailed statement demolishing Libri. Libri was indignant; if Liouville was so certain, then they ought to continue the discussion face-to-face. But Liouville refused. Instead, he presented his arguments at the next meeting once he had worked them out. Libri remained aggravated, now refusing to answer verbal arguments about technical matters "in the presence of a crowd of listeners who, in a delicate and difficult discussion, would be forced, for the most part, to follow only the form the discussion took," remaining ignorant of the content.[45] He insisted that the matter be settled in writing not only because Liouville constantly changed what he had said when it appeared in the *Comptes rendus*, but also "so that professional geometers, who are the only competent judges in this matter, could better comprehend what it is about."[46]

One front that developed in the war between Arago and Libri brought them into more direct confrontation: the history of science. Both of them had begun to devote increasing amounts of time to history during the 1830s, and to scientific biography in particular. By the end of the decade, their research was concentrated more on unearthing the records of the past than on directly advancing the sciences of the present. Arago, as perpetual secretary of the Academy, was responsible for writing the tributes to deceased members of the mathematical sciences section of the Academy—a duty he had taken up with more zeal than his predecessors—and he also began composing book-length treatises on scientific luminaries to include as the second half of the *Annuaires* published yearly by the Bureau des longitudes. Libri, for his part, had been a committed bibliophile and collector of manuscripts since at least 1823, when he was given access to the manuscript collections of the Archivio Mediceo by the Grand Duke.[47] He had for very long been pursuing a major, multivolume history of the mathematical sciences in Italy, and the first volume at last appeared in 1838.

Among his many biographical works, Arago had become particularly well known for his revisionist history of James Watt. Arago published his *Éloge historique de James Watt*[48] in 1839 following several years of research, and it was immediately translated into English (twice). Arago's guiding theme was that Watt's life exemplified the crucial role that technological invention and scientific discovery ought to play in the political and industrial progress of nations. Although the center of the book was the steam engine—its history and Watt's struggle to protect his intellectual property—Arago also wished to demonstrate that Watt was a savant in the strictest sense by assigning him credit for a purely scientific discovery: the composition of water.

Arago's reopening of the water question a half century after the fact set off a barrage of replies and counter-replies[49] that made the episode into a set piece in the history of scientific priority.[50] In 1834 Arago had traveled to England and Scotland to collect fresh sources for his memoir, and along the way he had come to believe that Watt had suffered a great injustice. By working with unpublished documents and consulting James Watt Jr., Arago came to believe that the received history of the discovery of the composition of water was the product of a set of vague and misleading claims that had been perpetrated by Henry Cavendish (the supposed discoverer), his supporter Charles Blagden (secretary of the Royal Society), and naïve historians who had failed to speak truth to power. According to Arago, not only was treachery likely to have been involved—separate copies of Cavendish's crucial paper had been predated by a year—but Blagden's analysis of the affair had relied on hearsay and had been suspiciously vague on matters of dating, just where precision was most important. Phrases such as "'*About the same time*' can prove nothing," noted Arago, "for questions of priority can depend on weeks, on days, on hours, on minutes."[51]

These two themes in Arago's history allowed him to bring ideas about invention and property to bear on conceptions of discovery. Arago reflected in detail on the ways in which British patent law had both helped and hindered Watt, both in his invention of the steam engine and in his ability to put it to profitable and public use. As a member of the French Chamber of Deputies, Arago took part in the legislative debates that eventually led to the revised French patent code of 1844. There, Arago objected strenuously and at length to the provision that banned purely scientific principles and methods from being patented. He at last succeeded in softening this restriction by appending the phrase "if no industrial application has been indicated."[52] Indeed, it was an inexplicable prejudice against ideas that had caused Watt such trouble with the British patent system: "men of genius, manufacturers of ideas seemed forced to remain strangers to material pleasures; naturally their history continued to resemble a legend of martyrs." It was as if men believed that ideas, unlike machines, "were made through no great fatigue or struggle!"[53]

Arago began to see his task as bringing the exacting methods of the law to the history of science: "The solution of a question of priority, when it is based, as the one that I have just presented is, on the most attentive examination of printed memoirs and the meticulous comparison of dates, takes on the character of a veritable demonstration." He also put forward a suggestion about the proper criteria for what constituted admissible historical evidence in settling priority disputes:

> It would not be asking too much, following the example of judges in matters of civil law, that historians of science never admit anything but written titles as valid titles to property; and, perhaps I should add, nothing but published titles. Then, and only then, would those quarrels, constantly arising, between national vanities, be done with.[54]

This focus on written and printed documents was not simply a matter of sound historical method, but of rewarding probity and openness. Arago was thus able to extend his celebration of Watt's moral character in a way that fit with his long-standing advocacy of scientific publicity. Cavendish was the retiring aristocrat to the plebeian James Watt. Where Watt was open and generous, and hastened to print his discoveries, Cavendish was secretive, and conspired with "typographers, compositors, printers" to falsify dates.[55]

While Arago was writing of steam and its composition, Libri was fast becoming—through his association with François Guizot—among the most powerful superintendents of the French state's manuscript heritage. Upon coming to power, Guizot had worked quickly to marshal historical research and documents in a bid to shore up the foundational narratives of the new King and the government. During the Restoration, he and other leaders of the liberal resistance had used history as a privileged terrain on which to engage in political critique. Where their revolutionary precursors had posited the possibility of a clean break with the past, these writers — including Guizot, Adolphe Thiers, and François Mignet — set themselves the intellectual task of recouping the strands of liberty that existed prior to the Revolution, and which would point the way to reconciling the positive consequences of the Revolution with order.[56] When these same men took up powerful political positions after the 1830 Revolution, they began to institutionalize this new focus on history. Guizot founded the Société de l'Histoire de France in 1833 to organize historical scholarship, and in 1834 he established the Comité des Travaux Historiques et Scientifiques with the aim of "overseeing research into and publication of hitherto unpublished documents pertaining to the history of France."[57] Libri was quickly enrolled in the massive historical projects Guizot undertook. He was named to the Comité in 1836 with the special task of dealing with documents related to the history of science, and in 1841 he was named by Guizot as the first director of a French commission entrusted with producing a detailed catalog of "all manuscripts in ancient and modern languages at present in the public libraries of the Départements."[58]

The final published volumes of Libri's monumental *Histoire des sciences mathématiques en Italie* appeared in 1841.[59] Libri, like Arago, declared himself interested in "investigating, in the writings of inventors, the first ideas that led to great discoveries." More specifically, however, Libri's aim was to establish, by exhaustive research, the foundational role that Italian savants played in the "the great scientific revolution" of the seventeenth century. His hero, whose accomplishments made up most of the fourth and final volume published in 1841, and from whom Libri even claimed ancestry, was Galileo: "This grand revolution is due to Galileo, the immortal genius who made and who prepared the way for so many beautiful discoveries."[60]

Like Arago, Libri was deeply committed to distributing credit where it was most due, but the two went about this in very different ways. For Libri, historical truth was to be found buried in tattered and forgotten manuscripts, in digging up old correspondences and friendships, in following the testimonies of contemporary witnesses. To give undue emphasis to print was deeply to distort the historical record. This was particularly true in the case of Galileo because, despite being open and generous about sharing his discoveries, Galileo had printed relatively few of them. The reasons for this were several. When he was young, he was too poor to afford it,[61] and once he was well known, printing brought its own troubles, thanks to both plagiarists and well-known political complications:

> Occupied more in making discoveries than in sending them to the printers, Galileo contented himself for very long with communicating them to his students and friends; so much so that, spreading everywhere in this way, they were often reproduced by plagiarists, who attempted to appropriate them. Later, when he contemplated finally collecting them together and publishing them, the inquisition arrested him and condemned him to silence.[62]

Besides, as Libri noted, there hadn't seemed much need to publish at the time: through his lectures, correspondence, and the circulation of manuscripts, Galileo's ideas spread throughout Europe rapidly, reaching just that portion of the public most equipped to appreciate and extend them.

But if manuscript and speech were the surest way for Galileo to ensure that his work came to the attention of just those best positioned to understand it, Galileo's practices made the historians' task all the more delicate. It meant digging up and cross-referencing the testimony of his students, of his friends, and of the academies. And it meant sorting through countless scholars' correspondence to dig up all the subtle clues that could prove just how many discoveries that had been attributed to others had their true source in something Galileo had communicated verbally or in manuscript: "When we treat questions of priority involving Galileo, we should never forget that before publishing his first work on *The Compass* in 1606, this great philosopher had been a professor at Pisa and Padua for 17 years and had communicated his ideas to thousands of students who spread themselves across Europe. This was a temptation that hardly anyone could resist: numerous plagiarists attempted to enrich themselves on his ideas."[63]

Using this perspective, Libri found theft and appropriation just about everywhere he looked. Most instruments, astronomical observations, physical laws, and theories associated with the revolution in natural philosophy of the seventeenth century could eventually be traced back, via some manuscript or letter, to Galileo. These included not only straightforward matters such as sunspots and thermometers,[64] but even the origins of the modern scientific method:

> It turns out, according to a letter written by Toby Mathiew [*sic*] to Bacon in 1619, that at least a year before the publication of the *Novum Organum*, the English philosopher would have known not only all of Galileo's printed works that had appeared by that time, but also that he would have had access to all of his unpublished and manuscript works as well.

Libri went on to explain that the "grand scientific revolution" that has often been attributed to Bacon was largely "due to Galileo."[65]

When Libri's volume on Galileo appeared in 1841, Arago was working on a long biography of the British astronomer William Herschel. Libri's book enraged him. Such cavalier use of personal testimonies and manuscript sources of questionable authenticity to establish discovery claims demanded a rebuttal. Using the flimsy excuse that Herschel too was quite interested in the sun, Arago inserted a thinly veiled forty-page critique of Libri's version of Galileo right in the center of his Herschel memoir and sent it to the printer. His specific target was the contested discovery of sunspots, which Libri had, among many other things, attributed to Galileo.

Where Libri's Galileo prudently kept the press at a distance, Arago's was unreasonably paranoid of print. Where Libri's Galileo was generous in spreading his ideas by manuscripts and demonstrations, Arago's was jealous of sharing his findings. Where Libri found historical truth in informal and manuscript communications, Arago insisted on the certainty and fixity that only print publication provided. Arago used the opportunity to put forward a general statement on historical method, intellectual property, and scientific sociability. "Among the moderns," Arago led off, "the discovery of sunspots has given rise to a raging and *confused* debate."[66] The problem, he suggested, was that there was no agreed-upon standard to decide what constituted discovery; there was no fixed measure of priority. Arago would now lay such a definitive principle down:

> There is but one rational and just manner in which to write the history of science: it is to rely exclusively, as I will in the following, on *publications* having a definite date; beyond that all is confusion and obscurity.[67]

The implications of the principle were as straightforward as they were powerful; Libri's rich world of manuscripts and private testimonials was of secondary
importance at best. They might be used to establish or challenge the authenticity of a particular printed document, but in themselves they had no value:
"Private communications just don't possess the necessary authenticity."[68]

With this principle in hand, Arago made short work of several of Libri's
priority claims on Galileo's behalf. He began with sunspots, a crucial—but
to Arago's mind simple—case. Fabricius published his *Narratio de maculis in
sole observatis et apparente earum cum sole conversione* with a dedicatory epistle
dated 13 June 1611. No publication, whether of Scheiner or Galileo, came
close. That basically settled it. But even if we took a broad view of publication, and included well-attested public demonstrations, Galileo still lost. Thus
we might choose to admit Galileo's famous demonstration in Cardinal Ban-
dini's garden, which was reported to have taken place in 1611. But we would
need to know the month. And no trustworthy witness could supply one. Guc-
chia spoke in June 1612 of a conversation with Galileo about sunspots more
than a year earlier. But this was vague, and even if we were generous and fixed
the date at May 1611, Fabricius still had the upper hand: with a dedication
written in June, and observations that clearly took place over at least two to
three months, his priority would still hold. And the claim of Fulgence Mican-
zio seventeen years after the fact that Galileo had shown the spots to Sarpi
using his first telescope in Venice in 1610 led to a technical contradiction. And
so on. None of Libri's claims from the manuscript record could stand up to
the careful historian's scrutiny: they were based on others' memories of events
and dates (often contradictory), attestations of friends (not credible), or later
commentators' extrapolations. The print record was the only way to reliably
settle disputes.[69]

Moreover, Arago showed, Galileo did care about priority as much as anyone else, but he was, *pace* Libri, *secretive*. How else to explain that his first remarks on the form of Saturn, and on the phases of Venus, were communicated
to the public in the form of puzzles and anagrams?[70] Arago had no time for
anagrams, and he deplored the new popularity of *paquets cachetés*. Just as the
latter were rising in popularity, he dismissed the widely held notion that one
could use them *pour prendre date*: "priority belongs incontestably to he who
was the first to put his observations before the public. . . . A sealed note can
only serve to protect someone's right to work in an area after someone else has
recently published on the same subject."[71]

When Arago later used the material from this attack to fashion a full biographical treatise on Galileo, he worked it up to read as a morality tale for those who, like Galileo, were "jealous of transmitting their work to posterity" by avoiding open publication. He retold an anecdote concerning the eighteenth-century Italian historian Giovanni Battista Nelli that had been related by Targioni Tozzetti revealing the extraordinary provenance of an important collection of Galilean manuscripts. In 1739, Nelli and a friend had gone to eat lunch at a tavern; along the way they purchased some *saucisson de Bologne* at a butcher shop, which they received wrapped in paper. When they sat down to eat, Nelli discovered that the paper wrapper was actually a letter in Galileo's hand. He quickly cleaned it and put it out of sight of his companion, and upon taking his leave headed straight back to the butcher shop.

> The butcher informed him that he often bought papers by weight just like that one from a servant whose identity he didn't know. Nelli purchased all of the papers that the merchant had, and after lying in wait for the unknown domestic a few days, he came into possession, for a certain sum, of all that remained of the precious treasures that Viviani had hidden 90 years earlier.[72]

The lesson: it does not pay to hoard knowledge, and Galileo, as great as he was, had paid a just price in the eyes of posterity. Thus, "savants might learn that the only sure medium for conserving their works is via a printer's *case*."[73]

Arago's moral lesson applied not just to the writing of history. In demolishing Libri's celebration of manuscript culture, Arago was responding to the criticisms of Libri, Biot, and others on the damage that rapid publication in journals such as the *Comptes rendus* was doing to the standards of scientific knowledge.

> What legitimate complaint can anyone make who, amorous of his discoveries as a miser is of his treasures, buries them away, keeping guard lest their existence even be suspected, afraid that some other experimenter will come along to develop and cultivate them. The public owes nothing to those who have rendered it no service. Oh! I do understand you! You wish to take the time to complete your work, to follow it through in all of its ramifications, to point out all of its useful applications! Feel free to do so, gentlemen, feel free; but it is at your risk and peril.[74]

Arago's principle was more than just a preference for publicity via the press: the point was that the act of making a discovery public was the only truly creditable act in science. No amount of evidence of prior discovery was relevant without publicity.

Others took note. Aragon's friend, Alexander von Humboldt, praised the Herschel book for its "profound knowledge of the history of discoveries"— a subject that had been "so badly done until now"—and he singled out the section "on the priority of discoveries" for particular praise.[75] Arago had thus invented a rational principle by which to allocate credit in the sciences. The debate was primed to move to the floor of the Academy itself.

Arago and Libri's first clash over Galileo at the Academy occurred in August 1843 over a matter of manuscripts and historical expertise. An Italian scholar, Eugenio Albèri, who had been helping edit a set of Galileo papers at the Palatina Library in Florence, had written to the Academy asking its support regarding a controversy raging over the nature of these papers. Albèri claimed that he had found in the papers the long-lost writings of Galileo on the satellites of Jupiter, which he noted Libri had claimed had been destroyed by the Inquisition. If Albèri were correct, it implied that Libri had gotten a crucial aspect of his narrative wrong because of insufficient attention to the manuscripts. Arago naturally leaned in to support Albèri's claim, and suggested that Libri had missed the importance of these papers because he was unable to understand astronomical data. Libri insisted that Arago did not understand the historical scene in Italy, and that Albèri had already been contradicted by all the relevant authorities. Arago promised to put "all the pieces of evidence in the trial . . . before the eyes of the French public, and all will be enabled to make an enlightened judgment."[76]

This was an early warning that their rival claims to historical expertise were ready to emerge into full-blown conflict. This occurred the following month, when a priority dispute arose over an instrument for measuring the fat content of milk that had been presented to the Academy by Albert Donné. The instrument maker Charles Dien charged, and Arago bore witness, that in fact Donné had received privileged access to a new invention of his own, and had stolen the idea by publishing a description of his milk-analyzer without citation.[77] This was what Libri was waiting for: he jumped into the fray, quoting verbatim Arago's principle of priority from his attack on Libri's *Histoire*:

There is but one rational and just manner in which to write the history of science: it is to rely exclusively, as I will in the following, on publications having a definite date; beyond that all is confusion and obscurity.[78]

Libri argued that to follow Arago's own principle would mean bestowing credit in this case on a plagiarist. The ensuing debate between Arago and Libri concerned where the limits lay to Arago's principle that formal publication was the only source of true certainty in matters of discovery. Libri argued that you could not impose limits to the principle without making it self-contradictory, and that to leave it unrestricted was to facilitate plagiarists, which was an absurd consequence. Arago claimed that his principle included within it the ability to call out plagiarists, since he had already stated the necessary conditions for its applicability:

A printed claim [*titre*] may be submitted to the same verifications as a banknote. Interested parties must have the right to object to its validity; contradictory claims must be debated with strict justice; this is a condition which, barring only the rarest exceptions, appears to me to imply the rejection of all posthumous complaints [79]

Very soon, the original question related to Donné's lactoscope faded from view. Dien and Donné, since they were not academicians, could not engage directly in the Academy's discussion anyway. Arago and Libri thus launched into a month-long debate about Galileo and the discovery of sunspots on the floor of the Academy.

Once he was provoked, Arago stated explicitly the attack he had left implicit in his Herschel biography: "Nothing is weaker, or more superficial, than the section of [Libri's] *Histoire des sciences* on the problem [of sunspots]." He not only attacked Libri's historical work, but again suggested that he was not up to the historical task because he had no expertise as an astronomer. Libri, deeply offended, prepared a lengthy reply to Arago's attacks that stretched to forty manuscript pages.[80] To Arago's claims about his lack of astronomical expertise, Libri responded by showing that Arago, in turn, had no true expertise as a historian. On a number of key points, Arago had naïvely used abridgements and translations of texts, rather than originals. He had shown himself to be unaware of several crucial sources, and worst of all, he showed that he did not understand how book publishing actually worked in the early seventeenth century. The weight of evidence showing that Galileo had demonstrated the existence of sunspots to several persons in Rome in April 1611 was astounding: "It would be difficult," he wrote, "to bring together a greater number of witnesses to attest to a single fact." On the other hand, Arago's insistence on trusting a single date on a print source was laughable:

> Where are the proofs? Where did Arago learn that the date of a book's preface is the date of publication, or that the date of printing can be taken for granted without discussion in matters of priority? Often and nearly always the date that the author gives to the writing, which is entirely arbitrary, precedes by a considerable time that of publication.

How was it that there was no documentary evidence of anyone mentioning Frabricius's book anytime near June 1611, when Arago supposed it to be published? "Those who have knowledge of bibliography" knew better than to trust such evidence: backdating was a notorious practice in the history of publishing.[81]

It does not appear that Arago responded to—or even saw—Libri's final response. But it didn't matter. Developed to help demolish an Italian rival with respect to a controversy in the history of science, and then hashed out over a priority claim on the Academy's floor, Arago's principle of priority by publication would continue to be deployed as he took it upon himself to arbitrate future disputes over priority and property. In fall 1846, a priority dispute would draw the attention not only of astronomers in Britain and France, but of most of the elite members of the physical sciences in those two countries.

The Geopolitics of Open Science

De Morgan's meditation on "Invention and Discovery" appeared on the eve of what he later called "the most remarkable case of simultaneous investigation which ever occurred."[82] Indeed, among the many spectacular Anglo-French confrontations over priority during the nineteenth century, arguably the most legendary concerned the discovery of a new planet—later named Neptune—in September 1846. The announcement that an observer at the Berliner Sternwarte had spotted a new planet in the vicinity of Uranus produced an immediate sensation across Europe. New planets were big news. Especially exciting and nearly unprecedented was that the discovery had been made by mathematical theory. Calculations based on discrepancies between observed and theoretical orbits had led to an accurate prediction that a hitherto unknown planet ought to exist and where it ought to be. The question of credit immediately focused on the individual who made the calculations and thereby spied the planet "at the tip of his pen" before it was glimpsed in the sky.[83] In a series of notes published in the *Comptes rendus* beginning in summer 1846, Urbain Le Verrier, a young astronomer connected with the Paris Observatory, had done just this.[84] Le Verrier had been studying unexplained perturbations in Uranus for several years, in part prompted by a suggestion by his director at the Paris Observatory, François Arago. Arago had also urged him to publish his results speedily, before waiting until he had a theory fully formed: Le Verrier's celebrated prediction in late August had been preceded by several previous notes establishing the anomaly and ruling out other possible causes. In late September, Le Verrier was ready with a specific predicted location: he wrote to Johann Galle in Berlin to look for the star. (The Berlin Sternwarte possessed a particularly large collection of relevant star maps, giving them a leg up on such a search.) Soon Galle was able to write back: "The planet whose position you indicated to me really does exist."[85]

When British astronomers heard of the discovery and of Le Verrier's immediately claiming credit for the prediction, some were stopped in their tracks. John Couch Adams, a recent First Wrangler in the Cambridge mathematical Tripos, had himself been predicting something rather similar for a year or more. The record needed to be set straight. John Herschel, George Airy, and others began to write letters and articles providing evidence that they felt would establish that Adams, and English astronomy, deserved at least partial credit for the prediction. In response, Arago stepped up to protect the rights of Le Verrier. The dispute was on.

The story of Neptune's discovery, and of the priority dispute that ensued, has been told several times. As we have already seen, priority disputes become the subject of historical accounts almost immediately, beginning with participants and contemporary observers. But even recent histories have continued to be geared toward answering the same question: who *in fact* deserves credit for the discovery?[86] Relatively few historians have considered what the controversy reveals about divergent, and shifting, conceptions of discovery and authorship in the physical sciences. Robert W. Smith, however, has shown that in the local scientific culture of Cambridge, where Adams lived and worked, print publication of discoveries might indeed have seemed secondary to personal knowledge. And recent work by James Secord has begun to explore the complex role of print media—and of newspapers in particular—in establishing the character of the Neptune discovery claim.[87] My aim is not to tell the history of these events again but rather to understand the conflict in the context of the debates over the nature of discovery, the man of science, and authorship that this chapter has been exploring.

Upon realizing their predicament following the spread of Le Verrier's announcement, British astronomers attempted to map out the case for co-discovery. They deployed several arguments.[88] First, Adams's detailed claim had been registered in the appropriate scientific institution: hadn't Adams "deposited in the two principal observatories of England, those of Greenwich and Cambridge, calculations of the heliocentric longitude, mass, longitude of perihelion, and eccentricity of the orbit?"[89] Second, the discovery was common knowledge among a sufficiently wide group of people: hadn't Adams's discoveries "been a subject of common conversation among [his personal friends] for the last two years"?[90] Third, there was a paper trail of detailed correspondence: hadn't Adams set forth with crystal-clear conviction his evidence in his letters to Airy, the Astronomer Royal?[91] Fourth, the certainty of imminent empirical confirmation of this claim had been proclaimed on the most public and distinguished of occasions: John Herschel, in his presidential address to the British Association just that summer, proclaimed of this nascent planet, "We see it as Columbus saw America from the shores of Spain. Its movements have been felt, trembling along the far-reaching line of our analysis, with a certainty hardly inferior to that of ocular demonstration." (Unfortunately, Herschel noted, the weeklies reporting his speech had failed him: "These expressions are not reported in any of the papers which profess to give an account of the proceedings, but I appeal to all present whether they were not used.")[92]

As Arago—who took charge of the debate on behalf of the French—noted, there was just one element missing: none of Adams's British defenders had made any "mention of any publication of Mr. Adams' work." And this was all that really counted. Its absence, as Arago proclaimed in his uncompromising analysis of the situation that he delivered to a lively audience at the Academy on 19 October, "was sufficient to put an end to all debate" in the matter. Restating verbatim his historical principle derived from his battles with Libri, he reduced to rubble even Herschel's modest suggestion that credit might be shared.

Figure 4.2 "Mr. Adams discovering the new planet in Mr. Leverrier's report." A French caricature that poked fun at the British claim to priority for discovery of the new planet. "Découvertes nouvelles, caricatures par Cham," *L'Illustration*, 7 November 1846, p. 156. Courtesy of Harvard University Libraries.

> There exists but one rational and just manner in which to write the history of
> science: it is to rely exclusively on publications having a certain date; outside
> of that, all is confusion and obscurity. Mr. Adams has not printed, even today,
> a single line of his research; he hasn't communicated them to any learned
> society: Mr. Adams therefore has no valid claim whatsoever to figure in the
> history of the discovery of the new planet.[93]

The argument was given some bite through a brilliant caricature in *L'Illustration* a few weeks later. It depicted an astonished Adams training his telescope
not on the sky but on the pages of Le Verrier's report from across the Channel
(Figure 4.2). Whereas Le Verrier worked his telescope with one hand and the
pages of his memoir with the other, Adams's arms flailed in midair, no printed
book or even manuscript in sight.

The self-assurance and agonistic character of the French response orches-
trated by Arago took the wind out of the sails of the British claim. Herschel
retreated, noting how those "Frenchmen fly at one like wild cats."[94] Airy sent
letters to Le Verrier and Arago all but implying surrender. The Royal Society
voted its Copley Medal (their highest honor) to Le Verrier. Some never forgave Airy and Challis for not following up on Adams's predictions. When Airy
died in 1892, and a movement to have him buried in Westminster Abbey lost
steam, its backers blamed continued ill feeling stemming from the Neptune

affair.[95] The direct confrontation between the French and British over Neptune thus died down relatively swiftly. Arago's publication principle received some press, but it is uncertain how many across the Channel actually saw its restatement in the *Comptes rendus*. But the idea received a great deal of attention in March 1847, due to Arago's English friend, Charles Babbage.

Although Babbage had not been directly involved in the incident, he seized on the controversy, seeing in it a characteristic failure of English institutions. As he was wont to do, Babbage became obsessed with the question and started a case file to gather precise evidence, both of the rules that applied and of the facts of the case. Regarding the former, his notes show: "Priority of pubn not of discovery the thing the world rewards." Regarding the latter, he acquired a detailed manuscript timeline of events, publications, and correspondence based on the sources available from 1834 through early 1847.[96] The occasion for his public statement was a dispute at the Astronomical Society over who ought to receive a medal for the planet's discovery. Babbage, offended by a point of order at the Society, wrote to the *Times* to deny publicly that Adams deserved any medal. His argument was uncompromising, and it followed Arago precisely: "the modern law relating to discoveries is, that they take their date from the time of their first publication to the world." He sent word directly to Arago of his article and of his support: "my own country from its neglect of science by the Government deserves to be outstripped and your country from its respect for sciences deserves to advance first in the race."[97]

Babbage had reason to be careful about his evidence, for the claim brought a flood of attacks. "B. is an ass & a vicious one but he cant kick or bite, luckily," was the private response of Richard Sheepshanks, the secretary of the Astronomical Society and an old foe. Sheepshanks's public response was more articulate; he went into great detail on the meaning and significance of publication. "The word *publication*, including every kind of communication, from a private letter up to a Royal Proclamation or Act of Parliament, has been the cause of much ambiguity." It was absurd, in the first instance, to focus on publishers rather than discoverers, since "a publisher is sometimes not the author of the thing published, and this may happen by fraud or by consent."[98] Next, since the word "publisher" was coming to be used to refer to the special case of printed documents, Sheepshanks suggested a new word, "evulgator," to recapture the older, broader term: "publication by printing, by open lectures, by transmission to learned societies, and the like." He admitted that in cases of confusion, "the *evulgator* ought to have a preference to the *first discoverer*." But in most cases the key point was simply to ask after the reliability of the evidence, not its particular form: " — 1st. Do the facts adduced contain clear evidence of a discovery? — 2ndly, Is the communication itself, and the date of the communication, quite certain?"[99]

Sheepshanks and several others vehemently denied the possibility of any law of priority altogether: "no one else has heard of the law; it is not supported by analogy, or usage, or common sense."[100] George Airy agreed: "I distinctly deny the existence, or the general reception, of such a law." The real question, Airy thought, was what kind of publication was most useful for prompting discovery. "Is it best to put the theory in type, and to leave it to produce its own effect by extensive dissemination? Or is it best to communicate it in manuscript to a few persons whose position is likely to enable them to undertake the search for the planet." Airy noted that from this perspective, Le Verrier's publications in the *Comptes rendus* "failed totally," and it was only his personal communication with the Berlin astronomers that had worked.[101] John Herschel argued that there could be no law of publication, since "the common sense of mankind will revolt, as it does against all legal definitions."[102]

The widespread British rejection of Babbage's version of Arago's law appears to align with a rejection of the legalistic style Arago had deployed for his historical narratives of discovery. As we have seen, Arago developed this style at the same time as he deployed a heroic narrative of an inventor struggling for pecuniary rights and rewards amid a complex, and sometimes hostile, legal system. If Arago's insistence on a law of publication derived from his experience with patent reform, then the British rejection of the argument might be seen as a rejection of this parallel.

There is some truth to this. Herschel, for example, was quick to reject agonistic narratives focused on first discovery. It was a shame, he thought, "to make rivals and competitors of two men who ought to be sworn brothers." Herschel encouraged several of his correspondents to adopt an alternative narrative that put aside "that ugly <u>word</u> 'priority,'" and instead celebrated simultaneous discovery as a form of cooperation.

> I think it is precisely one of the finest, most interesting & most valuable points in this discovery that it can be satisfactorily shown by evidence that whether

republished or not, the same result had been arrived at independently by two different geometers both starting from the ordinary recognized formulae of the planetary perturbations.[103]

Yet we have seen that British men of science were certainly not all averse to comparisons between patent law and discovery claims. De Morgan, who had distinguished legal matters involving "private rights and public peace" from "scientific history," nevertheless observed that "the cases which have occurred under [patent] law would be good study for those who write on discovery." Another individual who objected to Arago's and Babbage's priority principle, but who certainly considered the analogy to patents, was David Brewster.

Like Charles Babbage, his erstwhile partner in reform, Brewster was fascinated by Neptune. Though he remained on the sidelines during the height of the controversy, it seems he was only biding his time, for in May 1847 he published in the *North British Review* the most elaborate and complex analysis ever produced by a contemporary observer of the controversy. Brewster agreed with Babbage that British conduct in the affair was further evidence that their critiques of science in England seventeen years earlier were as valid as ever.[104] He was appalled. But he denied absolutely any claimed law of priority by publication.[105]

"Seeing . . . that there is no written code which regulates the rights of scientific discoverers," Brewster opened his argument by looking "in the Patent laws of Europe. . . . [for] those general principles which will guide us to a just decision of the case under our consideration."[106] Brewster was well placed to do so. He had been a vociferous critic of British patent legislation for years, ever since a disastrous experience inventing and patenting the kaleidoscope in 1817. He had subsequently made patent reform a central feature of his platform for reform in British science.[107] His aim had been to explore new ways to make the pursuit of natural philosophy a reliably remunerative activity. While a scientific discovery was not quite the same thing as a patentable invention, saleable spin-offs such as the kaleidoscope (based on various optical principles Brewster had developed) ought to be a reliable means of making a profit from philosophical activity, especially for those with limited financial resources to engage in litigation.

Patent case law, Brewster pointed out, did offer a precedent for a legal definition of publication, because evidence of prior publication of an invention could be used to invalidate patents. In this context, Brewster pointed out that publication clearly had nothing to do with printing, because disclosure to just two people *by whatever means* had been considered in English cases as sufficient to constitute publication.[108] Therefore, Brewster argued that while making a claim public was crucial, publication in *print* was not. By this criterion, Adams won hands down, for printing in fact had nothing to do with the legal definition of publication. Following Airy, Brewster argued that Adams had even done a superior job to Le Verrier in publicizing his prediction because he had foregone print publication and immediately informed precisely those persons (in England, at least) who were in a position to test the finding directly.

It is tempting to imagine that the divergence between Brewster and the French on the nature of priority and publishing was related to distinct legal cultures of invention. Laws relating to industrial property in Britain and France had developed within different legal systems and followed very different timelines. A truly integrated patent *system* would be passed in Britain only in 1852,[109] although by the late eighteenth century conventions for the granting of patents were perceived by many outside Britain as a model worth emulating. Indeed, the first modern patent law in France, passed in 1791 (following the abolition of all ancien régime privileges early in the Revolution), was based in large measure on British precedents.[110] The French system framed the patent as a contract between the inventor and the public. The state's role was to mediate and protect the natural rights of both these parties to exploit inventions.

Mario Biagioli has provocatively suggested that key conceptions of modern patent law emerged via the increasing importance of patent specifications, that is, on the insistence that inventors become literal authors describing their own inventions. First, by forcing inventors to articulate a precise claim to originality, the specification made inventions resemble ideas: it "put the 'intellectual' into 'intellectual property.'" Second, the publication of a specification that was detailed enough to allow others to reproduce the invention implied that a patent was a bargain struck between inventors and the state in which disclosure was exchanged for limited monopoly rights to exploit the invention.[111]

On this hypothesis, we might imagine that the insistence, at least in France, that publication was crucial to any claim to discovery arose as these new legal regimes were put into practice and revised in the new century. However, as helpful as Biagioli's periodization might be for a history of conceptions of intellectual property over the long term, we should be careful to avoid supposing any sharp epistemic break took place. In neither Britain nor France (nor even the United States) was the public status of patent specifications taken as a given during this period. In France, in fact, most specifications were effectively sealed until they had expired. (They therefore resembled something more akin to the *paquets cachetés* kept by the Academy than to publications.) British experts on the subject generally thought this aspect of the French system to be improper,[112] but in Britain itself there was little clarity on the matter. Many specifications found their way into the periodical press, though it was notoriously difficult to locate patents on any particular subject. (Patent specifications were distributed at random across three different offices in London, viewing each specification required payment of a separate fee, and taking summary notes on specifications — beyond title, date, and address of patentee — was reportedly forbidden by the clerks.)[113] When Charles Babbage drew comparisons between financial and moral rewards of discovery in 1830, it was precisely by *contrasting* the secrecy that tended to accompany inventions for profit with the necessary publicity that accompanied discovery claims.[114]

In fact, Arago's and Brewster's criticisms of, and suggestions for, their respective nations' industrial property regimes were quite similar. Both saw the tribulations undergone by Watt as demonstrating the insufficient attention of modern states to protecting philosophically oriented inventors. Both saw their respective nations' systems as impeding inventors by imposing exorbitant fees for the taking out of a patent. Both complained that patentees were put in a precarious and potentially ruinous position by there being no substantial prescreening of patent claims. (That is, patents were granted on the word of the applicant that they were indeed new and correctly described; the real test came later in litigation, when for a host of reasons a patent might be deemed invalid.) Both wished to expand, to varying degrees, the scope of what might be the subject of a patent, making it easier for savants in particular to take advantage of their property rights. Both understood the patent as a contract between inventor and public, in which public disclosure was exchanged for state protection of their property rights, but they both felt that the terms of the deal were stacked against inventors, thus discouraging not only disclosure of new inventions but also their very pursuit.[115]

Still, when it came to applying these ideas to scientific discovery, Brewster diverged from Arago in two crucial ways. First, he ultimately denied that discovery and invention (and the social identities of those who produce each of them) could be reduced to the same thing. Second, Brewster suggested that Arago played up the fiction of public disclosure, and the special role of the press, because of his privileged institutional position. Geopolitics mattered in questions of scientific publicity.

While Brewster obviously delighted in using legal tropes to discuss priority disputes, he denied that actual laws — which had given him so much trouble in the past — should apply to science. Thus, drawing on his knowledge of the history of discovery, Brewster imagined a fictional "jury which time has impannelled from all nations and from every period of modern science" to try the case of Neptune. It consisted of "historians of science, Montucla, Bossut, Priestley, Playfair, and Whewell, — and the distinguished philosophers Huygens, David Gregory, Muschenbroek, Smith, Robison, Hutton, and Dr. Young." Surveying how each of them had treated a set of historical priority disputes, he inferred how this jury would rule on Neptune if given the chance. According to Brewster, the jury's verdict would have been clear: "priority of discovery, even when that discovery has neither been communicated to friends nor published to the world, supersedes the claims of priority of publication." Legally speaking, Adams *had* published, but he need not have done so. Discoveries were fundamentally different kinds of things from inventions, and honor flowed from the act of discovery itself.[116] Brewster could cite as evidence his own career: in a race with Biot to work out a polarization theory of light several years earlier, he had taken to dating the pages of his laboratory notebooks to guard his priority claims by the day.[117]

This point was crucial for Brewster precisely because he was deeply concerned with marking out a unique social and professional identity for natural philosophers. However useful a good patent system might be for men of science, the honor that came from establishing priority in discovery was even more crucial for distinguishing men of science who were deserving of other forms of honor and patronage. British circumspection about the propriety of natural philosophers taking out patents in connection with their research would continue to be a distinguishing cultural feature throughout the century and beyond (even if they often did).[118] The situation in France for an elite savant such as Arago was quite distinct. The Academy of Sciences already provided mechanisms by which the state might directly recognize scientific achievement as such. The Academy of the ancien régime had once held the power of judging the utility of inventions on behalf of the state. The revolutionary patent laws had explicitly relieved the Academy, as a suspicious corporate group, of these powers.[119] Nevertheless, in part because patents were difficult and expensive to obtain, the Academy remained an important audience for inventors seeking an endorsement of their products. To blur invention and discovery was therefore, from Arago's perspective, potentially to raise the standing of elite savants, and to enlarge the sphere of influence of the Academy in the industrializing state.[120]

Second, while Arago framed his celebration of fast publication as a gesture of probity that made science more democratic and open, Brewster saw it as designed to favor exactly those who had privileged access to significant institutional support. Arago's rules for scientific priority, as we have seen, played up an analogy with the legal narrative that patents were an exchange of publicity for property rights. This is how Arago put it in a later treatise called "Sur la prise de possession des découvertes scientifiques": "truly valid claims to intellectual property are published claims. I insist on this point to condemn the negligence of those who, having made real discoveries, don't take care to enrich the public domain by printing them."[121] A savant gave up a new fact or theory to the public in exchange for recognition as first discoverer. According to Brewster, this argument for scientific publicity looked much less favorable to those on the peripheries of power:

> If priority of publication is to carry off the laurel from priority of invention or discovery, the philosopher must rush upon the world with his first conceptions — frequently the germs of great discoveries; and if the secret thus thrown to the wind does light upon good soil, the harvest will pass into an alien granary, should the seed have escaped from the grubs of science, or the parasitic monads that pick the brains of philosophers.

It was the "philosopher who inhabits a country where science is not endowed" who was "peculiarly exposed to danger" by early disclosure.[122] For Brewster, the struggling natural philosopher from Scotland, the insistence on immediate disclosure was a stratagem by which powerful institutions could coax every last discovery from isolated experimenters so that each might be adopted and exploited more efficiently by those with superior resources. This had been a racket that the Academy of Sciences had been running for over a century, and journals like the *Comptes rendus* were the modern and most efficient means by which to pursue it.

... the inventive philosopher, in the eager strife between *early* and *earlier* publication, must contrive new modes of despatch and diffusion. In order to appear in the *Comptes rendus*, or in *Poggendorfs Annalen*, or in the *Bibliothèque Universelle*, he must put in requisition the express railway train; or if he inhabits some mountainous region, where the post pays its angel visits, he must trust his despatches to the instinct of the carrier-pigeon, or to the sagacity of a balloon, which, with well regulated fuses, may drop its scientific budgets upon the seats of knowledge and of newspapers.

In the case where Arago's views won out, Brewster was prepared to celebrate the existence of periodicals which, whether due to low circulation or an obscure language, would allow men of science to claim their rights while evading "the robbers who would thus demand from him his intellectual property":

> Our Scottish philosophers would contribute their genius to *Johnny Groat's Journal*, in the far north, or to the Gaelic Magazine, which enlightens the Hebrides, while their Irish friends ... would embalm their discoveries in some Celtic periodical, which may sooner or later be civilizing the wilds of Connemara. In these dark-lanterns of knowledge, which exclude the vernacular eye, the discoveries of science will be as secure from depredation as if they were fossil tablets in strata not yet upheaved, or scrolls of disinterred MSS. among the ashes of Pompeii, or the lava of Herculaneum.[123]

Secret publications were as good as any other if all one wished to do was to establish a claim, since *true* scientific communication happened by other means anyway. Here then was a *dis*analogy between scientific discoveries and patented inventions: there was no reasonable way to restrict, even temporarily, a scientific discovery's adoption by others.

Brewster ended his polemic by lamenting the very real "injury done to Mr. Adams" by the loss of his priority claim. He then invoked the responsibility of the state to right the injustice: "it is only by an act of true liberality on the part of the Government; — it is only by a national recognition of his merits, that Mr. Adams can occupy his true place in the eyes of the civilized world."[124] Rewards, even when they took the form of merit rather than money, were about money as much as merit. The distinct institutional and political circumstances in which Arago and Brewster operated conditioned the different ways in which they perceived the nature of priority claims in the sciences. These differences were important, but underlying them was a shared sense that priority adjudication was not *simply* a matter of honor, but was central to just what natural philosophy was about. Getting priority right would do for the advancement of science what getting the patent system right might one day do for innovation in industry; and the two problems—together with other forms of intellectual property—were part of a broader, changing, field of possibilities for what it might mean to possess a discovery.

⟿

No consensus did emerge that priority claims were directly tied to authorship. De Morgan, for his part, had gone from asserting that priority generally depended on print publication to a denial that any such rule could ever exist: "Let the man who discovers be held the discoverer," he declared in 1860, having decided to revise and expand his treatise on "Invention and Discovery" in the aftermath of the Neptune affair. "The attempt to lay down a law of assignment of discovery has not succeeded: the law is quoted with respect until a dispute arises, and then all the facts of the case are discussed, which is tantamount to a downright refusal to obey the law."[125] But the very fact that De Morgan felt the need to deny such a law is an index of the shifting resources available to those wishing to claim scientific discoveries as their own.

Arago's words would continue to be invoked decades later in disputes over priority in the physical sciences, in France and beyond.[126] The next chapter will explore parallel questions about publication and priority in natural history, where even more ambitious programs to lay down laws yoking priority to publication arose. Here too, however, the matter remained a subject of controversy, and the resulting disorder became a prime motivation for those working to reform the system of scientific publishing later in the century.

It is not surprising that when conflicts over regulating credit for discovery arise, the nature of history should so often have become a privileged site of argumentation. The media and cultural practices through which knowledge claims are recorded and circulated—whether printed, written, oral, or even by human memory—help constitute (even if they do not determine) forms of life in the sciences and beyond.[127] Accounts of the nature of scientific discovery, central to so much twentieth-century philosophical and sociological reflection on the scientific enterprise, often relied for their historical data on discovery narratives whose authors had already done a great deal of work to pull out from the continuum of the past individuals, groups, and events, and to arrange these into causal sequences. The theories of history embedded in such accounts helped decide what counted as a discovery in the first place.[128] Thinking historically about ownership, credit, and identity in the sciences, therefore, requires that we attend to the formats and genres available to actors who have made claims to discovery or attributed such claims to others. This means attending not only to the affordances and limitations of particular media and formats but also to the narratives and values that are part and parcel of their use. By the same token, historians must attend to the ways in which their own media environments, which often share some genealogical relations to those of our actors, shape the historical narratives we produce.

5

What Is a Scientific Paper?

Approaching mid-century, the literary landscape of British science was strewn with the corpses of journals. "I have witnessed in my own recollection a failure of all the scientific journals almost that have been set on foot," declared Richard Taylor. "In the first place, Nicholson's Journal; in the next place, Thompson's Annals of Philosophy; The Royal Institution Quarterly Journal; Brewster's Edinburgh Journal; The Records of Science, and others." All these and more had gone under. The only ones that had any real staying power, according to Taylor, were his own. The *Philosophical Magazine* and the *Annals of Natural History*[1] could survive only because of the special circumstance that in his person were merged "editor, printer, and publisher."[2]

Taylor's insistence on the precarious status of scientific journals struck his examiners at the inquiry on the postal system where he was testifying as rather hard to believe. What about publications such as Charles Knight's *Penny Magazine* and the *Penny Cyclopædia*, he was asked. These seemed to "diffuse knowledge through the country" quite effectively and they were thriving, with print runs sometimes topping six figures. Perhaps the journals Taylor had listed failed for reasons of management rather than the inherent expenses of scientific publishing. Nonsense, Taylor responded.

I should hardly call those scientific works; they are compilations; they are hardly works of discovery, but rather entertaining miscellanies. I am not aware of a single new fact or original idea that has been communicated in them . . . They impart some knowledge to those who have none, but they would never be taken in or consulted by men of science.[3]

The examiners continued to push. Did he intend to "limit the extent of the circulation of scientific works to men of science"? Taylor explained that this was not his point. Rather, there was an important generic distinction that corresponded to a distinction among men: "if there were no men of science to make discoveries there would be no science to communicate to the ignorant."[4] Taylor had in mind a specific notion of what constituted a scientific journal, and entertainment or education had nothing to do with it.

This chapter concerns the central paradox underlying the rise of the modern scientific journal: the idea that there is a well-defined publishing genre that is the locus of legitimate scientific exchange arose against the backdrop of an increasingly fluid and diverse — not to mention vast — landscape of print media trading in natural knowledge. Likewise, though the skills and competencies that were brought to bear in settings of scientific work never stopped expanding, one particular identity took on a special role in defining a life in science: periodical authorship. As journals came to be central to the scientific enterprise — not simply for communicating discoveries but as a means of generating a scientific reputation — a great deal of work was invested in setting bounds to this emerging institution of science.

Historians have shown that the ease with which Taylor claimed to distinguish between those publications that advanced knowledge by publishing original discoveries and those that popularized it obscures a less tidy reality. The diversity of genres and formats in which scientific views and information circulated in mid-century Britain and throughout Europe was breathtaking. Besides books and treatises aimed at a wide variety of readers, articles on scientific topics found their way into a vast array of serial genres, from quarterly reviews, to encyclopedias, to intellectual weeklies such as the *Athenaeum* and the *Literary Gazette*, and even satirical papers such as *Punch*. A similar diversity of print media and genre held elsewhere. In France, scientific reporting had continued in daily papers such as *La Presse* and the *Journal des débats*. These became the breeding ground for a new group of scientific writers who focused on popularization [*vulgarisation*] and made it their particular task to mediate between science and society by their pens. Many of them were also prolific writers of monographs on a vast array of scientific and technical topics. In Britain especially, the line between such specialized periodicals and the wider landscape of serial publishing remained porous. Whether a given publication was an original contribution to knowledge, an attempt to diffuse knowledge already established, or a sensational tract intended solely to generate profit was sometimes impossible to decide, especially given that it might be all of these things at once.[5]

Conversely, transformations in the formats and genres associated with scientific societies and academies, from more expensive transactions to cheaper journals, and from longer memoirs to shorter papers, muddied the bounds between the commercial scientific press and the supposedly higher-status productions of these societies. As periodical authorship took on an increasingly central role as a marker of scientific identity, some men of science began to imagine projects by which to find a new means of defining the bounds of authoritative publishing. In natural history the order of knowledge became tied to the order of print in a particularly intimate way. There, the concept of priority — explored in the previous chapter for the physical sciences — took on a distinctive meaning as a principle through which to ground not simply honor of discovery but the stability of species names. Some, such as Hugh Edwin Strickland, who looked to harness the fixity of print as a means of controlling the proliferation of species names, were dismayed by the messy realities of scientific publishing. His efforts to map out and limit the media of natural history publishing in the 1840s mirrored countless endeavors at mid-century and beyond. These culminated in the Royal Society of London's *Catalogue of Scientific Papers*, a massive project that helped define and delimit the categories of the scientific paper and the scientific periodical.

Seriality at Mid-Century

The governmental inquiry for which Taylor was giving evidence in 1838 was set up to decide whether to establish a streamlined, inexpensive postal system for letter mail within Britain.[6] Since several witnesses were involved in publishing science, the hearings provide a captivating window into the conditions of producing science at the time. Besides Taylor, Dionysius Lardner, the editor of the *Cabinet Cyclopædia*, and Charles Knight, publisher of the *Penny Magazine* and many other works, testified. It was generally agreed by them that the expense and uncertainty of conveying letters, proof sheets, and publications was a great burden not only on purveyors of knowledge in print but on scientific writers and even readers. Each issue of a serial publication depended on intensive correspondence between editors and contributors, including soliciting contributions, submitting copy, and exchanging proofs. While there was a twopenny post operating within London and other metropolitan areas, postage further afield was far more expensive and uncertain, and because it was recipients who paid, senders were often disinclined to use the system unless they could be certain that what they had to write was going to be perceived by the addressee as worth the cost.

The inefficiency and expense of the general post was such that much of the exchange of materials in which men of science were engaged happened by other means. Individuals and societies often relied on booksellers' networks across Britain and the continent for the distribution not only of books and journals but of their correspondence. Those with connections to governmental officials made use of their special "franking" privilege to use the post for free to send their own letters and packages. Lardner testified that his own postage costs were low because he drew heavily on precisely such connections, but this was time consuming and unreliable, not to mention that it was an inherently unjust situation. What did it take to have access to such special services? "Some man connected or known to the aristocratic classes? — Yes." But also "any man who is a Fellow of the Royal Society, who lives among that class, is enabled to avail himself of those means of obtaining scientific communications." In certain towns, kindhearted "local treasurers" were unofficially appointed to receive and distribute philosophical commerce. Many items destined for foreign addresses continued to be sent through diplomatic channels, but these easily went astray. Just last year, months went by while "no English scientific journals were received by the French Institute, nor at Geneva."[7] Proceedings of societies, which had become important publications of scientific societies, were not easy to obtain, even for members. In London, locals could receive them by twopenny post, "but there are members residing in the country, who . . . do not get them at all, unless they have peculiar facilities."[8] Lardner invoked a recurring theme by contrasting the situation with that in France, where he had heard that the *Comptes rendus* "is distributed all over France, I believe gratuitously, by the post."[9]

These delays hurt men of science because it meant that British discoveries might be ignored on the continent and possibly scooped. "The papers which appear in the English scientific journals, are, as soon as those journals reach the Continent, translated and published in the journals of France, Italy, and Germany, and also in America," explained Taylor. Although the editors of the various journals had entered into agreements to exchange their publications, the cost of postage made it more expensive to receive these than to purchase them of booksellers in London.[10]

These obstacles also limited who could participate. Many of those involved in scientific work were "men in very straitened circumstances"; for them, the financial burdens of publishing and correspondence were particularly troublesome. Some were lost to science simply because they could not afford to send their discoveries through the mail. Relieved of this burden, "men of science would communicate by post as freely as they converse together when they have the means of meeting."[11]

Several witnesses saw the postal system as one among many hardships under which benevolent authors, editors, and publishers labored in the face of government restrictions on the free circulation of knowledge. These had come to be known as taxes on knowledge; postage was, according to Lardner, "a tax on the bread of the mind."[12] Besides the expense and uncertainty of the mail, there were other taxes, some of which had been instituted to generate revenue for the state — usually during wartime — and others which had been designed to function as indirect censorship. Among the former were excise taxes on paper, established in 1712 but increased (for certain kinds of paper) during the Napoleonic Wars. The latter tended to target serial genres that focused on political news. The "stamp duty" on daily papers focused on current (generally political) events — so-called because it was enforced by the obligation to use papers provided by the Stamp Office — had been raised to new levels in 1819 to quell the increasingly popular radical press.[13]

The lobby against the taxes on knowledge had become vigorous by the mid-1830s, but it was only gradually and over several decades that these costs of doing business for publishers and correspondents diminished. The lobby to establish the Penny Post was successful. Its launch in 1840 not only eased communications between editors, writers, and correspondents, but changed the very genre of letters in England by putting less pressure on correspondents to make each letter sent worth the recipient's investment. The number of letters exchanged increased a great deal, but the average letter likely said much less.[14] In 1847, the establishment of a cheap rate for posting books made distribution of periodicals cheaper as well. The stamp duty on newspapers was also reduced in 1836, and was abolished in 1855. Paper duties were halved in 1836, but were abolished only in 1861.

While these developments made life easier for those running the sorts of publications favored by Taylor, they did even more to spur the accelerated expansion of cheaper weekly and daily papers, allowing new titles to enter the market and reducing the subscription prices of others. Cheap periodicals such as the *London Journal* and *Reynold's Miscellany* with massive print runs began flooding the market in the 1840s, so that by 1850 these publications outpaced the production not only of more expensive periodicals but even the cheap weeklies begun in the 1820s and 1830s aimed at public instruction.[15] Journals like Taylor's were utterly dwarfed by the increasingly vast array of general publications with larger print runs in which science found some place. The many publications of Charles Knight in London and of the Chambers in Edinburgh are particularly well known, but there were many others.[16]

Aimed more squarely at an intellectual class of readers, the *Athenaeum* and *Literary Gazette* were weeklies that had become well-established sites for circulating scientific information. They were perhaps most well known for their extensive coverage of the annual British Association meetings. Charles Dickens's satirical reports of the fictional "Mudfog Association for the Advancement of Everything" were aimed not only at the meetings but at the new genre of science reporting itself. Dickens's fictional correspondent "Box"— who fancied himself the lynchpin of the meeting "inasmuch as the notion of an exclusive and authentic report originated with us"—risked life and limb to report not only the ridiculous scientific claims in each section but such mundane details as the lodgings and menus of "Professors Snore, Doze, and Wheezy."[17] But the *Athenaeum* also included a great deal of intelligence on scientific and technical topics, including summaries of proceedings of societies, meteorological reports, and letters from men of science on matters of contemporary interest and controversy. Its importance in this regard was so well established that when *Nature* was founded in 1869, the botanist Joseph Hooker saw it as an imprudent attempt to take over a role long played by the *Athenaeum*. Researchers took these publications very seriously. For example, the physicist William Thomson believed that the best way to ensure that his work would be translated and published in Paris was by having abstracts appear in the *Athenaeum*. The downside, as George Stokes pointed out, was that its editor would cut anything that "the great mass of readers" would not understand; "they are sure to cut out the symbols . . . and there is danger that in so doing they may cut out all the pith of it."[18]

Even dailies such as the *Times* included a great deal of correspondence from active men of science on topics they took to be of more general concern. Characteristically, it was in the *Athenaeum* and in the *Times* that much of the controversy over the discovery of Neptune in 1846 and its aftermath was argued out among the men of science involved. Illustrated weeklies as diverse as the *Illustrated London News* and the satirical *Punch* were substantive sources of scientific news.[19]

Seriality characterized not only journals that were so named but many other popular mid-century genres, including encyclopedias and other reference works. Encyclopedias were normally published in series, and although ostensibly planned as an organized whole with a preset order (alphabetical or thematic), it was common to treat them much like journals, with editors and authors finding ways to mobilize the alphabet to carry on a polemic or announce a new discovery.[20] Moreover, publications such as the *Encyclopædia Metropolitana*, the *English Cyclopaedia*, and the *Encyclopedia Britannica* employed many active scientific investigators as writers, and the articles that appeared in them often went beyond overviews of a given subject and were fully engaged in contemporary debates.[21] Even monographs could be published in parts, thus spreading out the risk of a work that might not sell.[22]

Most self-declared scientific journals remained hybrids, geared toward varied audiences and aims, from popularization to the circulation and even archiving of new discovery claims. Those that announced themselves as popular, such as Edward Newman's *Zoologist: A Popular Miscellany of Natural History* and his *Phytologist: A Popular Botanical Miscellany*, opened their pages to the communication of new facts as much as to educational or entertaining articles. Conversely, even those journals largely focused on communicating new discoveries were careful to include more entertaining and readable matter as well. The conductor of the *Magazine of Natural History*, J. C. Loudon, felt the need to defend this hybrid policy: "Had our Journal been appropriated exclusively to subjects of deep research, and only open to the communications of experienced Naturalists, it might have taken a higher stand as a philosophical work, but it would not have been productive of the general good that it was

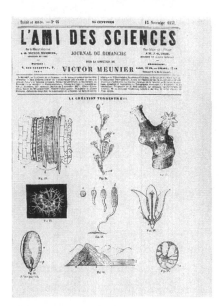 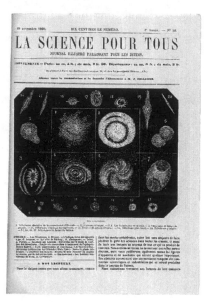

Figure 5.1   Two Paris illustrated popular science weeklies around 1860. *L'Ami des sciences,*
15 November 1857; and *La Science pour tous,* no. 52 (29 November 1860). Writing popular
science was becoming a profession in its own right at this time. Courtesy of Harvard
University Libraries and University of Kansas Libraries, respectively.

our object to promote."[23] Even Richard Taylor made sure his journals could
draw as a broad readership as possible and was willing to reject papers based
not simply on the quality of research but on their ability to attract readers.[24]

The situation in France provides a partial contrast. There, a cadre of writers
actively took on the title of *vulgarisateur* (popularizer) as a professional identity.[25] Many, like François-Vincent Raspail and Jacques-Frédéric Saigey in the
1820s, had come to Paris from the provinces and taken up journalism when
they found a path toward a research career blocked. But they were quicker to
abandon much hope of keeping up their research and made *vulgarisation* into
their specialty. Figures such as François Moigno, Victor Meunier, and Louis
Figuier got started by writing *feuilletons scientifiques* (usually for dailies such as
La Presse) and moved on to editing periodicals on their own along with writing other scientific works targeted at students and others. Many of the new
periodicals foregrounded illustrations, including weeklies such as *La Science
pour tous* (founded 1855) and *L'Ami des sciences* (1855–62) (Figure 5.1). These
took after illustrated magazines such as the *Magasin pittoresque* (founded

1833) and *L'Illustration* (founded 1843), which also regularly included articles on scientific topics.[26]

While these new periodicals dedicated to science could contain a great deal of current scientific intelligence, they defined their task as mediating between scientific elites and broader audiences. In the *Revue synthétique* (1842–44) and *L'Ami des sciences*, Meunier merged strands of utopian socialism with scientific popularization, envisioning his job as synthesizing scientific and social progress: "To popularize [*vulgariser*] science is to use everyday language to recount the efforts of science to constitute a new social order."[27] Moigno and Figuier did not yoke science so intimately to a particular political vision, though both also understood their job was not simply to transmit science but to reframe and critique it.

It was true in France as well that the genres in which scientific information and news might be found were as diverse as the expanding landscape of publishing itself. These continued to include monographs, encyclopedias, and other serialized reference works, as well as a wide variety of serials. The constant barrage of new titles was satirized by the caricaturist Jean-Jacques Grandville in 1844 in a depiction of a pump that could instantly flood large cities with prospectuses and periodicals whose contents largely recycled old ideas in slightly new form (Figure 5.2). Still, in Paris it appears to have been easier for publishers to sustain journals that were targeted more specifically at readers and institutions focused on following the research front. Like the long-running *Annales de chimie et de physique*, the *Annales des sciences naturelles* and the *Journal de pharmacie* were run by editorial boards of prominent savants that published hundreds of memoirs per year. More specialized periodicals directed by individual savants, such as Joseph Liouville's *Journal de mathématiques pures et appliquées*, also survived. Unlike in Britain, many of these publications were able to eke out an existence because the state, through the Ministry of Public Instruction, was willing to subsidize editors and publishers and thus kept independent publications with small readerships such as Liouville's journal afloat.[28]

These specialized journals were increasingly perceived as similar in kind to the memoir series published by academies, a fact that some lamented. It was to recapture the glory days of the *Annales de chimie* of decades past that Moigno claimed he had launched his science journal *Cosmos*. Once, he reminisced, the *Annales de chimie* was truly geared toward communication and critique, whereas each issue was now just "a heavy pile of memoirs printed one after

Figure 5.2 Jean-Jacques Grandville's imagined pump that spewed forth publicity recycled out of older ideas, capable of inundating a large city in an instant. *Un autre monde* (1844), p. 277. NC1499.G66 A42 1844x. Houghton Library, Harvard University.

another without order, intelligence, or life." Liouville's revival of Joseph-Diez Gergonne's mathematical journal in 1836 exemplified the change in orientation. He differentiated his editorial style from predecessors by promising only "to publish the memoirs addressed to us, rarely giving accounts of that which has been published in other collections," and by also avoiding involvement in rancorous debates.[29]

The interminable collections that Moigno bemoaned could also describe the shorter papers in the *Comptes rendus hebdomadaires*. Modeled on journalists' accounts of academic meetings, the Academy of Sciences' weekly had begun as a representation of what happened each week at the Academy. But reports of discussions were phased out, and by mid-century their weekly serial was primarily a collection of research papers (albeit short ones).[30] In 1853, early in the Second Empire, the Academy revoked journalists' privileged access to the papers following each meeting, making it far more difficult for other journals to produce detailed accounts of papers presented, and gave their own publication (delivered each Sunday after Monday meetings) priority over others.[31] The generic shift away from reports on meetings to independent papers matched a broader shift in the French press from a style that derived from representations of oral eloquence in legislative and other meetings, to new stylistic canons that privileged journals as constituting a public forum wholly untethered from oral performance.[32]

The *Comptes rendus* was deemed so crucial to the Academy that it was kept afloat despite putting the Academy into constant financial peril.[33] In 1856 the Minister of Public Instruction reprimanded the Academy for having been long delinquent in publishing its annual report of activities. It was via these reports that France's premier scientific body was meant to render itself accountable to the public—but for two decades it had focused instead on putting out its weekly journal. The Academy defended itself, saying that now "the *Comptes rendus* are a necessary publication" and that all foreign academies had launched imitations of it. The Academy worked differently now: academicians no longer really pursued projects as a collective body; everyone simply pursued their particular "work of research and discovery" on their own. The Academy was best understood as a collection of scientific authors, and the weekly *Comptes rendus* was by far the best representation of the new form of scientific activity.[34]

As the academician became just one author among others, his identity as a judge diminished accordingly. Although commissions were still appointed to report on manuscripts, few reports actually got written (outside of prize commissions). Even vehement critics of the Academy such as Victor Meunier came to recognize that this was not so much a moral failing of academicians—as François-Vincent Raspail had believed—as a structural inevitability: "if out of ten reports only one gets produced, the limit of what is possible has probably been obtained." In order to fulfill the job of judge and rapporteur, an academician "would have to renounce his personal work and resign himself to no longer doing anything original."[35] This was all to the good, for *rapporteurs* had a terrible track record anyway: "How many things, declared absurd at the moment of their birth, contribute today to the glory of their authors and to the shame of their judges!" Meunier admitted that this was not always due to the incompetence of those judges: "where is the criterion by which to recognize, in an instant, the just claims of that which is new?" Academies set up to be universal audiences for scientific claims were the wrong sort of institution. A "good organization of society in science" required that every idea, from its birth, be "exposed to the full light of publicity, discussion, control." Errors would be eradicated through the slower, but more democratic, process of judgment inherent in the circulation of ideas. The best that could be hoped for from the Academy was that it would contribute to such publicity, both for original works of its members and others.[36]

In Britain, even if Richard Taylor insisted on distinguishing between his own journals and those of mere popular interest, it was still more common to draw a line between all of these journals run by a commercial firm and the publications of societies. But this distinction was eroding. Men of science who helped edit journals such as Taylor's increasingly brought to them a sensitivity about demarcating original content from reprints and reviews. William Jardine, who helped Taylor (and then William Francis) edit the *Annals and Magazine of Natural History*, understood his job to be in part to advise Taylor on just such subtle but important points, noting for example that a paper ought to "have come among Bibliographical notices—it is not an original paper but would have done very well though rather long as a <u>review</u>."[37] Jardine's sensitivity to recognizing original content, even in journals, was not unusual. Although journal editors such as Taylor did not use formal referee systems, he did check with a variety of helpers when specialized opinions seemed advisable. The recent spread of referee systems used by scientific societies in the 1830s provided a ready vocabulary for this task.[38] As journals like Taylor's became repositories of memoirs, new publications appeared—including T. H. Huxley's *Natural History Review*, the *Chemical News* and, most famously, *Nature*—which were in a certain sense a reinvention of the sorts of publications that William Nicholson and Alexander Tilloch had launched at the turn of the century, focused more on circulating news and serving as a clearinghouse of correspondence.

As in France, by mid-century many of the proceedings publications of societies had evolved from their origins as summaries of meetings to develop their own identities as independent publications. In many cases, the abstracts of longer papers that these publications had initially contained came to be treated as independent papers, even if they were sometimes prepared by editors rather than authors.[39] The Royal Astronomical Society had taken to using its *Monthly Notices* for the publication of "matter of merely temporary interest" quite early, and by 1847 it formalized this arrangement.[40] In the aftermath of the Neptune affair in 1846, both this Society and the Royal Society took steps to ensure that new discovery claims that they received were put into print as swiftly as possible.[41]

A major generic transformation was gradually taking place by which the preference for longer memoirs associated with transactions was giving way to a preference for shorter papers. One way to follow this transition is via the feedback between authors and referees in the systems of prepublication review that many societies had by then adopted. At both the Royal Society of London and the Geological Society, among referees' most common criticism of memoirs was that they were longer than they needed to be, and that the abstract was sufficient or even superior, recommending excision of anything that was not strictly an original claim or fact. Here is an early example: "the abstract which has been read from the minutes of the Society, containing all the information which is to be found in the body of the paper, would by most readers be preferred I believe to the original."[42] In 1834 Henry de la Beche testified that the utility of the abstracts published by the Geological Society was coming to be perceived as a problem, because "for the most part they contain the cream of the papers read, many of which afterwards appear in the 'Transactions', and that therefore they hurt the sale of the latter."[43] Although these proceedings had their origins in commercial journals dedicated to diffusing knowledge to broader audiences, this focus on condensation could have the effect of further limiting the reading audience for such papers, since much of the contextual matter regarding methods and historical framing was being re moved or relegated to citations in footnotes.

Societies began also to question the utility of quarto transactions, becoming less likely to insist on their luxurious format and expense. After 1830, even the Royal Society of London regularly looked for ways to save money on the printing of the *Philosophical Transactions*, altering fonts and margins over the years to squeeze more words onto each page.[44] Other societies went further. In 1844 the Geological Society of London entered into a collaboration with the publishers Longman & Co. and transformed their *Proceedings* into a *Quarterly Journal*, eventually allowing their *Transactions* to lapse entirely.[45] It would be focused on original papers, but like other commercial journals it would include miscellaneous news from other sources, providing readers with a statement of the "progress of geological inquiry." The stipulations were that the *Journal*'s content would be controlled by the Society's officers, who would provide editorial services at no cost, save for work on the miscellaneous (news) portions. Longman's would keep any profit, and advertisements were allowed (but there were to be none on the cover). This gave rise to a trend. Within the year, the Royal Horticultural Society transformed its quarto *Transactions* into an octavo *Journal*, and soon the Chemical Society of London launched their own *Journal*. The Zoological Society of London flirted with producing "a Quarterly Illustrated Journal in 8vo of an intermediate character of a more popular form" too, but settled on expanding their *Proceedings* and adding illustrations.[46] In 1855, the Linnean Society transformed its proceedings into two *Journals*—one each for zoology and botany—in which original papers were given pride of place by pulling them out from the flow of meeting minutes.[47]

By the 1860s, British science was awash in monthly or quarterly octavo journals, and it was becoming difficult to distinguish those published by societies from those put out by a commercial publisher (Figure 5.3). While early proceedings publications, which were generally unillustrated, were cheap enough that they were given away free to Fellows, the transition to journals usually meant that societies were selling these more expensive illustrated periodicals through booksellers. This brought them into direct competition with commercial publishers, and in some cases, the trend changed the very nature of society membership. At the Chemical and Astronomical Societies, for example, members were coming to be seen largely as subscribers to their periodical, with little to no expectation that they would attend meetings or otherwise participate in the life of the Society.[48]

Other transformations in the social life of science were contributing to the shift toward periodical authorship. In the wake of the *Comptes rendus*, Jean-Baptiste Biot had worried that the turn to writing short notes and away from larger works might lead to the breakdown of long-standing modes of collaboration. But emerging institutions found new ways to adapt the journal to their purposes. With the rise of new laboratory spaces that combined teaching and research, periodical authorship became a key ingredient of a new kind of collaborative institution. At Giessen in the 1830s, the organic chemist Justus von Liebig had turned his pharmaceutical institute into a research and training mecca in which practical instruction in laboratory techniques led regularly to research projects that produced original results. Liebig ran his own journal, the *Annalen der Pharmacie* (later the *Annalen der Chemie und Pharmacie*), and he encouraged students to publish their results, telling one student that "nothing arouses more enthusiasm in young people than seeing their names in print." He pointed to the older French custom whereby students' own re-

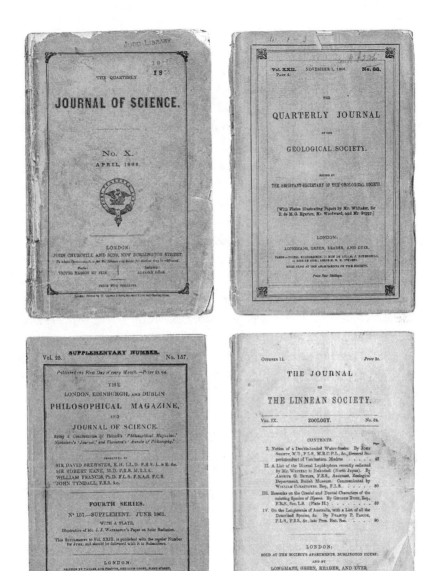

Figure 5.3 By the 1860s, quarterly and monthly scientific journals in octavo format were as likely to be published by a learned society as they were by a commercial publisher. Pictured here are issues of the *Quarterly Journal of Science*, no. 10 (April 1866); the *Quarterly Journal of the Geological Society* 22, no. 4 (November 1866); the *Philosophical Magazine* 23, no. 157 (June 1862, Supplement); and the *Journal of the Linnean Society* (*Zoology*) 9, no. 34 (11 October 1866). Author's collection.

search was subsumed in the larger work of a supervising professor as a great "discouragement to young people."[49] The focus of Liebig's laboratory on the analysis of organic compounds lent itself extremely well to this morselization of collective research into individual projects fit for brief essays. Liebig himself was said to write his articles "while the crucible containing the results is still in the fire, sending the lines to the print shop just after having burnt his fingers lifting them out."[50] The so-called Giessen Model was well known by mid-century, and attempts at imitation spread across Europe, not only via Liebig's own students (who came to Giessen from as far away as the United States) but also eventually in other fields. The result was an important subclass of scientific periodicals whose identity was tied directly to a particular research group and — perhaps even more significantly — a mode of collaboration predicated on individual contributions fit for the genre of the short scientific paper.[51]

In 1830, the dream of Babbage and Augustus Granville had been that contributions to the *Philosophical Transactions* would become a standard by which to judge a life in science. The actual considerations in play when the Council decided whom to support for election would continue to depend on personal knowledge, but many candidates did begin to perceive that being an author of papers was especially significant. In 1851 a young T. H. Huxley was told that publishing in the *Philosophical Transactions* was a key qualification for getting into the Society. Some Fellows, such as Michael Faraday, decided to support only candidates who had published in the *Transactions*.[52] The Philosophical Club, an exclusive group of Fellows that dined together, limited its membership to those who had published in the transactions of one of the major societies.[53] In 1840 the Royal Society introduced a standard form to be filled out by applications to the Fellowship. It included space following "The Discoverer of . . . ," "The Author of . . . ," "The Inventor or Improver of . . . ," "Distinguished for his acquaintance with the science of . . . ," and "Eminent as a . . ." Most of these certificates focused on authorship, often colonizing the space given to the other categories to give information on separate works and memoirs published in transactions (Figure 5.4 is an example). In 1853, the Society began requesting that copies of the publications listed be submitted as well.[54]

Before mid-century it was exceedingly rare for candidates to mention journals in these lists, although in the 1850s they began to do so, and the Society sometimes indicated to authors that journal publications might indeed be relevant to admission.[55] In France, a tradition dating to the beginning of the century had savants looking to gain election to a prestigious academy or teach-

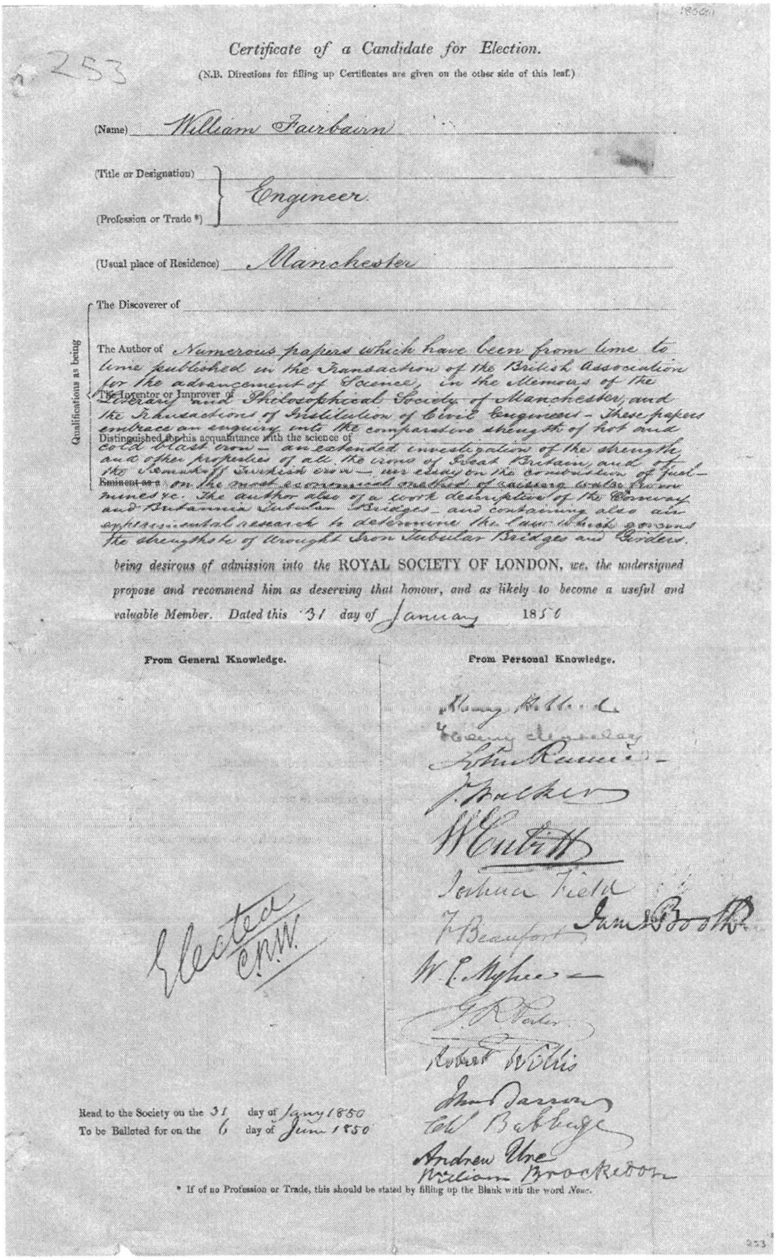
Certificate of a Candidate for Election.

(N.B. Directions for filling up Certificates are given on the other side of this leaf.)

(Name) William Fairbairn

(Title or Designation) (Profession or Trade *) Engineer

(Usual place of Residence) Manchester

Qualifications as being

The Discoverer of

The Author of Numerous papers which have been from time to time published in the Transaction of the British Association for the advancement of Science, in the Memoirs of the Literary and Philosophical Society of Manchester, and the Transactions of Institution of Civil Engineers – these papers embrace an enquiry into the comparative strength of hot and cold blast iron – an extended investigation of the strength and other properties of all the irons of Great Britain, and of the Lemakoff Turkish iron – an essay on the construction of fuel – on the most economical method of raising water from mines &c. The author also of a work descriptive of the Conway and Britannia Tubular Bridges – and containing also an experimental research to determine the law which governs the strengths of Wrought Iron Tubular Bridges and Girders.

The Inventor or Improver of

Distinguished for his acquaintance with the science of

Eminent as a

being desirous of admission into the ROYAL SOCIETY OF LONDON, we, the undersigned propose and recommend him as deserving that honour, and as likely to become a useful and valuable Member. Dated this 31 day of January 1850

From General Knowledge.	From Personal Knowledge.
Elected CRW.	Henry Holland
	Henry Moseley
	John Rennie
	J. Walker
	W. Cubitt
	Joshua Field
	F. Beaufort
	Jam. Booth
	W. C. Mylne
	G. R. Porter
	Robert Willis
	Chas. Darwin
	Chs. Babbage
	Andrew Ure
	William Brockedon

Read to the Society on the 31 day of Jany 1850

To be Balloted for on the 6 day of June 1850

* If of no Profession or Trade, this should be stated by filling up the Blank with the word None.

Figure 5.4 The election certificate of the engineer William Fairbairn (1850). This is typical of mid-century certificates, where qualifications connected to authorship began to dominate the form and colonize the space provided for other forms of qualification. Royal Society of London, EC/1850/11.

ing position produce self-published pamphlets — known as *Notice des travaux* or *Exposé des titres* — that laid out their claims.[56] Early on these included diverse kinds of information, including books published, a description of their significance, as well as memoirs submitted to an Academy, especially if there existed a favorable report. But they also included much else besides, including social status, discovery claims, inventions, acquaintances, and a general commitment to science.[57] They did not, however, tend to include a simple list of publications, and works published in journals were either ignored or referred to as a group. A paper was more likely to be referred to as having been presented orally — "read at the Institute," for example — than as having been published by a journal. Gradually, however, descriptive paragraphs took a backseat to lists in which the basic unit was the publication, and over time more of these were listed as having been published in commercial journals. But the increasing visibility of publication lists also owes a great deal to another kind of publication that flourished especially after mid-century. Societies, publishers, and editors began to produce massive catalogs of science in print, providing persuasive tools for distinguishing not only who counted as a scientific author, but what counted as a scientific publication.

Printing, Names, and Natural History

> Scientific natural history has now become as much a matter of literary research as of physical observation.
> —HUGH EDWIN STRICKLAND to CHARLES DARWIN, 31 January 1849[58]

When in 1847 David Brewster decried the notion that priority claims were connected fundamentally with authorship, he was especially piqued by the claim that this was a law of science. "If it is a modern law," Brewster challenged, "it will be no difficult task to point out its date, and to tell us by what parliament of science it was enacted."[59] In fact, a group of zoologists had recently come together under the aegis of the British Association to lay down just such a law. The rules they enacted to regulate naming practices in natural history helped spur a debate about the order of natural historical knowledge that lasted well into the next century, one that brought to the fore questions about the nature and limits of periodical publishing in science more generally.

Within those fields focused on the discovery and naming of natural objects, "priority" had taken on a meaning and significance that surpassed other branches of knowledge. Here it was not simply a matter of apportioning credit or honor for discovering new objects of knowledge but of the fundamental ordering of that knowledge. Debates about natural systems of classification were notoriously fraught and prone to disagreement, but there remained hope that there could at least be agreement about how to assign them names. The conviction that these two activities — classification and naming — could be more or less separated made it possible to imagine that stability of names might serve as a preliminary step toward later theoretical progress.

During the 1820s and 1830s, long-standing uncertainty about names developed into a pressing source of controversy. Imperial London was the principal destination of a continuous stream of natural objects in want of description and classification. The flood of new information seems to provide a natural explanation for such dissent, but this was not in itself that new. Gordon McOuat has shown how principles of naming in zoology became the locus of struggles over legitimate consensus in zoological communities, dividing not only distinct subdomains and theoretical orientations, but also political commitments, pitting radicals against conservatives.[60] The British Museum's vast holdings in zoology were under the charge of John Edward Gray, once a prominent radical who was a sworn enemy of a group of more conservative zoologists. To shield zoology from radical influence and establishing order, the latter had to look elsewhere. They looked to the printing press.

In 1835, Hugh E. Strickland, a young geologist and ornithologist, graduate of Oxford and son of a Baronet, launched an ambitious campaign to bring order to "the present anarchical state of things" in zoology.[61] The field was being inundated, he felt, by radical practitioners who fixed on taxonomical innovation as a means of reform. Strickland hoped to neutralize reformists' seeming attempts to bring chaos to zoological knowledge by shielding actual species names from their radical plans. With Gray, who had his own ideas about names in charge at the British Museum, Strickland wanted a "code of laws for naturalists."[62] Appealing to a Lockean conventional philosophy of language, Strickland argued that what mattered about species designations was not so much their *accuracy*, but rather their *fixity*. The surest basis for such a program of fixity was *priority* — as it was expressed in the printed record: "With respect to established species, priority seems to be the universal law for the adoption of specific names."[63]

The idea that the first name applied to a new species ought to be the one that sticks was not new; it had been laid down in general terms by Linnaeus himself. Even in France, often maligned by the British as the ideological source for reckless innovation in search of a chimerical "natural system," the general idea that priority of names ought to be respected was common.[64] But what Strickland wanted was to elevate a general guideline about how naturalists should behave into a law that would constrain their behavior (on penalty of exclusion from the community). The legitimacy of zoology as a universal activity depended on it: "the so-called commonwealth of science is becoming daily divided into independent states, kept asunder by diversities of language as well as by geographical limits." English zoologists were kept from communicating with their French counterparts because "their *scientific* language is almost as foreign to him as their *vernacular*."[65]

Strickland campaigned for years and at last got the question of zoological nomenclature onto the agenda at the British Association meeting in 1841. Strickland's nomenclature committee included several of the most prestigious names of the British zoological establishment, including J. O. Westwood, Charles Darwin, J. S. Henslow, and Richard Owen. Their report the following year enshrined the "Law of Priority" as the "fundamental maxim" of right conduct in natural history: "The name originally given by the founder of a group or the describer of a species should be permanently retained, to the exclusion of all subsequent synonyms."[66] After noting certain exceptions to the fundamental maxim, the committee specified precisely what did and did not constitute a legitimate claim to priority in naming a species. In order to secure the desired permanence of nomenclature, the committee insisted that priority depended on printing:

> To constitute *publication*, nothing short of the insertion of the above particulars *in a printed book* can be held sufficient. . . . MS names have no authority. . . . Nor can any unpublished descriptions, however exact . . . claim any right of priority till published, and then only from the date of their publication.[67]

Although the committee "conceded that mere insertion in a printed book is sufficient for *publication*," they strongly recommended that authors submit their new species definitions to "such periodical or other works as are likely to obtain an immediate and extensive circulation."[68] This focus on a specific medium as a condition of the legality of any name had only rarely been a part of previous guides to naming practice; nor was it clearly stated in Strickland's earlier writings on the subject.[69]

The Strickland Rules received a lot of press and were discussed by zoologists not only in Britain but on the continent too.[70] The extent to which the rules actually influenced naming practices was harder to gauge. Despite much support, they faced criticism from important figures such as Louis Agassiz and Gray, who controlled the British Museum. The latter managed to keep the draft rules from being officially discussed at the British Association meeting, and thus from any official endorsement by British zoologists.[71]

Among the obstacles that Strickland and his allies encountered was that naming was connected with various different problems of order. Notwithstanding Strickland's philosophy of language, names of zoological species and genera were far from entirely arbitrary designations. Some of what made names useful was that they did reflect beliefs about the true relations of genera and species. The binomial system of naming came from Linnaeus, and while he had understood his own system to be more or less artificial, it remained connected to very real judgments about similarity and difference that were not easily pulled apart from theoretical classification. Even Strickland admitted that some names were simply too misleading to be allowed to stick.

A second obstacle was that many naturalists perceived the right to name a new species or genus to be central to establishing a reputation in natural history. Indeed, many earlier attempts to lay out principles of nomenclature had highlighted the requirement that names bestow honors in a just manner. Strickland's committee tended to disparage considerations of fame as a hindrance to consensus. Charles Darwin was particularly agitated about the use of personal names in nomenclature—which he called "the greatest curse to natural history"—and he contemplated launching a crusade to ban the practice. "I do not think more credit is due to a man for defining a species than to a carpenter for making a box," he complained to his friend Joseph Hooker. "But I am foolish & rabid against species mongers . . . they act as if they had actually made the species, & it was their own property."[72] As Darwin gathered allies for his campaign, Hooker convinced him to stand down: "the swarm of snobs with various qualifications & claims for fame & who seek fame alone is still very great & by Jove old Darwin they will be down on you like Sikhs if you do not look out . . ."[73] Any attempt to wrest the priority of naming from its association with personal honor and property was unlikely to succeed—priority was doomed to be pulled in multiple directions.

Finally, Strickland's insistence that the fixity of printing was the way to guarantee the fixity of species names was not necessarily a solution to the problem of order at all. The first name put into print might not only be bad, but it might also be obscure. Giving preference to print as the public record of natural historical knowledge meant ignoring other, potentially more relevant, modes of circulation. Though Darwin had helped draft the original rules, he himself came to doubt the wisdom of strict adherence to print priority on these grounds: "I am in a perfect maze of doubt on nomenclature," he admitted to Strickland in 1849. He worried about the many cases in which well-described names in widespread circulation turned out to have been anticipated by some otherwise forgotten printed text. When that happened, Darwin argued, "I believe if I were to follow strict rule of priority more harm wd be done than good."[74]

Other candidate procedures for fixing names were available, including maintaining a council of elite naturalists who would have "the sole privilege to alter, or increase, when necessary, the nomenclature of natural history."[75] J. E. Gray understood it to be his prerogative as Keeper of the British Museum's vast zoological collection to dictate zoological names, a task he pursued by maintaining loose-sheet catalogs from which he regularly published books and memoirs.[76] Among elite botanists, the conviction that those who maintained public repositories had ultimate authority to control names was especially common. When zoologists looked into revising and ratifying Strickland's Code in the 1860s they attempted to bring elite botanists on board. Most declined. Both Joseph Hooker and George Bentham, who jointly administered Kew Gardens, were skeptical that general rules could be reliably applied without reference to an authoritative individual or body. Hooker insisted that there was harmony among botanists on such issues, precisely because there was a clear hierarchy: "Our leading systematists following one plan & being themselves dutifully followed by the masses." Bentham thought rules would not go far unless they were supplemented by regular meetings of "men whose name should have sufficient weight and authority . . ."[77]

Nevertheless, by the 1860s botanists were also applying themselves to developing a code. Here, the scope of what constituted publication was a vexed question; many agreed that a strict restriction to print was inappropriate in a field in which collections of specimens held such a central place. From Massachusetts, Asa Gray urged this point: "May not the affixing of a name to a sufficient specimen in distributed collections (a common way in botany) more surely identify the genus or species than might a brief published description?"[78] The rules drafted by the Swiss botanist Alphonse de Candolle and approved at a meeting of French botanists in 1867 agreed that such labels were sufficient. More controversial was whether public readings of papers at societies and academies should count. French botanists engaged in a heated discussion on "the sense in which the word *communication* should be understood."[79] Even among zoologists, Alfred Russel Wallace noted, it was still "a disputed point whether the date of reading a paper publickly [*sic*] is that of publication as regards claiming priority."[80] George Bentham pointed out that in biology, unlike other fields, priority was not simply about establishing "independence of observation or discovery," but of naming, which was "a more complicated question." Bentham, then president of the Linnean Society, denied the admissibility of public reading, "because it does not give fixity; the author himself does not feel bound by it and . . . may alter his names before or during the printing." But printed documents brought their own difficulties, since the dates they carried were so often inaccurate. He noted with approval the recent practice of the Academy of Natural Sciences in Philadelphia, which had begun publishing its journal with a list of the dates on which each separate paper was actually distributed.[81] Later in the century, commissions of zoological nomenclature doubled down on the point that "print publication is essential," drafting new rules that explicitly ruled out readings at meetings or manuscripts as legitimate media for priority claims in natural history.[82]

But there was a tension at the heart of any plan to lodge the fate of zoological knowledge in print, for it put zoologists at the mercy of publishers and printers, where considerations of profit might overrule science. It also raised the fundamental problem of drawing a line around what constituted a legitimate print publication. Natural history descriptions might be found not only in reputable monographs and periodicals but in a wide spectrum of more general and even popular publications. How to unearth a true first description? In response to Darwin's concerns about bad or obscure descriptions, Strickland admitted that the rules would likely need to be supplemented by new regulatory mechanisms to discipline authors and publishers.

> Let a committee (say of the Brit Association) be appointed to prepare a sort of *Class List* of the various modern works in which new species are described,

arranged in order of merit. The lowest class would contain the worst examples of the kind, and their authors would thus be exposed to the obloquy which they deserve . . .[83]

Strickland was wont to vent his frustration over the commercialism of much of what passed for zoological publishing. When the *Zoologist*, a new monthly, launched in 1843, Strickland complained that, despite some useful articles, it "studies popularity too much and Science too little."[84]

Strickland and his allies—including his father-in-law and naturalist William Jardine—made numerous efforts to shore up the state of zoological publishing. In 1843, they helped set up a collective of natural historical authors and readers—the Ray Society—to mitigate the influence of commercial publishing by printing natural historical works through subscription. "The object," Strickland explained to Charles Lucien Bonaparte, "is to save from oblivion such works of this kind as would probably entail loss to the authors if they published them at their own cost."[85] Relying on printed descriptions, after all, made sense only if natural historical works were able to make it into print. Jardine blamed the commercial motives of publishers for the fact that works that ought to have been published as books were split up, scattered in random places, and "forced to a periodical" if they were printed at all. Astronomers had received large grants from the British Association to print "dry but excellent" star catalogs, Jardine complained, but zoologists hadn't had as much luck. He cited the sad case of John V. Thompson, a zoologist who could not find a publisher or subscribers for a book, and was "obliged to publish his discoveries in two or three periodicals and Transactions . . ."[86]

But not everyone agreed that publishing in periodicals was such a bad fate, or even that natural historical works were being forced into the shadows by the nasty publishing market. "If this is the case," John Edward Gray protested, "why are the scientific societies languishing for papers and why are there not more important papers in Taylors Annals?" Jardine, meanwhile, attributed the shoddy character of scientific periodicals to the profit motive, noting that on the continent governmental subsidies allowed specialized scientific publications to flourish.[87] Just how one interpreted the status of scientific journals in Britain depended on the circumstances. In the mid-1850s, Charles Darwin claimed that there was no appropriate venue available in which he could publish his theory on species. But just a few years later when T. H. Huxley was mulling over editing the *Natural History Review*, Darwin warned him to make sure it was a true *review* rather than a "journal for original contributions in Nat. History." Of the latter, he noted, "there is now superabundance."[88]

After promising not to compete with publishers or societies, the Ray Society was launched, gathering over 600 members within a year and publishing volumes for its subscribers at a quick pace.[89] Strickland quickly set to his next project: charting the vast world of zoological publications. He lobbied the Society to invest in a series of bibliographical works, the most important of which would be a grand catalog of all zoological publications. A central record of zoological works might go a long way to avoiding the "great inconvenience" caused "by substituting a name just *dug up* out of some obscure provincial periodical" for those in general circulation.[90] His plan to do this efficiently was that the Society should acquire, expand, and print the personal bibliographical collection of the Swiss zoologist Louis Agassiz, which had become renowned for its reputed size. Strickland eventually got approval to go ahead with the project after promising to do the editorial work for free. The task turned out to be far more difficult than he had expected and forced him away from research for nearly a decade.[91] Agassiz's records turned out to be a mess—they contained inaccurate citations, titles, and many works that seemed to have no connection to zoology at all. Strickland noted that had he started from scratch, he would have "excluded all merely popular and elementary essays." There were other zoological catalogs, but what made this one special and so difficult to produce—was that it included not only separate works but the contents of journals and transactions.[92] Not only were many of these characterized by "extreme rarity and limited circulation," Strickland pointed out, but it was precisely periodicals that were the worst offenders in mingling the popular and the scientific: "Our popular 'Magazines' of Natural History teem with trifling notices, often anonymous, sometimes brief and indefinite, sometimes wordy and inflated, but which do not contain a single fact of scientific importance." But the obscurity and hybrid nature of scientific periodicals was precisely the reason that a catalog was needed, for it could also begin the work of sorting truly scientific publications from mere magazines and popularizations.[93]

Scientific Papers and Bounding

The first volume of Strickland's *Bibliographia Zoologiae et Geologiae* appeared in 1848. It was one among several comparable efforts beginning around mid-century. All of these were dwarfed, however, by the Royal Society of London's *Catalogue of Scientific Papers*, a massive project beginning in the 1850s to produce a complete list—organized by author—of all the scientific papers that had appeared in scientific periodicals since 1800. The *Catalogue* was the most resource-intensive special project that the Royal Society undertook during the nineteenth century. Its construction reshaped the past according to a novel vision of the bounds of scientific publishing,[94] projecting backward to 1800 a vision of the history of science in which the edifice of knowledge was built up out of individual papers each connected unambiguously to an author. It did as much as anything else to cement the idea that scientific publishing involved special kinds of authors and special kinds of texts, both easily demarcated from the wider publishing landscape in which much scientific communication nevertheless remained embedded. And it encouraged men of science to embrace specific authorial habits in the future. The making of the *Catalogue* provides a fascinating window into the process by which the scientific journal came to be defined and delimited. Catalogs are powerful agents in processes of canon-formation.[95] But because they are tools used not only by historical actors but also by the historians who study them, their role as historical objects is easily missed.[96]

The idea for a catalog of scientific memoirs started in the United States, with a young engineer who worked for the US Coast and Geodetic Survey. The idea found its way to England, where the British Association passed it on to the Royal Society.[97] In the process, what began as a project to catalog publications of learned societies in the physical sciences became something much bigger: a catalog of all serial works of science—including not only transactions and proceedings but scientific journals and ephemerides.[98] Building such a catalog was nearly unprecedented. It meant treating independent scientific journals—normally published by commercial publishers for a profit—on a par with the publications of scientific societies, valuable not simply for their ability to circulate knowledge quickly but as contributions to the permanent record of science. The contents of journals were not normally the sort of thing anyone would choose to catalog. It meant drawing a boundary around a group of publications that were not all easily assimilated. Some objected that such imprudence would cripple the venture. It was one thing to deal with the memoirs appearing in transactions, carefully vetted, substantial works that were intended by their authors as permanent contributions to knowledge, suggested James David Forbes. But it would be a waste of time to include every article, for example, in the French Academy of Sciences' *Comptes rendus*, "or even in the better class of scientific periodicals." William Thomson agreed, urging "condemnation on the planners of the Scientific catalogue."[99]

The Council of the Royal Society, an organization actively looking to maintain its relevance in a quickly changing scientific landscape, jumped at the *Catalogue* project. Since the 1830s, it had pursued multiple reforms that focused especially on its reading publics, and its library had followed an acquisition strategy such that its periodical holdings had come "to constitute its chief distinction."[100] It had also started its own scientific journal, the *Proceedings*. But this measure had only gone so far. The proliferation of specialized scientific societies in London and beyond made it implausible to suppose that the Royal Society's publications served as a clearinghouse of scientific discoveries in England or even London. In Paris, by comparison, the weekly *Comptes rendus* of the Academy of Sciences had quickly become well known across Europe. Although a great deal of scientific exchange occurred beyond the ken of the Academy's meetings, it was possible to imagine (even if falsely) that the *Comptes rendus* was not simply an account of the Academy's doings but of the best of French science. Although attempts at imitations were widespread, efforts to found something like it in London had foundered.[101]

If the Council of the Royal Society wished to extend its reach over the world of natural knowledge, it needed a different strategy. The *Catalogue* was it.[102] But the scope of this task turned out to be enormous. A section of the library was transformed into a bibliographical workshop, pumping out bibliographical slips by the thousands. Work on the ground to assemble the contents was directed by Walter White, the assistant (and later head) librarian. Soon the Society hired (his distant relative) Henry White to focus exclusively on *Catalogue* work. Under their direction, an expanding troupe of "boys" produced individual slips containing references to papers from the Society's vast collection of periodicals. Using carbon and copy paper, they prepared four slips per entry, expecting that they would gradually arrange the material according to several ordering schemes.[103] When they found gaps in a series, the Library Committee in many cases purchased the missing volumes. Not content with the contents of the Royal Society's library, in 1864 White's bibliographical workforce fanned out across London in pursuit of catalog-worthy periodicals to be found in other libraries, including the British Museum, and the Chemical, Linnean, and Medical & Chirurgical Societies.[104]

The decision to make the *Catalogue* into a complete listing not only of papers in transactions but also in scientific journals did not resolve the question of what, specifically, would actually get into the catalog. Neither "scientific journal" nor "scientific paper" were natural kinds. Moreover, since the role of periodicals in science had changed a great deal since the beginning of the century, any decisions that made sense for the present would not necessarily translate easily to the past. Officially, the idea was to include papers relating "to all branches of knowledge for the promotion of which the Royal Society was instituted, excluding matter of a purely technical or professional character."[105] But the directions from the Committee were more specific than that. The non-Society periodicals they had in mind were those "containing memoirs, published by individuals": the sorts of things to be found in transactions. News and reviews, especially if anonymous, were to be left out—so that there needed to be some means of selecting out original contributions toward improving natural knowledge.[106]

Since the work would consist largely of indexing the contents of particular periodicals, the most efficient way to proceed would be to decide which journals would count. While it was relatively straightforward to distinguish society publications from commercial journals, it was more difficult to differentiate independent scientific journals that contributed to the progress of knowledge from those that contained only news, synthesis, education, or entertainment. But since the bulk of the indexing would be done with relatively unskilled labor, it was in the selection of periodicals that the Council and Catalogue Committee played the largest role in shaping the *Catalogue*'s contents. If all else failed, the Committee pointed out that simply settling on "a perfect list of the scientific periodicals" would itself be a worthy outcome, setting implicit limits to the task of searching the literature.[107] The recorders amassed a core set of publications by trawling the catalogs and shelves of important libraries in London and picking out those publications that seemed to fit the bill. Periodicals deemed ambiguous were forwarded first to the Library Committee and then if necessary to the Royal Society's Council for a final verdict.[108]

By 1864, the clerks had indexed most of the serials in the Royal Society Library. They printed a large broadside list of these serials and dispatched it to societies, academies, and known experts on scientific publishing for feedback (Figure 5.5). The list contained 453 items and was dominated by publications of societies, which made up nearly two-thirds of the titles listed. In terms of

Figure 5.5 Preliminary list of journals and transactions to be indexed for the *Catalogue of Scientific Papers*, distributed by the Royal Society for feedback in 1864. Royal Society of London, MM/14/184.

geographical distribution, the list was made up mostly of serials from Western and Central Europe. The remainder were scattered throughout the rest of Europe, North America, and the British colonies.[109]

Responses flowed in from across Europe and America, usually with praise for the project along with scores of publications that they had missed. Johann Christian Poggendorff, longtime editor of the *Annalen der Physik* in Berlin and author of a biobibliographical dictionary for the exact sciences, supported the project but recommended separating out the exact sciences from the rest. The Austrian mineralogist Wilhelm von Haidinger successfully lobbied the Society to include an entire genre — the yearly *Programms* published by German and Austrian Gymnasiums and which often included research papers of instructors — and he offered to do the indexing himself. Joseph Henry, whose report from the Smithsonian helped spark the project, was scandalized by how meager the Royal Society's list of periodicals was. He complained that the number of scientific serials excluded from the list was larger than the number that were included.[110] By the time the *Catalogue* went to print, the list of serials had tripled in size, reaching 1,378.

In principle, the Society wanted to include worthy papers even when they had appeared in serials that fell out of the bounds of properly "scientific" publications. Borderline cases — "Medical and Technological journals" in particular — were to be scanned "for the sake of a few strictly Scientific papers scattered through them."[111] But there were strict limits. Social respectability and scientific specialization seem to have been the most operative criteria. Proceedings publications — despite sometimes consisting largely of abstracts — were indexed extensively, while periodicals such as the *Mechanics' Magazine* — which did contain many articles that might be deemed contributions to scientific knowledge — were entirely excluded. The *Athenaeum*, an important venue for reports on scientific meetings and lectures, was ignored, as were the quarterly review journals that often employed prominent scientific investigators to write on scientific topics. No encyclopedia articles were included, and newspaper reports were left out entirely.

Because the model for the *Catalogue*'s contents was the memoirs published by learned societies, ingenuity was required to transform many of the journals in the list into the sorts of things commensurate with these publications. Throughout the period, but particularly early in the century, independent journals often relied on reprinting, summaries, and translations for much of their content. It was thus not straightforward to identify what should count as a genuine original memoir. Publication dates were not always helpful in this regard: it happened regularly that a memoir published by an academy or society was preceded by an earlier summary in an independent journal that might appear without the author's knowledge. The Committee ultimately chose to include reprints and translations along with the originals under a single entry. The *Catalogue* provides an interesting (if imperfect) window into the extent to which papers were reprinted, excerpted, and translated over the course of the century. In the case of an author such as Jöns Jacob Berzelius, who had an international reputation but wrote in a language with relatively few readers, the majority of his entries listed in the *Catalogue* included references to multiple periodicals (as many as ten) published across Europe.

Shorter versions of a paper were generally listed along with more extended versions, even if the former was an abstract or a summary prepared by an editor or journalist rather than the original author. In the case of certain periodicals, this was more often the case than not. Consider the example of the Paris journal *L'Institut*. Founded in 1832, when many Paris dailies and weeklies employed journalists to report on academic meetings, it consisted largely of reports of academic meetings with summaries of papers (often written by a third party). The indexers nevertheless chose to include many of these extracts and journalists' accounts alongside memoirs. The same went for the reports written by French academicians on papers and inventions presented at meetings. From one point of view, these were simply commentaries and judgments and thus not original contributions. On the other hand, they were viewed by many as excellent abstracts of papers—more valuable than many of the original papers themselves. Many of these also found their way into the *Catalogue*.

In cases such as these, the problem of attributing authorship became a perilous task: was the report writer the rightful author? Or the author of the paper being reported? Or perhaps it was the academician who actually presented the work at the meeting. Authorship proved too restrictive a concept to contain these possibilities: the Committee attempted to set rules of attribution, but these proved difficult to follow in practice.

The problem of authorship is all the more significant in that this became the organizing principle for the whole enterprise. The temptation to produce a subject classification, or even an alphabetical title keyword index, loomed large, but the Committee ultimately decided to focus on the author. Augustus De Morgan—so respected for his bibliographical acumen that he was a member of the Catalogue Committee despite not being a Fellow of the

Society—had long warned of "heavy and numerous difficulties" entailed by subject classification.[112] Names were the most rational basis for cataloguing books; this was the only ordering principle that De Morgan felt would generate universal assent.

But when it came to periodicals, attributing authorship was often problematic. The difficulties were not limited to proceedings and academic reports; they extended to the many journals that published original papers anonymously or pseudonymously. The list of such journals reads like a who's who of what some called the "higher class" of scientific journals, including the *Philosophical Magazine*, the *Quarterly Journal of Science*, the *Bibliothèque universelle*, the *Edinburgh Journal of Science*, the *Annales de mathématiques pures et appliquées*, the *Annalen der Physik*, and the *American Journal of Science*. Although anonymous attribution of original research papers was especially common earlier in the century, the practice had by no means disappeared by midcentury.[113] The *Catalogue*'s editors publicly expressed their exasperation that many of the journals with which they were forced to deal did not correspond to their vision: "None but those who have been engaged in a task of this kind can form any idea of the difficulty occasioned by such omissions."[114]

The editors did what they could to incorporate anonymous and pseudonymous papers into their system: "No pains have been spared," they wrote, "to assign the Memoirs to their respective Authors."[115] But this was a Herculean task; some authors, such as the natural philosopher Thomas Young early in the century, had used pseudonyms for the majority of their contributions to periodicals.[116] In ambiguous cases, such as articles signed by initials, they were often forced simply to make educated guesses. But assigning authorship in this way proved perilous, especially when dealing with publications outside England. A French statistician later uncovered various cases of misattribution of French authors in the *Catalogue*.[117]

The catalogers reached out to editors, societies, publishers, and suspected authors themselves to uncover clues about authorship. They asked authors to confirm their contributions, sending them proofs for corrections, and even asking them to send in their own publication lists.[118] Such requests prompted writers to reflect on their status as authors. Some zoologists had gone through a similar process during the publication of the *Bibliographia Zoologiæ et Geologiæ*. Strickland had also reached out to authors for help after his first volume was criticized for inaccuracies and omissions. Murchison wrote to him that after updating the proof, he felt "quite appalled at the list under which I am buried," and he went on to reflect on what he still wished to accomplish as a geologist. Charles Darwin was likewise surprised by the "awfully long" list that had resulted, and recommended deleting less important items.[119] The Society's requests for lists starting in the 1860s were greeted with similar reflections. David Brewster reported taking "much pleasure" in helping get his list exactly right. Some requested separate copies or proofs of the list for their personal use.[120] The editors of the *Catalogue* hoped that it might change authorial practice itself, "inducing contributors to Scientific Journals to give their names in full."[121]

Many authors expressed confusion and even frustration over just what kinds of publications were supposed to be included. James David Forbes tried to add his encyclopedia articles, reviews, and separate works, but he was informed that none of these were admissible. John Herschel also expressed confusion over what counted: "I do not know what are the limitations under which the citations are made further than that they . . . exclude books & separate publications." He noted in particular "whether an Encyclopaedia is to be considered a <u>Book</u> more than a Vol of Memoirs <u>by various authors</u> I do not know."[122] Herschel had written extensively for the monumental *Encyclopaedia Metropolitana* put together by Samuel Taylor Coleridge, on topics such as Light, Sound, and Physical Astronomy—Coleridge had singled them out as "enlarging the boundaries of our scientific knowledge."[123] Herschel understandably wanted them counted among his scientific contributions. Herschel was one author who did maintain meticulous lists of his publications, but he never differentiated between specialized venues and venues such as the *Athenaeum* and the *Times*.[124] In subsequent decades, authors who published in genres outside the bounds of the *Catalogue* continued to complain that it misrepresented their output. The French chemist Marcellin Berthelot pointed out "very important omissions," providing a list of his books in the hope that the editors might make an exception.[125]

After about a decade of work, the Royal Society began printing in 1866, having convinced the government to foot the bill.[126] The first series, covering the years 1800–1863, took up six volumes of about 1,000 pages each. Over the next sixty years, four more series would be published to cover the remainder of the century. As they readied volumes for the press, the Committee was eager to learn how the *Catalogue* would be received. It soon became evident that they were right to focus so much on issues of authorship.

Reviewers were quick to hail the *Catalogue* (Figure 5.6) as a grand,

Berzelius, *Jöns Jacob.* **243.** Om basisk fosforsyrad Kalkjord. Stockholm, Öfversigt, I., 1844, pp. 136–138; Journ. de Pharm. VII., 1845, pp. 367–369; Liebig, Annal. LIII., 1845, pp. 286–288.

—— **244.** Bidrag till några salters historia. Stockholm, Öfversigt, I., 1844, pp. 203–210.

—— **245.** On the hypothesis of Mr. PROUT with regard to Atomic Weights. [1844.] Silliman, Journ. XLVIII., 1845, pp. 369–372.

—— **246.** Åsigter rörande den organiska sammansättningen. Stockholm, Acad. Handl. 1845, pp. 331–360; Poggend. Annal. LXVIII., 1846, pp. 161–187; Taylor, Scientific Mem. IV., 1846, pp. 661–682.

—— **247.** Att skilja Iridium från Osmium och Rutenium. Stockholm, Öfversigt, II., 1845, pp. 145–149.

—— **248.** Ueber die Classification der Mineralien. (*Repr.*) Erdm. Journ. Prak. Chem. XXXIX., 1846, pp. 297–311.

—— **249.** Om Talkjordshydrat, motgift för Arsenik. Stockholm, Öfversigt, III., 1846, pp. 231–233.

—— **250.** Om Bomullskrutet. Stockholm, Öfversigt, III., 1846, pp. 283–291.

—— **251.** Ueber die Bildung eines wissenschaftlichen Systems in der Mineralogie. (*Transl.*) Poggend. Annal. LXXI., 1847, pp. 465–476.

—— **252.** Om Allophansyrans rationella sammansättning. Skand. Naturf. Förhandl. V., 1847, pp. 347–349.

—— **253.** Om knallsyror. Stockholm, Öfversigt, IV., 1847, pp. 1–2.

—— **254.** Om œnanthsyra. Stockholm, Öfversigt, IV., 1847, pp. 3–4.

—— **255.** Kolsvaflad etyloxid, Xanthogensyra Zeise. Stockholm, Öfversigt, IV., 1847, pp. 49–51.

—— **256.** Organiska saltbaser af animaliskt ursprung. Stockholm, Öfversigt, IV., 1847, pp. 119–120.

—— **257.** Om Allophansyra. Stockholm, Öfversigt, IV., 1847, pp. 151–153.

—— **258.** Sur la découverte de l'acide lactique dans l'économie animale. (*Transl.*) Journ. de Pharm. XIII., 1848, pp. 477–480; Phil. Mag. XXXIII., 1848, pp. 128–133.

Berzelius, *Jöns Jacob,* et *Pierre Louis* **Dulong.** Nouvelles déterminations des proportions de l'eau, et de la densité de quelques fluides élastiques. Annal. de Chimie, XV., 1820, pp. 386–395; Schweigger, Journ. XXIX., 1820, pp. 83–84; Thomson, Ann. Phil. II., 1821, pp. 48–50; Tilloch, Phil. Mag. LVIII., 1821, pp. 203–208.

Berzelius, *Jöns Jacob,* och *J. G.* **Gahn.** Undersökning af några i grannskapet af Fahlun funna Fossilier. Hisinger, Afhandl. Fysik, IV., 1815, pp. 148–216; Annal. de Chimie, IL, 1816, pp. 411–422; III., 26–39, 140–161; IV., 1817, pp. 243–245; Annal. des Mines, I., 1816, pp. 463–484; Schweigger, Journ. XVI., 1816, pp. 241–305.

—— —— **2.** Analys af ett fossilt Salt, från Fahlu grufva, och Insjö sänkning. Hisinger, Afhandl. Fysik, IV., 1815, pp. 307–317.

—— —— **3.** Tantalmetallens egenskaper, halten af syre i dess oxid, dennes mättnings-capacitet och kemiska egenskaper. Hisinger, Afhandl. Fysik, IV., 1815, pp. 252–262; Schweigger, Journ. XVI., 1816, pp. 437–447.

—— —— **4.** Undersökning af några i trakten kring Fahlun funna Fossilier, och af deras Lagerställen. Hisinger, Afhandl. Fysik, V., 1818, pp. 1–27; Oken, Isis, 1819, col. 391–409; Schweigger, Journ. XXI., 1817, pp. 25–43; Thomson, Ann. Phil. IX., 1817, pp. 452–460.

Berzelius, *Jöns Jacob,* och *L.* **Hedenberg.** Misslyckade försök att erhålla svafvelbunden qväfgas. Hisinger, Afhandl. Fysik, II., 1807, pp. 99–102; Schweigger, Journ. II., 1811, pp 158–162.

Berzelius, *Jöns Jacob,* och *Wilhelm* **Hisinger.** Expériences galvaniques. (*Transl.*) Annal. de Chimie, LI., 1804, pp. 167–173.

—— —— **2.** Försök med Elektriska Stapelns verkan på Salter och på några af deras baser. Hisinger, Afhandl. Fysik, I., 1806, pp. 1–38; Gilbert, Annal. XXVII., 1807, pp. 269–324.

—— —— **3.** Undersökning af Cerium, en ny metall, funnen i Bastnäs Tungsten. Hisinger, Afhandl. Fysik, I., 1806, pp. 58–84; Annal. de Chimie, L., 1804, pp. 245–271; Nicholson, Journ. IX., 1804, pp. 290–300; X., 10–12; Tilloch, Phil. Mag. XX., 1805, pp. 154–158.

—— —— **4.** Undersökning af Spinell från Åkers Kalkstensbrott i Södermanland. Hisinger, Afhandl. Fysik, I., 1806, pp. 99–105.

—— —— **5.** Undersökning af rosenröd syrsatt Manganes från Långbanshyttan i Wermeland. Hisinger, Afhandl. Fysik, I., 1806, pp. 105–110.

—— —— **6.** Undersökning af Pyrophysalith, et nytt Stenslag från Finbo i Dalarna. Hisinger, Afhandl. Fysik, I., 1806, pp. 111–118; Annal. de Chimie, LVIII., 1806, pp. 113–121; Nicholson, Journ. XIX., 1808, pp. 33–37.

—— —— **7.** Undersökning af en grönagtig Stenart, från Glanshammar i Nerike. Hisinger, Afhandl. Fysik, II., 1807, pp. 203–205.

—— —— **8.** Undersökning af Orsten (Lapis Suillus). Hisinger, Afhandl. Fysik, III., 1810, pp. 379–388.

Figure 5.6 Page 340 of the first volume of the Royal Society's *Catalogue of Scientific Papers* (London: Eyre and Spottiswoode, 1867). Biodiversity Heritage Library.

monumental work.[127] According to a review in the *Athenaeum*, the *Catalogue* quashed any lingering notion that the Royal Society had become an "effete body." *Chambers's Journal* reported in December 1867 that it would revolutionize literary research: "Any student desirous to know what has been written on any scientific subject since the year 1800 will have only to look into the great *Catalogue of Scientific Papers*."[128] Reviewers were also quick to reflect on what science looked like through the lens of the *Catalogue*. Two things stood out: science was growing out of control, and it was not the sort of thing that any individual could be expected to follow. An expansive review in the *Athenaeum* warned that "[t]he sciences are breaking down under their own weight." Because "the mass of publications containing original investigation has increased so much, and is increasing so much faster," no one could be expected to be an infallible authority, even in any one branch.[129] "Those who are not aware of the state of things go with confidence to men of name with the question whether a little matter . . . be original or not." But this was sheer naiveté: authentic knowledge was lodged in printed sources, and by this measure it was growing at a rate with which no individual could be expected to keep pace.

Although each individual's knowledge of important periodicals might remain small, the *Catalogue* demonstrated just how large the world of serials had become. Readers who thought that the number of scientific serials was small, perhaps a few dozen, would "be surprised to hear that the number of black birds baked in this scientific pie is about *fourteen hundred*":

> Looking roughly at the number of entries in a page of the Catalogue, we surmise that there will be not far from 200,000 scientific communications registered in the whole work, being 2,500 for each year which is contained in 1800–1863. What a coral-island science will be![130]

Here was an awe-inspiring image: science as coral, a massive edifice built up invisibly over time by the constant addition of tiny polyp-like papers. Representations of science as serial accretion would become massively popular in the following decades, epitomized by the positivism of Ernst Mach and Henri Poincaré. This was not so much an underlying reality at last revealed by the *Catalogue*; rather, in raising the profile of the scientific paper, the *Catalogue* offered a medium for this epistemic imaginary.

But it was not self-evident that science — even science as it was represented in serials — really did look quite the way the *Catalogue* suggested. Like Herschel and Forbes, some commentators were mystified by the logic of the *Catalogue*. What purpose could there be in simply grouping together the publications of learned societies with a selection of commercial journals? Of course, monographs and larger works — according to some the true bastions of scientific progress — were missing. But that was not all. The principles of selection used by the editors had distorted "the progress and history of discovery both in Physical and Natural Science" according to one Royal Society Fellow writing in the *Athenaeum*.[131] Why had they excluded "the many short, but frequently important communications sent to journals not professedly scientific, such as are to be found in the columns of the *Times*, and more especially in the *Athenaeum* itself?" And then there were the many "important books of 'Voyages and Travels,'" which often contained "Papers relating to almost every branch of Science." Not to mention that much good science was published in government documents: "Scientific Reports published in the Proceedings of Royal Commissions and Parliamentary Committees, which are little known to the world at large, although frequently very valuable." These kinds of documents were not only central to scientific communication, but they were "far more difficult to discover in the mass of miscellaneous matter with which they are surrounded, and in which they may almost be said to be buried, than those which have appeared in the specially scientific journals." Another commentator noted that with more care, the editors might have been able to include "many valuable memoirs now in great part unknown, or, at all events, forgotten." According to these criticisms, if the *Catalogue* was truly meant to be an aid to scientific research, then it had missed the mark.[132]

But some might have argued that there was value in just such forgetting. By ignoring the diversity of formats in which science got into print, the *Catalogue* put bounds on what should be preserved, simplifying the record and pushing scientific authors to embrace certain publications and scorn others. While evidence suggests that many did use the *Catalogue* as a research tool,[133] it found other kinds of uses. Nearly all early reviewers, for example, used the *Catalogue* to become auditors of scientific productivity. Reviewers scanned each new volume's pages looking to discover the most prolific authors, and nearly always reported what they found. "Brewster numbers 299 in this century; Cauchy, who belonged entirely to this century, 478; Challis, 190; Cayley, 308,"[134] read a typical report. Some reviews, such as one in the new journal *Nature*, read like long bibliographical lists themselves.[135] The *Catalogue's*

	I. Bd.	II. Bd.	III. Bd.	Zusammen
Oesterreich	4	2	6	12
Deutschland	30	22	39	91
Frankreich und Belgien	38	49	27	114
Grossbritannien und N.-Amerika	24	25	35	84
Deutsche in Russland	2	4	2	8
Italien	8	4	—	12
Dänemark	—	2	1	3
Schweden	1	1	—	2
Niederlande	2	1	2	5
	109	110	112	331

Figure 5.7 Data derived by Wilhelm Haidinger from the *Catalogue of Scientific Papers'* first three volumes, showing the number of contributors with over fifty publications organized by region. *Archiv der Mathematik und Physik* 51 (1870): 14. Courtesy of Harvard University Libraries.

layout seemed to encourage such score-keeping; it not only grouped together papers under authors' names but numbered them for each author. In later decades, as new series of the *Catalogue* appeared covering later years, the total publication count for each author from the previous series was carried over, such that observers could easily keep track of the running totals.[136] Some reviews, for example in the *Wiener Zeitung*, allowed that "the mere comparison of numbers is no basis for final judgments of value, which lie rather in the content of each communication." But who, in truth, could resist? He assured the reader that a "quick glance is no less stimulating." Another reviewer noticed that many of the articles cataloged — especially of the biggest contributors — were exceedingly short, but no matter: "in pure science a few words may represent a week of hard thought."[137]

Although the *Catalogue* made it easiest to compare the publication counts of individual authors, enterprising statisticians found ways around this. The running tallies of entries published in each series were a source of quantitative evidence about the progress of science more generally. For example, in late century the *Catalogue* was used to support the claim by the Austrian social critic Max Nordau that clerical claims that science was in retreat were baseless. Here was objective proof "by figures that science does not lose, but continually gains ground."[138] In 1870, the Austrian mineralogist and editor Wilhelm von Haidinger "took a statistical look into this grand work" and derived a rubric by which the *Catalogue* could be used to compare productivity by geopolitical region. Counting each author who had published at least fifty papers, he organized these into a table according to region (Figure 5.7).[139]

But counts such as Haidinger's likely revealed more about the varied uses that authors and editors made of periodicals across time, geography, and subject matter than they did about the comparative scientific productivity of individuals and nations. They also, of course, revealed something about the choices that the catalogers themselves made about inclusion and attribution. To see why the French, for example, did particularly well according to Haidinger's measure, consider the most prolific French author (and second most prolific overall) in the first series, Augustin-Louis Cauchy. The mathematician had become notorious for his use of the Paris *Comptes rendus* after it was launched in 1835. Rather than submitting short summaries of larger works, Cauchy chose to use the new weekly to publish his research the way a writer of fiction might publish a serialized novel, putting two- to five-page installments into print on a weekly basis. (For academicians, the only publication limit imposed was number of pages per week.)[140] Some French savants, such as Jean-Baptiste Biot, condemned Cauchy for his lack of authorial decorum.[141] Yet Cauchy's publishing habits were only the most extreme instance of a general trend. (Ironically, Biot himself was the second-largest contributor to the *Comptes rendus* after Cauchy during this period, having published 121 notes over twenty-five years.) Outside France many condemned the French for their obsession with publishing short notes staking out priority claims without confirming their findings.[142] But in London, Philip Sclater, John Gould, and John Edward Gray (the most prolific author in the first series) also produced massive publication lists, due largely to their numerous publications in the *Proceedings of the Zoological Society*. In Berlin, meanwhile, Christian Ehrenberg and Heinrich Rose used the Prussian Academy's *Monatsbericht* in a similar fashion.

But the *Comptes rendus* of Paris dominated the *Catalogue* like no other publication. Although the *Catalogue* listed over 1,400 periodicals, about one out of every eighteen entries in the first series contained a reference to the *Comptes rendus* (despite its having existed for fewer than half the years covered by the first series). This rate was much larger than that of the next most cited source, the *Annales de chimie*. Next followed a long list of commercial journals: Poggendorff's *Annalen der Physik und Chemie*, the *Philosophical Magazine*, the *Journal für praktische Chemie*, Liebig's *Annalen der Chemie und Pharmacie*, the *Journal de pharmacie*, the *Astronomische Nachrichten*, Silliman's *American Journal of Science and Arts*, and the *Annales des sciences naturelles*. The *Comptes rendus* was alone among society publications in the top ten, although several proceedings journals—including those of the Zoological Society of London, the British Association, the Société Géologique de France, and the Akademie der Wissenschaften in Vienna—made the top twenty. Although the *Catalogue* had its origins in an idea to index the memoirs contained in transactions and academic collections, these made up only ten percent of the *Catalogue*'s references, while independent journals accounted for fifty-eight percent of them. When all was said and done, the *Catalogue*'s representation of the world of science in print inclined unmistakably toward the commercial press.

Other biographical genres came to resemble the *Catalogue*'s lists as well. By the 1870s, many of the *Notices des travaux* that French savants compiled to support their candidacy for membership in the academies and for university chairs included an exhaustive, enumerated list of their periodical publications.[143] Similarly, election certificates for the Royal Society began to consist largely of publication lists. Accounts of authors' publications had dominated these documents since the 1850s, but they were still integrated into narratives in the form of sentences and paragraphs sometimes describing the substance of their contributions. During the 1860s candidates began to do away with these syntactic connections, leaving a bare list of titles of papers and periodicals (sometimes adding more bibliographical information, such as year, volume, and page number). By the mid-1870s, nearly all regular election certificates were made up of bibliographical lists.

The attraction of using the *Catalogue* to compare the productivity of individuals and groups went alongside what might have been its most enduring role: as an aid in assessing and narrating a scientific life by authorship. "The cataloguer is the *vates sacer* of these heroes," wrote one reviewer. "Without him they are lost in the bulk of periodicals which are not at hand . . ."[144] In the 1860s, Michael Foster and Charles Lyell had imagined the fruits of a life in science to be grand treatises, but the availability of authors' publication lists suggested a different vision that repositioned periodical authorship at the heart of a scientific life. In the 1870s, obituary notices and appreciations of researchers across Europe began routinely to invoke the *Catalogue*. Proofs of a productive scientific life often cited the numerical count of a subject's papers appearing in the *Catalogue*. When Justus von Liebig died in 1873, a eulogist wrote: "the mere list of Liebig's contributions to Science covers nearly eleven large quarto pages of print, and embraces 317 titles." When the Belgian statistician Adolphe Quetelet passed away in 1874, a remembrance in *Nature* also turned, appropriately, to numbers: "The many-sidedness and fertility of his mind may be seen from his scientific memoirs enumerated in the Royal Society's *Catalogue of Scientific Papers*, amounting at the close of 1863 to 220."[145] The *Catalogue* also provided a ready-made itinerary by which to follow the unfolding of a career. When Wilhelm von Haidinger—who had vigorously promoted the *Catalogue* himself—passed away in 1871, two detailed obituaries followed the *Catalogue* closely in narrating his life. Franz Ritter von Hauer and Eduard Döll both referred to the periods and changing foci of Haidinger's research career by grouping together subsets of papers in the *Catalogue* and giving paper counts in each case.[146] Some saw the *Catalogue* as a successor to Johann Poggendorff's 1863 *Biographisch-literarisches Handwörterbuch zur Geschichte der exacten Wissenschaften*,[147] which had included short biographical information and bibliographical information about researchers in the exact science. But biographies here had been whittled down to a list of papers, and the biographer was simply an indexer carrying out a set of instructions.[148]

—&—

Histories of modern science in Europe and the United States have long lived in the shadow of the *Catalogue of Scientific Papers* and its representation of the bounds of scientific identity and authorship. But the *Catalogue* is far from a neutral witness in the history of scientific publishing. It is not simply a product of its time but helped shape imaginaries not only of the scientific literature but of the identity of researchers. Cutting across transactions, proceedings, and selected high-status specialized journals—but excluding nearly everything else—nothing did more to fix a conception of scientific periodicals as collections of original papers attributed to individual authors. This was a work of canon-making writ large. It became a key part of the bibliographical apparatus used by historians of science, not only for biographers but also in efforts to study the history of science quantitatively. The *Catalogue* and similar lists did not simply make such research easier, but by privileging a certain representation of the course of a scientific life, they helped shape underlying assumptions about the most valuable fruits of this calling.

The use of publication counts as an objective measure of productivity that could be pursued across time and geography would only continue to gain ground. In the mid-twentieth century, the *Catalogue* remained a key resource for those interested in such topics as "Men's Creative Production Rate at Different Ages and in Different Countries." More generally, "the number of scientific papers a man publishes" has remained a favored measure among those interested in "generating, in an objective fashion" ranked lists of scientific productivity.[149] But by attending to the ambiguities, sleights of hand, and elisions in the construction of these monuments to science productivity, we can begin to glimpse the ways in which received boundaries between experts and non-experts — and the literary practices through which these boundaries are reproduced on a day-to-day basis — were erected in the first place.

6

Access Fantasies at the Fin de Siècle

No one professes to be an advocate of closed or secret science. But demands that science be more open, or more public, have been attached to a dizzying range of concrete ideas about what that might mean. Today, the fight for open science is fought on many, sometimes conflicting, fronts. They include opening research up to nonprofessionals, putting data into online databases as it is produced, and making scientific papers free and accessible to working scientists across the globe. These visions of scientific exchange constitute different kinds of access fantasies. They are claims about the way in which access to science — conceived variously as a set of practices, membership in a community, or a body of knowledge — ought to be facilitated and secured: who ought to have access, how that access is maintained/delivered, and just what is to be accessed. These visions are not necessarily illusory, but they exist as imaginaries nonetheless. And these imaginaries are political, in the sense of politics as the activity through which individuals and groups articulate, negotiate, and implement claims upon one another and upon collectives. To articulate an access fantasy in science is to articulate a specific claim about the politics of knowledge, and it is often to invoke, even if implicitly, a criterion of the legitimacy of scientific communities.

Throughout the period covered by this book, the expanding role of the periodical press in scientific exchanges was driven in no small measure by actors claiming to expand the relevant publics of scientific knowledge. Founders of commercial scientific journals around 1800 said they were spurred to action by the expensive and imposing character of the publications of academies and societies. Others put forward periodical print as offering an alternative form of scientific sociability to those more exclusive institutions. Societies and academies in turn explained their embrace of proceedings publications as a response to expanding conceptions of the place of science in civil society. Those who insisted on print publicity as a condition of claiming the honor of a discovery pointed out that print was accessible to all. These arguments, diverse as they were, all took part in a particular kind of access fantasy. The public that they conceived for scientific knowledge was relatively broad and undifferentiated; even if in practice it remained quite restricted, in principle it was boundless. And what they aimed at producing was access to "knowledge" in a rather abstract sense of that term.

But in the final decades of the nineteenth century, a different kind of access fantasy rose to prominence. In this vision, the relevant public was more specific: it was made up of active researchers, trained in some branch of science, and likely trying to make a career out of their devotion to science. Those engaged in science were now referred to as workers—*travailleurs* in France—and the knowledge those workers needed was embodied in a very specific medium: the scientific paper. Finally, new media technologies and infrastructures were being imagined as essential to this kind of access. Index card subscription services that functioned like specialized clipping bureaus proliferated alongside proposals for centralized repositories and distribution agencies designed in part to overcome the disorderly and unregulated marketplace of scientific publishing.

In a word, the scientific literature had come into being. In this key concept was embodied a powerful image of the collective knowledge of experts and of the orderly progress of knowledge. It was in principle open to all, but in practice what mattered was that it was accessible to those who were scientific practitioners—authors—themselves. The scientific literature was the terrain for political battles over access to and control over knowledge. If science was to serve a special role in the modern democratic polity, then it ought to embody modern democratic values. But it could play this role only if it was defended from unscrupulous publishers and unqualified authors.

The moment the scientific literature appeared as an object of widespread attention, observers perceived it as being in need of repair. It proved to be an unwieldy entity, weighted down by varied functions and meanings. A British Library bibliographer named Frank Campbell worried that "the development of Periodical Literature has been such as to constitute a very considerable danger to the progress of knowledge."[1] Could the same medium that was used for the "*momentary* dissemination of knowledge" be used at the same time as "the *permanent* record of it"? Besides that, periodicals focused on science — particularly those associated with societies and academies — had also associated with the vetting and validation of knowledge claims. Finally, scientific authors looked to periodicals to build their reputations and careers by amassing publications, using lists of their papers as calling cards establishing their expertise. Could a format under the control of an assortment of publishers, editors, societies, and academies be trusted to all the uses to which it was now being put? These issues went beyond problems of access, and yet the problem of how to organize communications to make them simultaneously more equitable and more efficient caught the attention of savants across Europe.

Following the Franco-Prussian War, savants in both France and Britain argued that the rising industrial power of the German Empire demonstrated that science in their respective nations needed better organization and support from the state. Their arguments began with the technological innovations science made possible, but they also encompassed moral and cultural considerations. In the first instance, this meant reforming and expanding scientific education, but it also meant guaranteeing open and efficient communications. Some observers, such as the French astronomer Henri Saint-Claire Deville and the British astronomer Norman Lockyer, argued that it was the communal character of the scientific enterprise that made it special; it ought to be a model of social cohesion. The scientific literature was a key medium through which savants had come to imagine social exchange. Lockyer pointed out that the very success of educational reform had changed the nature of scientific sociability by producing a new class of scientific worker: "Students are turned out by the score . . . who have taken part in original research. Such persons constitute a class which has only lately come into existence":

> Whether or no they are to spend their lives in a dull routine of teaching or testing, falling gradually further and further behind the times, or whether they are to aid or even to follow the advance of knowledge depends largely upon the facilities for acquiring information which are afforded to them. They leave the University, or the University College, with its well-stocked library, and

forthwith their touch or want of touch with the outer world depends almost entirely on the periodic literature of the science to which they have devoted themselves.

For Lockyer and many others, ensuring that the scientific polity truly encompassed all such workers meant that the scientific literature would need to be put on a more secure footing.

In France, activists and entrepreneurs located at the margins of the Paris academic elite pursued several enterprises to alleviate the potential alienation experienced by savants spread across the nation. One was to educate young savants in the most effective use of the scientific literature. Another was to develop new media technologies that could bring scientific papers within the reach of all scientific workers. The 1880s and 1890s saw the emergence of several attempts by young, politically minded entrepreneurs to bring about a revolution in scientific information. They launched small but ambitious organizations dedicated to collecting, disseminating, and organizing information about scientific papers in the name of decentralization and democracy. It was the new Association française pour l'avancement des sciences that became the most consistent patron of bibliographical reform in French science, setting itself in opposition to the role played by the Academy of Sciences and its *Comptes rendus*. Opening and equalizing access to the scientific literature was becoming a form of social activism.

In Britain, these problems were perceived in terms of systemic organizational inefficiencies as much as they were in terms of informational inequalities. Lockyer's *Nature* claimed to be "the accredited organ of Science among the English-speaking peoples," a forum for news of all that was new. But even Lockyer saw this as far from sufficient. "Is not some integration of the body scientific needed?" wondered the physiologist Michael Foster. "Does each of us work economically for the body's ends?" The literature had become bloated with sewage, "which has to be got rid of before the stream can again become free from impurity."[2] Worries such as these led to ambitious schemes through which scientific societies would wrest control of scientific venues from publishers.

La Decentralisation Scientifique

In early September 1870, Bismarck's troops were closing in on Paris. Although the war between France and Prussia had begun only about a month earlier, the Prussians had caught French troops off-guard. The embarrassing mismatch — almost immediately attributed by many in France to the superiority of German education, industry, and science — would cast a long shadow for decades to come. The emperor Louis Napoleon had surrendered on 2 September, but a coup d'état led by Léon Gambetta and others immediately dissolved the Empire, and the newly minted Third Republic renewed the declaration of war. All that was left was the capital. Prussian command, with an eye to the international reputation of a new unified German Empire, chose to avoid a direct assault on Paris and simply to wait until its inhabitants capitulated. They expected this to be a matter of a few weeks at most. But with the prospect of a last stand at Paris, civic and military officials had moved swiftly to fortify Paris with new defenses and supplies. Herds of cattle and sheep had been brought within the city limits, and fortifications around the city were redoubled. Thus outfitted, and with some daring and desperation as well, the siege of Paris would drag on much longer than the Prussians had anticipated. But if a nutritional crisis was postponed, Parisians faced another pressing shortage: information. Military command, under Gambetta, had abandoned Paris and regrouped at a distance in Tours. Attempting to slip anything or anyone past the Prussian blockade was prohibitively dangerous, and the telegraph cables that normally kept Paris connected not only with other European cities but with its provincial centers had been cut.[3]

The single means of escape was by air, which in 1870 meant by hot air balloon. Gambetta himself — determined to carry on the battle — had escaped Paris by this means in the early stages of the siege. But there was a catch. Since no one had yet invented a reliable method of steering these vessels, people and information could get out (as long as balloons were still to be had), but there was no way to get them back in. The metropolis whose political and intellectual life had so dominated the nation was in the unprecedented position of information isolation.

For the members of the Academy of Sciences, the situation was especially alien. While savants in the provinces had complained for decades of their intellectual isolation, now it was the Parisian elites who were forced to fend for themselves. They continued to publish the *Comptes rendus*, but it came to resemble the proceedings of a provincial academy, circulating within the bounds of the city and focusing on matters of immediate, practical interest. The chemist Marcellin Berthelot abandoned his more esoteric studies of the foundations of organic chemistry and led several committees investigating the chemistry of explosives. Papers were read on the purification of fats from diverse sources for nutritional purposes, the preservation and transport of meats, the mechanical efficiency of techniques for milling wheat, and the healing of bullet wounds.[4]

But it was a technical breakthrough originating outside the Academy that would ultimately reconnect Paris with the newly formed republic. The only means remaining for getting messages back into Paris had been by carrier pigeon. But since there was a finite number of pigeons trained to return to Paris, they were by no means guaranteed to complete their journey, and their maximum payload was about a gram, their utility seemed limited. Compression was called for. René Dagron — a photographer who had recently made his name by pioneering precision microphotography as an aesthetic medium — offered his services. The idea of taking a massive quantity of text and, through photographic techniques, reducing it to a small fraction of its size for replication and transport had been speculated about before, but here was incentive to put the idea to test. Dagron showed that he could shrink a page of a large-format newspaper to the size of one square millimeter. He was hired and dispatched via balloon to set up a microphotography base outside Paris. Although his balloon did not make it past Prussian lines and he spent a week evading capture, he eventually set up operations at Tours. In Paris, meanwhile, the central telegraph office was refitted to project incoming microtexts onto the wall (Figure 6.1). When the pigeons with their cargo began arriving in Paris, messages were transcribed and sent for delivery by the regular Parisian post. All told, there were about fifty-nine successful avian deliveries, bringing to Paris nearly 100,000 official and private letters.[5]

In the new year, the Prussians grew impatient and decided to bombard Paris after all, and the war soon came to an end. Following the siege, the members of the Paris Academy, too, began to rethink the nature of intellectual cooperation and organization in distinctly national terms: "It is being said on all sides, and with justice, that it is by science that we have been defeated." So admitted Henri Saint-Claire Deville, as he addressed the Academy in March 1871 on the appalling state of "scientific organization" in France. For too long the initiative of those "grand and glorious scientific bodies" scattered through-

Figure 6.1 Depiction of the projection and transcription of microscopic dispatches at the central telegraphy office during the Siege of Paris, in Gaston Tissandier, *Les merveilles de la photographie* (Paris: Hachette, 1874), 239. Courtesy of the BnF.

out the French nation had been smothered by the stiff hand of the state. The catastrophic defeat showed that now was the time to "intervene actively and directly in the affairs of the nation."[6] Observers pointed out that the Academy itself had been complicit insofar as it had benefited from and helped carry out the state's dictatorial approach to managing French scientific life.[7] Among the most important implements of centralization had been its own journal, the *Comptes rendus hebdomadaires.* Saint-Claire Deville's manifesto caused a stir not only in France but across the Channel. British men of science were keen to work out what the implications of the war between France and Prussia might be for themselves. Norman Lockyer, who had launched *Nature* just two years earlier, saw much in Saint-Claire Deville's manifesto that applied just as well in Britain, even if their predicament was of quite a different kind. Instead of faulty organization, British men of science, according to Lockyer, "have no organisation whatever."[8]

The historian Ernest Renan also diagnosed the political and intellectual defeat of France as a problem of central organization. Invoking an analogy from biology, he argued that "an animal in which organization is highly centralized cannot survive the amputation of one of its appendages."[9] By mid-

1871, when Renan published *La réforme intellectuelle et morale,* the idea was widespread. In the later years of the Second Empire, liberal republican elites, Proudhonist socialists, and others had been urging the need for decentralization across politics, education, and administration. The push for local self-management that many savants came to see as crucial to their own professional self-interest had also been a key platform of the Comité Central founded on the initiative of the Socialist International[10] at the beginning of the war, and which had its apotheosis in the Paris Commune. But now the cataclysm of defeat had given savants — both the elite and those on the periphery — the occasion to push the case that a reform in the social organization of knowledge could transform society itself.

Renan, the Academy, and nearly everyone looking to renovate intellectual and political life looked especially to educational reform. The Prussians had triumphed because of their superior educational organization; the competition between cities and states fostered a robust network of universities that had produced a formidable intellectual and technical elite.[11] "Nearly everything the First Empire did in this regard was regrettable. Public instruction cannot be handed down directly from a central authority."[12] "Administrative centralization," the chemist Jean-Baptiste Dumas affirmed, "applied to the university, has ruined higher education."[13]

Complaints about the state of French science and learning had become fashionable during the 1860s. Chemists especially — including Adolphe Wurtz and Louis Pasteur — had lobbied hard to reform higher education on the model of German universities.[14] The minister of public instruction in Louis Napoleon's government, Victor Duruy, championed the cause himself. Wurtz had been commissioned to investigate German university laboratories in particular, and his visits culminated in a report published in 1870 that detailed the facilities, organization, and architecture of the lavish chemical laboratories that had been built in German universities over the past two decades. The movement also led to the founding of the École pratique des hautes études in 1868, a new administrative unit that was supposed to be the locus for fostering a more intimate relationship between teaching and research in French institutions of higher education. Louis Napoleon had himself paid lip service to the need for educational and scientific reforms, although his government put far more financial resources toward the lavish new Opéra and Haussmann's redesign of Paris.

With the establishment of the Paris Commune in March 1871, which proved as bloody as the siege itself, the nascent campaign to revitalize French science from the top down never really got off the ground. The initiative came from those further from power, many of whom aimed at breaking the power of the elite Paris academies and faculties. Several national specialist societies were founded in the years after the war, including societies dedicated to physics, mathematics, and zoology. The Association française pour l'avancement des sciences — modeled on the British Association — was assembled in the months following the siege. Its leaders came from a wide range of backgrounds, and included not only savants, but engineers and industrialists; many had roots in the provinces, and most were not members of the Paris Academy.[15] With the motto "Pour la science, pour la patrie," there was no doubt about the postwar context of its founding. The president's speech at the first annual meeting, by Armand de Quatrefages, developed an extended analogy between the "*travailleur scientifique*" and soldier: "In our time more than ever, the intellectual and the scientific terrain also have their battles, their victories, and their laurels." Vengeance against Germany was more likely to come from science.[16] The point was not simply that science had a central role to play in the nation's revitalization and its eventual revenge against Germany, but that savants as a corporate group of workers should take on some of the qualities of military discipline, with its values of self-sacrifice, rational organization, and collective honor. The equation between scientific and military organization, central to the mission of the new Association française, was powerful enough that the more awkward aspects of the comparison — military organizations did after all possess a strong central command above a rigid hierarchy — were hushed over.

Young researchers lamented the seeming indifference of the academic elite to the international march of knowledge. "I tell this to you in complete confidence," the young mathematician Gaston Darboux wrote to his friend Jules Houël in 1870, "most members of the Academy pay no attention at all to work published outside of France." Not a single mathematician there, he claimed, even knew how to read German. The indifference had the trickle-down effect of making scientific information harder to come by for everyone else. "Here in Paris, there is not a single library where one can consult all the important collections," Darboux went on. "If I want to read a journal, I have to subscribe to it myself." Hardest to get a hold of was the newest material, he said, because librarians, wary of letting unbound issues of periodicals get into the hands of readers, allowed them sit for up to a year until sewn into annual volumes.[17] On the eve of the war with Prussia, Darboux had convinced the

Ministry of Public Instruction to fund a new periodical he hoped would help deliver French mathematicians from their isolation. The *Bulletin des sciences mathématiques* was supposed to be a revival of an older kind of periodical, he said, one aimed at encyclopedic coverage of all that was new in a given field of knowledge. He cited the *Bulletin universel des sciences* from the 1820s, where the radical François-Vincent Raspail and other hired hands had summarized the newest discoveries.[18]

The 1860s had been a golden age for the emerging profession of the scientific *vulgarisateur* who made it his mission to bring scientific news to a broad reading public by using everyday language. But this was something different. In summer 1871, the editor of the *Revue des cours scientifique*, a journal that focused on printing university lectures, renamed it the *Revue scientifique*. Émile Alglave explained that the war with Prussia had prompted him to refocus the *Revue* on building unity among French savants and providing a medium by which to lobby the state on behalf of savants as a corporate group. The vogue for *vulgarisation* was no longer what was called for. In attempting to bring science to so many readers, much of its content had been so watered down as to lose its value. His aim was to reach only those who were "capable of the effort" involved in actually cultivating science. A better term was *populariser*, by which he meant the consolidation of a true *peuple scientifique*.[19]

When the young physiologist Charles Richet took over the journal's editorship in 1880, he reiterated this discomfort with *vulgarisation*; if anything, the *Revue* was "vulgarisation pour les savants." Although Richet was brought up among the intellectual elites of Paris—his father Albert was director of surgery at the Paris Faculty of Medicine, and he trained under both Étienne-Jules Marey and Marcellin Berthelot—he cultivated an identity as a champion of marginalized knowledge and knowers. His own interests were incredibly diverse; besides his laboratory research, he published novels, maintained broader philosophical interests, and became committed to the pacifist movements just then gaining steam in Paris. His research spanned topics at the far edge of respectability, including split personalities, hypnotism, and automatic writing.[20] Richet imagined the *Revue scientifique* as a vehicle for binding together the diverse world of savants just as his own work crisscrossed the increasingly variegated landscape of knowledge. The *Revue* would be one solution to the problem that "scientific and technical journals, and the bulletins of our countless learned societies have become so numerous that it is impossible to remain absolutely *au courant*."[21] Richet explained that "savants from all the sciences form a large family, and our hope is that this journal will be their journal."[22]

If late nineteenth-century savants formed a family, however, it was a strange one. Scattered across distances and fields, most of its members were strangers to one another. The glue that connected them together, Richet posited, was scientific papers. He used his platform not simply to supply readers with information but also to give younger savants a basic training in how to use the literature in their own work. This meant, for example, knowing how to accomplish a diligent search of the literature. He explained that each author was responsible for consulting not only classic books and introductions but the vast expanse of specialized periodicals. He talked about how to cite sources correctly. If it demanded a great deal of labor, forcing one to consult countless volumes for just a few citations, Richet promised the reward was equally great: "you have furnished the current state of knowledge on a given subject, so that savants after you will be able to start from your memoir."[23] Conversely, failure to read diligently would cause chaos and lead to disputes over priority and credit. These were moral matters, he emphasized, for they demonstrated one's *sincérité*.

Among the young savants who might have come across Richet's instructions was the young mathematician Henri Poincaré. He soon learned firsthand the perils of navigating the scientific literature for those on the periphery. In 1881 he began to publish work on the theory of what he called "Fuchsian functions," gradually realizing that their study led to striking analogies with what were by then famous non-Euclidean geometries. These quickly generated a lot of commentary in the European mathematical world, but not all of it was admiring. The German mathematician Felix Klein, editor of the *Mathematische Annalen* at Göttingen, unleashed a series of attacks on Poincaré's papers, which he claimed mangled several key concepts and which clearly demonstrated that he had failed to keep up with the literature. In correspondence he chided Poincaré for his willful ignorance, telling him that he ought to have begun his research "knowing the whole bibliography." Among colleagues in Leipzig he was less diplomatic, fuming "that in France probably no one even knew that the *Mathematische Annalen* existed at all." Not only did French mathematicians fail to do their duty in keeping up with the literature, but they also erred in "publishing overly quickly," putting claims into the public domain without adequate justification. Parisian mathematicians used the *Comptes rendus* to stake claims "in very short Notes on 'methods' in which it was impossible to tell whether 'you were capable of actually making use of them.'"[24] For his part, Poincaré scrambled to find the references Klein had sent him. But having recently moved to Caen for a teaching job in the north-west of France, he did not have access to many of the relevant periodicals in the libraries there.

Poincaré did manage to emerge relatively unscathed from his conflict with Klein, and he soon made it back to Paris with its better-stocked libraries, first with a junior teaching position in mathematical analysis, and then with the prestigious chair in mathematical physics at the Sorbonne. In 1885, he launched an innovative new project to create a literature information service for mathematicians using index cards. The circular advertising his plan pro-claimed that "not one of us have not been forced to engage in arduous bib-liographical research, and we can imagine a moment when it will become impossible to embark on any research whatsoever without some new instru-ments in hand with which to work."[25] If successful, Poincaré promised that his bibliographical system would save others from the humiliation that Klein had earlier heaped on him.

Taking advantage of the 1889 International Congress in Paris, Poincaré convened an international conference on mathematical bibliography to ham-mer out the details of his scheme. Individual cards would correspond not to documents, but rather to subjects. Each could contain as many as ten refer-ences to papers falling under the same three-figure classification symbol. While later purveyors saw in these cards a methodological revolution, Poincaré's edi-torial team reported that they chose the card format over books simply so that publication of their *Répertoire* could proceed as quickly and continuously as possible. Sets of cards would be published as soon as enough references had accumulated to fill them. The user — with the classification scheme in hand — could arrange them into the correct order as they were received.

Poincaré's mathematical bibliography operated for decades under the management of a succession of secretaries — Charles-Ange Laisant, Maurice d'Ocagne, and Georges Humbert. Unlike Poincaré, they were relatively minor figures in French mathematics, and they were each committed to improving and equalizing access to scientific information. They had all been trained at the École polytechnique and had begun their professional careers as military engineers. Humbert and d'Ocagne were Catholics who experienced profes-sional obstacles because of the polarized political atmosphere of the Third Re-public. Charles-Ange Laisant, who became the *Répertoire*'s most consist publicist and organizer, had transitioned from mathematics and engineering to politics. He had been elected to the Chamber of Deputies and edited a radical journal called the *Petit parisien*. Eventually, however, he came back to mathematics, bringing to it a keen interest in the politics of organization.[26]

Laisant came to see scientific exchange as a model of political solidarity. He also founded a periodical titled the *Intermédiaire des mathématiciens*, a question-and-answer service. Readers were encouraged to pose questions to their peers as well as to send in responses to others. Laisant explained that his publication was "not geared to one particular category of mathematician, but to all who work, from the most illustrious to the most modest."[27] It was precisely the exemplary cohesion among mathematicians, Laisant argued, that made the *Intermédiaire* both useful and conceivable: "These exchanges of views, these reciprocal communications are to our eyes a great benefit. Science is the great pacifier, the most noble and powerful agent of civilization."[28] Later, as president of the Association française in 1903, Laisant spoke on the "Social Role of Science": Science was coming to exercise a moral action on minds through what he called "the universal law of solidarity."[29] He explained that the law of solidarity combined with "modern media of communication" strengthened all kinds of social and political relations within France and beyond.[30]

Laisant's language of solidarity had become part of mainstream political parlance during the 1890s. Politicians and intellectuals looking to revitalize French republicanism had embraced *Solidarisme* as something approaching an official philosophy of the Republic, emphasizing the necessary bond between the individual and the collective, the duties of each citizen toward society, and the necessity of cooperative action in all aspects of life.[31] In political practice, *Solidarisme* provided a new intellectual platform for republicans and moderate socialists to join forces during the 1890s, holding more extreme factions at bay. The "Radicals" (most of whom had ceased to be very radical) that came to dominate the chamber in the last years of the century passed several laws explicitly "in the name of solidarity." Legislation was enacted to encourage the formation of mutual assistance societies, medical aid was guaranteed to the destitute, and state pensions were provided to several classes of workers. These years arguably marked the beginning of the French welfare state.[32]

Laisant seized on new information technologies as a means of bringing *Solidarisme* to science, leveling the hierarchies that kept the less privileged on the outside of the circles in which much scientific information still circulated.

Laisant and his colleagues were not alone. The mathematical bibliography project turned out to be the first of many similar efforts to bring access to the scientific literature to those on the periphery. Some of these were connected with a learned society, but many were private ventures, the vision of some entrepreneur who perceived some gap in the marketplace for scientific information. Many attempted to reach beyond traditional books or periodicals by adapting for specialized fields innovations that had been pioneered in the wider press. They looked to employ tools such as card catalogs, press clipping services, and private circulating libraries as the basic elements out of which a new system for distributing scientific information might be fashioned.

The journalist Charles Limousin coined a term, *éphémérographie*, to designate information services dedicated to the periodical literature. During the 1890s, several enterprising journalist-savants laid plans in Paris to try their hand at this new field of scientific journalism. "Working alongside professional librarians, what we need are *professional specialized bibliographers*, actual savants, whose job it is to index and to classify the works collected [*dépouillés*] from periodicals by ordinary clerks or assistant librarians."[33] Baudouin, Herbert Haviland Field, and Limousin were each young savants who attempted to do just that. All of them viewed the problem of access to the scientific literature as central to the problem of decentralizing scientific organization.

In 1894, Baudouin founded the Institut de Bibliographie Scientifique shortly after completing his medical studies in Paris.[34] He stated that his goal was to build up a "vast organization . . . for enhancing, in a way never before seen, the bibliographical research work of savants."[35] At the center of his vision was the *fiche*: the standardized card would form the basic building block of a massive repository of information about scientific papers. Through judicious deployment of the fiche, Baudouin explained, "take any subject whatsoever (for example: the treatment of pneumonia by cold baths) and it is possible to obtain its complete Bibliography in a matter of seconds, that is, the totality of works published on the question."[36]

As a young student, Baudouin was deeply impressed by the *Index Medicus*, a current bibliography for the medical sciences that had been founded by the American doctor and bibliographer John Shaw Billings in 1879.[37] Billings's publication had quickly achieved fame not only in the United States but in Europe as a model for a periodical index of specialized literature. By the 1890s, however, it was verging on financial ruin. Baudouin experienced his eureka moment when he realized he could not only pick up where Billings might be forced to leave off, but he could do better by exchanging the book format of the *Index Medicus* for what amounted to a combination of a vast card catalog and customized indexing service. He soon expanded to the biological sciences, hoping one day to treat the natural sciences as a whole.[38] Like Poincaré's *Répertoire* of mathematics, his principal aim was not to produce general summaries of progress in a given field, but to provide subscribers timely and targeted information on any given subject. The Institut dispatched copies of all *fiches bibliographiques* deemed relevant to a subscriber (based on their stated interests) as they were indexed by the clerks. This system would save subscribers the cost and trouble of amassing references to every single medical paper, and it would allow—at least in principle—a quicker turnaround time for getting information to researchers: "Using the system of index cards arranged by subject, every question is updated, each evening, once the indexing of weekly or even daily publications has been finished and the cards made."[39]

For those with less ready access to a large library, Baudouin also offered *fiches analytiques*: these were cards containing abstracts of articles that his staff (Figure 6.2) produced on demand. He also began to build up a circulating library; inspired by Mudie's Select Library of London,[40] Baudouin offered to lend out any book in the Institut's collection by post.[41] He had provincial subscribers especially in mind, but even in Paris, lending libraries for specialized publications were virtually unheard of. He also advertised a translation service and a clipping and copying service. Baudouin hoped to provide a complete bibliographical package:

> Once having subscribed to the ensemble of services of the MUSÉE DE BIBLIOGRAPHIE, you will be enabled to carry out work on any subject whatsoever without needing to possess a single book in your own library. Let us suppose, for example, that you must write on "the Fauna of the Gironde." The *Répertoire sur Fiches* service will allow you to acquire by mail all the necessary bibliographical indications. The *Bibliothèque* will then send the books that you need. If you don't read German, English, or other foreign languages, the *Fiches analytiques* service will send you a short précis of these foreign works.[42]

The roster of services Baudouin offered seemed always to be expanding. Figure 6.3 shows a price list of the services available in 1896. He soon offered a *service de bibliothéconomie* as well, whose staff would assemble private libraries for

Figure 6.2 Clerks producing analyses of scientific papers at Baudouin's Institut de bibliographie scientifique. *Bibliographie scientifique* 3, no. 1 (January 1897): 7. Courtesy of the BnF.

INSTITUT INTERNATIONAL DE BIBLIOGRAPHIE SCIENTIFIQUE

FONDATEUR: Dr MARCEL BAUDOUIN

PARIS, 14, Boulevard Saint-Germain, 14, PARIS

TARIFS GÉNÉRAUX   SERVICES PRINCIPAUX

I. — ABONNEMENTS

PAYS	I. — BIBLIOTHÈQUE CIRCULANTE	II. - FICHES CIRCULANTES	FICHES Bibliographiques	FICHES Analytiques	V. — Renseignements Bibliographiques.
1° FRANCE ET COLONIES	France et Algérie........ 20 fr. Recouvré par la poste..... 20 50	France et Colonies....... 10 fr. Recouvré par la poste...... 10 50	10 fr. 10 50	10 fr. 10 50	France et Colonies........ 5 fr. Recouvré par la poste..... 5 50
2° ÉTRANGER	1° Pays compris dans l'Union postale, possédant ou service de *Colis postaux*.. 40 fr. 2° Autres pays........... 50	1° Pays compris dans l'Union postale............... 20 fr. 2° Autres pays............ 25	20 fr. 25	20 fr. 25	1° Pays compris dans l'Union postale........... 10 fr. 2° Autres pays............ 15

II. — FRAIS DE PRÊTS ET D'ENVOIS

PAYS	CHAQUE LIVRE PRÊTÉ. - CATÉGORIES			FICHES CIRCULANTES	CHAQUE FICHE communiquée		CHAQUE RENSEIGNEMENT COMMUNIQUÉ	
	C.	B.	A.					
1° FRANCE	0 fr. 25	0 fr. 30	1 fr.	1° FRANCE	CINQ centimes	50 cent.	1° FRANCE	50 Centimes
2° ÉTRANGER	0 fr. 50	1 fr.	2 fr.	2° ÉTRANGER	DIX centimes	UN franc	2° ÉTRANGER	UN franc.

Figure 6.3 List of services (with prices) offered by Baudouin's Institut de bibliographie scientifique. *Bibliographie scientifique* 2, no. 1 (January 1896): 20. Courtesy of the BnF.

savants or doctors by furnishing them with an appropriate set of starter publications as well as library furniture, and would also renovate existing libraries
by reclassifying their contents. Finally, following the model of news agencies
such as the *Agence Havas*, Baudouin also hoped to become a central agency for
distributing scientific news to journal editors.[43]

Another attempt at a scholarly indexing service was launched by Charles
Limousin himself. His effort was both broader and less comprehensive than
Baudouin's. Rather than starting from one specific discipline with a plan to
expand out later, he indexed whichever periodicals he could reliably get his
hands on. Limousin intended his *Bulletin des sommaires des journaux scientifiques, littéraires, financiers* (1888) to pick up where generalist indexing journals left off and to provide subscribers a bibliographical guide to the contents
of more specialized publications. There was nothing very new in such publications, but Limousin's ambition was to found a service that would provide subscribers personalized references on cards. His enterprise was inspired directly
by the press clipping bureaus that had recently flourished across Europe. He
collaborated with one such service, the *Argus de la presse*, as a means of sharing
users. Limousin formatted his *Bulletin* so that users could cut up its pages and
paste them onto individual reference cards. (He published a single-sided edition for this purpose.) He also lobbied the French Ministry of Public Instruction to support his project on a larger scale and to fund a bureau that would
regularly clip and index all current scholarly journals so as to make specialized
knowledge in print accessible to interested citizens.[44]

Both Limousin and Baudouin were motivated by what Baudouin called "*la
décentralisation scientifique*," — a concept he claimed to be among the first to
champion.[45] Baudouin was particularly interested in equalizing access to science for those who were at a geographical distance from Paris. "A savant who
has achieved a notable experimental result, or a critic who has undertaken a
synthetic study on some scientific subject, must know what other researchers
have seen, found, or written before them on the same question. But how, in
a city with no important scientific library for them to use, can he acquire the
information that he needs?" He hoped "to put into the hands of all those who
take an interest in nature the working elements that until now could only be
acquired in the largest of cities."[46] Baudouin had grown up in the Atlantic
coastal village of Saint-Gilles-Croix-de-Vie. On arriving in Paris in 1883 to
complete his medical education, he had immersed himself in editorial work
for commercial medical journals, and by 1892 he was the secretary of the As-

sociation de la presse médicale française. Even after many years in Paris, he remained fiercely loyal to his provincial roots. Later, after getting out of the information business, he returned to live in his native village and became an expert on its archaeological riches.[47]

Limousin's motives, like those of Charles-Ange Laisant, were focused as much on social inequality as geographical distance. Limousin had been trained as a typographer and was a disciple—like his father Antoine—of Pierre-Joseph Proudhon, the legendary anti-state socialist. In 1865 he had become a secretary of the Paris Branch of the International Workers Association and then became an important publicist for socialist causes, editing several publications and working toward the great goal of workers' emancipation.[48] During the Second Empire, his activities as the founder and editor of the *Tribune ouvrière* had landed him in prison when the state shut it down, ostensibly for a story he ran about a workers' strike. He later got better at knowing what he could get away with. In 1880, he founded the *Revue du mouvement social*, a monthly dedicated to topics relevant to the grand cause. Over time, however, he gradually shifted away from the political focus of the *Revue* toward the task of informing his readers about developments in a range of contemporary issues in the sciences and the arts. In 1887, he abandoned the *Revue* and replaced it with the *Bulletin* to accomplish this goal in a more systematic way. Adapting his Proudhonist brand of decentralized socialism, Limousin aimed to bring the democratizing possibilities of the mass press to knowledge workers, not by engaging in *vulgarisation*, but by making specialized research practically accessible to fellow workers wherever they might be: "In the many country sides of Brittany or Burgundy, the Alps or the Pyrenees, as well as lands in between, there are workers, thinkers, artists desiring to stay current on the work of their colleagues, and to communicate to their colleagues the results of their own researches and meditations."[49]

Another bibliographical venture closely linked to Paris was the Concilium Bibliographicum. Although it was spearheaded by an American, and it eventually operated out of Zurich, it was planned out in Paris in conversation with zoologists. Herbert Haviland Field had earned his doctorate working on the embryology of amphibians in the laboratory of E. L. Mark at Harvard University in 1891. While working on the degree, far from the European centers of power in biology, Field had become fascinated with problems of scientific information. A culture of working with classified index cards had spread throughout America's libraries during the mid-nineteenth century, and it was particularly strong in Mark's Harvard laboratory. Mark understood his principal obligation to his students to be to train them in two skills: one was modern methods of microscopic anatomy, and the other was modern methods of bibliography. Among other things, "students were encouraged," he recalled, "to form the habit of making out their bibliographic references on separate cards of standard size, and advised always to carry about with them blank cards for this purpose . . ."[50] Field, a polyglot known for his prodigious memory, was thus especially well placed to grasp the potential uses of this information technology. Upon completion of his doctoral thesis, he set off for Europe to make the customary rounds in eminent European laboratories, and it is there that he began to plot his vision of a bibliographical empire.

In 1894, after stints in Freiburg and Leipzig, Field settled in Paris to carry out research under the zoologist Alphonse Milne-Edwards. There, he took part in the burgeoning life of the Société zoologique. Under the leadership of Raphaël Blanchard, the Société had made a name for itself by hosting the first International Zoology Congress (1889), where it began lobbying zoologists to adopt an international agreement on rules of nomenclature.[51] The Law of Priority, that the name of a species depends on the first description of it in print, was the centerpiece of Blanchard's program. The close link between print and knowledge that this implied led Blanchard and his colleagues to begin investigating the vehicles of transmission and registration that could give such an agreement a concrete, material foundation. The members of the Société were thus well situated to appreciate Field's bibliographical vision when he presented it to them in 1894. He promised that his enterprise would provide the material substrate of nomenclatural order. For example, the crucial problem of dating species descriptions could at last be solved in a straightforward way: the date that would count would simply be the date on which his central bibliographical bureau received a copy of the publication in question. His enterprise could come to be a *dépôt légal* of species.[52] The Société's role in fostering international contacts among zoological organizations was a boon as Field negotiated with European and American learned societies, governments, and bibliographers to generate consensus on the utility of his plan. The lobbying of Blanchard and E.-L. Bouvier helped give bibliography a prominent place at the 1895 International Zoological Congress in Leiden. This conference saw the establishment of the first International Commission on Zoological Nomenclature (and the Code that remains the basis for current zoological naming conventions) but, thanks to the efforts of Blanchard, it also saw the founding of a Commission for International Bibliography to oversee an international bureau of zoological bibliography that would be run by Field.[53]

The Concilium ultimately set up operations in Zurich after the Swiss government offered to fund the central office.[54] By that time, Field had already gathered a powerful group of supporters. Besides the Société zoologique, he had established relations with the Zoological Station in Naples, as well as with Julius Victor Carus, founding editor of both the *Zoologischer Anzeiger* and the *Zoologischer Jahresbericht*. Field's office employed both clerks and trained zoologists. It was the job of the latter to classify the massive quantity of periodicals that streamed into the Zurich Bureau from authors, editors, and scientific societies. A decade into its existence, in 1906, the Concilium was doing very well; it had distributed over two million cards to users, and had indexed 68,000 zoological articles.[55]

Limousin's grand plan never really took off as he envisioned it, but both Baudouin and Field had greater success. Field's enterprise, in particular, flourished until his own untimely ill health and passing after World War I.[56] But all of these enterprises speak to a new kind of information activism that was thriving across Europe by the 1890s. It took a particular form in France, where scientific life was often perceived as being under the stifling control of a Parisian state-funded elite at the academies and the Parisian institutions of higher education. Britain saw a similar flourishing of information activism, but it took quite a distinct form. Here, few independent entrepreneurs emerged to intervene in the distribution of scientific information. Rather, members of the elite itself, most of them based in London and Oxbridge, took up the cause of access by attempting to rebuild a communications infrastructure that they worried had gone off the rails.

The Machinery of Scientific Periodicals

In May 1892, a pamphlet appeared in London titled *On the Organisation of Science (being an essay towards systematisation)*. Its pseudonymous author, A Free Lance, called for a radical reform of the "machinery of scientific periodicals":

> However vicious and ultimately injurious may be that system of centralisation and administrative despotism which is so strikingly displayed by, e.g., our French neighbours . . . yet one can scarcely suppress at times the regret that we have not here in England some means of authoritatively centralising

and organising that part of the machinery of science with which it is my purpose to deal.

The daily life of scientific investigators was rife with communications breakdowns. Free Lance argued that what ailed the machinery of science was precisely what ailed the machinery of the British government, which had become "a network of absurdities involving pecuniary waste and administrative inefficiency."[57] His target was in part the proliferation of scientific societies, but it was also a complaint about the influence of market forces in scientific publishing. One supporter summarized his proposed solution: "in each country each subdivision of science should have its one central and accredited journal in which all papers on that subject worthy of publication should be published."[58]

The pamphlet caused a stir. It was reviewed extensively in the British scientific press, and a growing murmur of discontent over the state of scientific publishing quickly grew into a sustained roar. The scientific literature was the subject of an ever-expanding laundry list of unflattering depictions—in such terms as chaos and disorder,[59] uncontrolled dispersion and growth,[60] wasted and inaccessible resources, and undisciplined authorship[61]—for the next several years.

Some articulated the problem as a change in the conditions and practices of scientific reading. At the beginning of the century, when scientific journals were often framed by their projectors as anthologies of all that was new and interesting, reading one might very well mean passing through each page of an issue. While a few scientific serials such as *Nature* might be read in this way, this was no longer what observers really expected scientific journals were for. To Charles Darwin in 1865, the perceived shift in reading habits had been "an old subject of grief": "I have often thought that science would progress more if there was more reading. How few read any long & laborious papers."[62] But to the electrical engineer James Swinburne, this had simply become a natural condition of scientific life. Professional researchers no longer read "for general information in an indolent way." Scientific readers wanted "to know either all that has been done on a given subject, or whether some discovery has been hit upon before."[63] Skimming through volumes and flipping through indexes was the more common way in which one engaged with scientific texts. And while one might read in the privacy of one's own home, it was increasingly likely that one would do so in public places where interruptions were frequent. In part this was because access to the specialized periodical literature was mostly to be had through libraries and institutional collections. But it was also because reading often happened when one had a moment to do so on a train or at a station.[64] Access to the scientific literature meant being able to find the papers one was looking for without having to read any more than was necessary. "There is," announced Michael Foster in 1894, "a very pressing need of some easy machinery by means of which an inquirer may discover the existence and learn the exact date and position in literature of the papers which have been published on the subject upon which he is working."[65]

Talk of rebuilding the scientific literature spread to many fields of knowledge, but practitioners of different branches emphasized different issues and possible solutions. Free Lance tipped his hand about his own disciplinary identity when he referenced "the wilderness of periodical literature" confronting those who visited the Linnean Society's reading room.[66] With its many annuals, catalogs, and nomenclators — H. E. Strickland's *Bibliographia Zoologiae* had since been followed by comparable efforts — arguably no branch of science had developed as many guides to tracking the literature as zoology or botany, and yet natural historians were particularly disturbed about the state of publishing. The great evil remained synonymy: species were constantly being renamed by new discoverers, not because of dishonesty but simply because they had no knowledge of the previous discovery.[67]

To be a working zoologist meant having access to the literature as one worked. This likely required having access to a suitable library with the right periodicals and indexes, which varied a great deal according to geographical and professional circumstances. Otherwise there was the danger of being accused of negligence even if one's own work was deemed new, for editors and referees came increasingly to insist that authors cite their precursors. As in France, such demands took on moral overtones. One tract on the subject referred to "the moral obligation which rests with every one who has to deal with questions involving the study of work done by those who have preceded him."[68] The marine biologist T. R. R. Stebbing argued passionately not only that keeping up with the literature and citing others was essential, but that it required "individual self-denial for the general welfare." Stebbing was so moved by these issues that he set himself up as a watchdog of publication malpractices in marine biology, rooting out invalid names to keep the literature pure.[69] Some condemned such vigilantism, ridiculing Stebbing as a "most pious priority purist," among those misguided souls who have "made the hunting of original names an end in itself, and have believed themselves to be advancing science when they were only adding to its confusion." Others, meanwhile, applauded his efforts as "a lesson to some of his timid and short-sighted fellow workers."[70]

Despite publications such as the *Catalogue of Scientific Papers* and *The Zoological Record* providing informal (but unenforceable) pointers to the kinds of publications to be preferred for scientific authorship, there remained "no limit to the media of publication which may be lawfully employed" to record new zoological information.[71] Most continued to agree that papers read at meetings could not count, "because both the author and the society have it in their power to alter name or description after the reading of the memoir."[72] A British Association Committee in 1896 recommended further restrictions: "New names should not be proposed in irrelevant footnotes or anonymous paragraphs." Names appearing in a "student's textbook, or in a short review of a work by another author," also had to go. Some suggested a central publication registry: "no specific name should be recognised unless it be entered by the author at some central office, together with a properly published copy of the work in which the description appears. The name would then be checked, dated, and placed at once in the Index."[73] But the fact remained that all it took to start one's own publishing venue was access to a press.

There was also a growing acknowledgment that print publication was far from a straightforward criterion for establishing exact dates. Many periodicals, especially those published by scientific societies, sent issues to the press gradually over a period of months, sometimes even years. In France, one author suggested that the header on each page of scientific publications should include the date of printing alongside the page number to remove all ambiguities.[74] But it wasn't clear that the date of printing was the right criterion at all. As an American ornithologist pointed out, "publication implies distribution, and has no necessary relation to the date of printing."[75] And yet scientific societies subjected memoirs submitted to a long process of reading and vetting before they went to press. As one editor noted in elaborating on the frustrations of a zoological author:

> [A]n author's paper is first sent in to the secretary, then read in open meeting and discussed, then (it is probable) made still more public by some printed abstract . . . After the author's hopes have thus been raised; after his results have been proclaimed to contemporary, perhaps to competing, workers; after

a lapse of time, which in the case of some work may be fatal to the author's claim of priority: it may be that, after all this, his paper is refused ultimate publication.[76]

And yet dates of reading remained problematic, since a paper might go through substantive revisions between reading and publishing. There was no solution to this dilemma. The confusion surrounding dating a discovery was not simply a matter of bookkeeping: it was a consequence of a commitment to a theory of discovery that could be pinned to down to an individual act at a moment in time. The scientific literature had given material form to the ambition that these acts might all be recorded with precision — but in practice it just didn't work out.

Physicists were also quick to respond to Free Lance's call to save scientific communication, although their situation was quite distinct from natural history. The scattering of physical papers across a wide range of periodical publications was notorious. Unlike many other fields, the principal specialist society representing physicists in Britain, the Physical Society of London (1874–), was only a couple of decades old.[77] Not only had it not done much to sponsor catalogs and handbooks, but its journal was not a very prestigious venue.[78] Norman Lockyer lamented that physicists published "in the Philosophical Transactions and Proceedings of the Royal Society, in the *Philosophical Magazine*, in the Proceedings of the Physical Society, in the Reports of the British Association, and in the Transactions of the Cambridge and the Manchester Philosophical Societies." There were also "the principal Scotch and Irish societies" and various other independent journals.[79] Moreover, the most dominant publication in physics was the *Philosophical Magazine*, a journal that was under the sole control of a commercial publisher, Taylor & Francis.

British physicists, even those in London, felt as if they were living in a scientific backwater, bereft of information. Lacking access to continental journals, some reported relying on summaries of papers that appeared in local periodicals as a substitute for reading papers in publications that rarely crossed the Channel.[80] In 1883, the physicist Silvanus P. Thompson had lamented the plight of a "would-be reader of original memoirs and researches, who is compelled to journey from one shore of England to the other in order to consult the *Edinburgh Transactions*, the *Cambridge Transactions*, the *Comptes rendus*, the volumes of *Poggendorff's Annalen*, and those of the *Annales de chimie et de physique*, or the memoirs of any one of the five great Academies of the European Continent."[81] Another observer summed up the situation as follows: "an English physicist . . . has no simple means of following the progress of his own special study."[82]

The obstacles facing scientific readers had their inverse in authors' worries that their latest publications would go unread entirely. Chief among the supposed virtues of periodicals had always been that they disseminated one's work at a distance. But British authors expressed fierce skepticism that venues for scientific publishing accomplished this well at all: "There is no complaint more frequently heard abroad than that important papers of English scientific men are almost inaccessible to the foreigner."[83] One reason for this was that many believed that transactions and journals of societies were the best places to publish science, and yet it was notorious that the circulation of these publications "is almost negligible, and the majority of the papers are never heard of by outsiders." Local societies, James Swinburne noted, "shroud valuable papers of all sorts in their transactions, and bury them in public libraries."[84] Free Lance had called for their abolition: "all these desultory provincial publications *must cease* . . . The various provincial philosophical and 'royal' societies . . . should *cease to exist as such*."[85] Foster cast scorn on all those "Transactions of the Club of the Lovers of Natural Knowledge in Weissnichtwo" with their memoirs "on the Physiology of This or That."[86] William Ramsay, in an otherwise friendly address to the new People's Palace Chemical Society, begged his audience not to publish its own journal, for more journals only meant more sorrow.[87]

Oliver Lodge complained that "mere printing in a half-known local journal is not proper publication at all; it is 'printing for private circulation.'"[88] This was hardly exaggeration. The expansion of scientific periodicals had been accompanied by the rise of a vast network of private exchange of printed papers through the separate copies authors obtained. Some, such as the chemist John Young Buchanan, thought that this was the only way authors could expect to find readers: "Unless the author distributes lavishly separate copies of his paper in every quarter where he considers it important that it should be read, it will pass unnoticed."[89] Separate copies were made available to authors, sometimes for a fee, by both learned societies and independent journals.

Despite their obvious utility, separate copies could be viewed as problematic from several points of view. First, precisely because they were deemed to be such a significant part of the circulation of knowledge, many editors and publishers were concerned that they hurt their sales and subscriptions.[90]

Many journals put an upper limit on the number that could be requested, and insisted on a temporary embargo on their distribution.[91] Second, they were viewed as causing bibliographical chaos because they were often cited by authors as independent pamphlets. Many committees looking to reform scientific publishing focused especially on this issue. The publishing reform committee of the Association française pour l'avancement des sciences, for example, lamented that "separate copies appear as completely distinct pamphlets, often not even bearing any indication of the collection where the work has been inserted and thus confounding researchers."[92]

Third, the practical problems of distribution and citation highlighted by the phenomenon of separate copies gave the lie to any image of scientific information as open to all. Access to separate copies remained largely the purview of small elites, since "the bulk of such copies usually find their way to men of established scientific position."[93] Taken to the extreme, the prominent role played by separate copies reduced the role of publishers and societies to their operating as suppliers of papers for a vast economy of private exchange that was the real locus of scientific communication. For those committed to a vision of scientific communication as an equitable system, such a state of affairs bore an uncomfortable resemblance to earlier epochs prior to the availability of journals and indexes. The literature search was for outsiders.

Physicists convened a special session at the British Association meeting at Nottingham in September 1893 to discuss the physics literature and its problems.[94] Their debate focused on several possible plans: one was to create a comprehensive index of physical papers, another was a cooperative abstracting service, and still another plan envisaged taking central control of the publication of physical papers entirely.[95]

One partial model was chemistry. The Chemical Society of London had been founded in 1841, just after the rise of proceedings journals, and it had very much been a product of that age. From early on its Council pursued the goal of turning its periodical into "a sort of Compte Rendu of all the work done in chemical science throughout the country."[96] Its frequency was increased as fast as finances would bear, and by 1862 it was published as a monthly journal. Henry Armstrong, president of the Chemical Society in 1894, boasted that "our Society has now so thoroughly established its position . . . that it has a right to expect—indeed, demand, the support of all chemists in the kingdom."[97]

The Society's ability to publish frequently was perceived as being linked to the size of its membership, which it constantly worked to expand. Unlike many previous models of what belonging to a learned society should be, membership in the Chemical Society implied little more than being a subscriber to its journal. There was no expectation of attendance at meetings, no requirement that members be active researchers, as long as one paid one's dues. Armstrong believed that the mere possession of the Society's journal might be enough to exercise a salutary influence on pharmacists, doctors, and other professionals: "Contact with virtue cannot fail to promote higher actions and aspirations. Thus, if every pharmacist were a Fellow of our Society ... and but placed our Journal month by month on his counter," then, argued Armstrong, "might he and his assistants derive inspiration at least from the contemplation of its covers."[98]

Since the 1850s, the Society had also endeavored to publish abstracts of important papers that appeared in other venues. In 1871, on the initiative of then president Alexander Williamson, a more ambitious service was launched with the aim of providing abstracts of *all* chemical papers to its Fellows on a monthly basis. The enormous cost of the operation—it required hiring scores of young chemists to produce the abstracts—was at first met by special grants and donations, but by 1876 the steadily growing membership of the Society allowed their ordinary funds to keep the operation going.[99] As a budding chemist, Armstrong had been on the first team of abstractors,[100] and he had come to see in this activity an ideal training for young chemists. He was involved in editing the Society's publications for decades, and he became deeply impressed with the central role that efficient genres and procedures of publication played in binding investigators together in the collective search for truth. As president of the Society between 1893 and 1895, Armstrong expounded elaborate codes of conduct with respect to scientific publishing and authorship.[101]

Armstrong dwelt on the role that specialized societies could play in shaping literary habits and styles once they established themselves as obligatory passage points for authors. "Learned societies allow far too much freedom of individual action," he complained, "and ... allow the gravest literary malpractices to pass unnoticed." Sharper control of publishing would be a help to both authors and readers. Armstrong wanted to rein in authors who produced papers that were "ill-arranged and intolerably diffuse, being full of unnecessary detail." The problem was not simply one of cost and space. Such papers "are also difficult to read and understand, and eminently uninteresting." Scientific publications should be short, readable, even entertaining, and they should emphatically not be "mere transcripts of the laboratory notes."[102]

Armstrong's stylistic ideal was the abstracts he had been trained to write as a young chemist. He felt that these were often "superior to the originals." Abstracts allowed a writer's main points to shine through in all their clarity, unencumbered by unnecessary and unreadable accounts of data or standard procedures. Armstrong did not believe that the data that accompanied most scientific papers were of sufficient detail to be of much realistic use to readers — their real role was more symbolic, evidence of an author's "bona fides," and therefore expendable. Once in charge of the abstracting service himself, Armstrong had pushed his chemical abstracters to shorten even further the abstracts that they produced; he admitted with some pride that his "staff are certainly called on to school themselves very severely."[103] But if the Chemical Society had created a pocket of order and discipline in British science, Armstrong worried that the disarray of neighboring fields and of the Royal Society threatened that order.

For Armstrong and others, the *Philosophical Transactions* exemplified a key impediment to efficient communication in science. It was not simply that its volumes were large, infrequent, and expensive (complaints that were by then nearly a century old). Authors published in the *Transactions* not primarily to communicate their findings but rather to build their reputations. More generally, the scientific literature had come to constitute a "machinery of honors," according to Free Lance. If attaching honor to publishing papers had begun as a salutary social innovation in science, it was now partially to blame for the glut of worthless papers becoming a "distinct obstacle to a better organisation."[104] (Ironically, therefore, publications such as the *Catalogue of Scientific Papers* that had been built to solve the problem of locating papers may have contributed to the urge to publish by giving prominence to authors' publication lists.)

Deciding where to publish involved a potential trade-off between spreading one's discoveries or advertising one's self. Everyone agreed that the most prestigious scientific publication in Britain was the *Philosophical Transactions*, but paradoxically its contents were about as likely to be read as those of the most obscure local society: "Those who are not Fellows generally know nothing about them until they find them by chance."[105] Many complained that most of the material in any given issue of the *Transactions* was neither relevant nor intelligible to them, and its lavish quartos were not only expensive but a positive nuisance in an age when much reading occurred while on the go: "We must have the papers at our individual disposal, and in a far more handy and less expensive form than that of the Phil. Trans.; such ponderous tomes cannot be carried about, and an ordinary brief abstract of such a paper . . . is of little use."[106] But those looking for professional advancement—an increasingly relevant concern by the late nineteenth century—were in no position to forego "the honours of large type and quarto pages in the Transactions."[107] Why? "The reason is very simple," said Free Lance, "*kudos*." Authors chose to go to the Royal Society because "they know that that is a better advertisement for themselves." (His next book, *Towards Utopia*, imagined a society that was not only without intellectual property, but in which all works were published anonymously and biographical dictionaries did not exist.)[108] Armstrong lamented the "tendency to regard the Royal Society as a good stage on which to advertise. I have heard it said that young fellows at the Universities have been advised to send their papers to the Royal rather than to any other Society from this point of view." Allowing the prestige of publications to decide such matters meant sacrificing the interests of science. Decisions that should be made on "purely impersonal grounds" were thus left to "men who have no individual judgment on scientific questions."[109]

According to Armstrong, "if the Royal Society were to-morrow to become defunct as a *publishing society*, the scientific world would be not a wit the poorer."[110] Lockyer suggested that as things stood, "the Royal Society is an obstacle to the realisation of a satisfactory scheme for the publication of English physical papers."[111] Once upon a time the Royal Society had plausibly been a central repository of scientific information, so that "men knew where to look for all that was new."[112] When new specialized societies multiplied early in the century, they were pressured to pledge not to interfere with the Royal Society's rights to publish science, and to serve "as a feeder to that great original trunk of scientific information."[113] Now, however, the "original trunk" had become the parasite. Some felt that the Royal Society should reinvest its energies (and funds) in the role it had begun to fill when it launched the *Catalogue of Scientific Papers* and become a third-party manager of scientific information, a "Universal Bibliographer," and its library (Figure 6.4) a central archive and bibliographical workshop for science.

Michael Foster was the senior secretary at the Royal Society, but he also admitted that the development of scientific societies, "and of several independent scientific periodicals, has entirely changed the circumstances under

Figure 6.4 The principal library of the Royal Society of London at Burlington House. *Popular Science Monthly* 81 (1906): 479. Courtesy of Harvard University Libraries.

which the Society's publications and meetings were instituted."[114] Foster was in a position to act. He convened a committee at the Royal Society with representatives of the major specialized societies to consider the feasibility of putting into action some plan to coordinate their activities.[115] At the first meeting of the "Procedure Committee," Foster presented an outline of his agenda. Then they met in November 1893 to discuss several proposals that had been submitted by committee members and others.[116] The committee took into account every aspect of the Society's protocols: the organization of meetings, the format and distribution (and even the existence) of both its *Proceedings* and the *Transactions*, alternative publication schemes (especially of abstracts and catalogs), the procedure for vetting manuscripts, and the role of sectional committees (corresponding to the various branches of knowledge). The central issue, however, was Foster's proposal to create an alliance between the publishing wing of the Royal Society and the principal specialized societies based in London. He wanted to see each such society put in charge of publications committees that would be responsible for refereeing papers submitted to the main Society. The *Philosophical Transactions* would be split into several distinct parts corresponding to each of these specialized sections: "Phil. Trans., Chemistry and Phil. Trans., Physiology," and so on. (Papers that fit into none of the sections would be published in a section reserved for general papers.) The key to the proposal was that it contained a financial incentive for each committee to maintain high standards. Each branch would be allocated a certain share of a publication fund. If the cost of publishing the papers approved by that section came in below that amount, the surplus went to the specialized society itself; conversely, any deficit from publishing too much would be paid by the society. Such a plan would at last bring a halt to the unchecked expansion of the literary waste menacing scientific workers.

Foster's plan received support from Norman Lockyer, Arthur Rücker, and several others. But others were skeptical of a plan that would tether British science as a whole to the Royal Society. Henry Armstrong was sympathetic in principle to any plan to impose literary discipline on authors but wary of the possibility that chemists would be subject to such control. He compared the situation to colonial governance. Borrowing a line from the former Home Secretary, he argued that "the prosperity of the whole is best secured by making every part prosperous . . . there is no conflict between the interests of the Colony and the Empire . . ."[117] Armstrong suggested an alternative plan that would decouple the distribution of papers entirely from reward and validation. Since the tag "*Philosophical Transactions*" played an undeniable role in professional advancement, he suggested transforming what he saw as the all-but-moribund periodical into a kind of brand name instead. Once it was decided that a paper was "of sufficient importance" to be published in the *Transactions*, it should be sent to the appropriate specialist Society for publication, and printed with the special designation "Accepted for the Royal Society's 'Transactions.'"[118]

If some were concerned about the freedom of the major specialized societies, the upshot of these plans spelled a dark fate for minor and regional societies. Some, such as the Norfolk and Norwich Naturalists' Society, seemed willing to cooperate to some extent: "we may at times refuse papers of great originality and value . . . such papers, though the Society be honoured by their reception, are out of place in our records if they have no reference or application to the county of Norfolk."[119] But others bristled at the notion that "all independent—i.e., non-metropolitan—associations . . . obliterate their separate identity and become branches of a central organisation." Another warned that "one Science one Journal would also mean one Science one Editor, and the reign of a powerful but small centralised bureaucracy."[120]

Especially conspicuous by their relative absence in these discussions were the publishers of commercial journals. It was implicit in most of the discussions that these journals should either be taken control of by some scientific society, shut down, or relegated to the periphery of respectable science. Their proprietors thus had little incentive to participate in any of these schemes. In physics, the major journal was a commercial publication: "The *Philosophical Magazine* was the personal property of Dr. Francis, and even in the interests of science it was unreasonable to try to evict a man from his own property." Lord Rayleigh pointed out that the establishment of a central publishing organ might bring with it the "great difficulty of censorship and the fear of a dominating clique."[121]

The aim of wresting control of scientific publishing from the marketplace of periodicals could indeed be interpreted as an illegitimate power grab by those with institutional power. Worse, the primary mechanism of inclusion and exclusion was itself coming under suspicion. Since mid-century, many of the larger metropolitan societies and specialized societies in London had developed referee systems to make decisions about what they should publish. But in 1892 Lord Rayleigh caused a sensation when he came across a paper rejected long ago by the Royal Society and which seemed to anticipate James Clerk Maxwell's revolutionary kinetic theory of gases by over a decade. Rayleigh had been led by surprisingly prescient remarks in a paper by an author unknown to him named John James Waterston to go digging in the archives for an explanation. He was astonished to find a dusty manuscript by Waterston on the "Physics of Media that are Composed of Free and Perfectly Elastic Molecules," which laid out in near-perfect detail the essentials of the kinetic theory well before Maxwell's own work. He also found the savage referee report that had consigned it to oblivion: "the paper is nothing but nonsense, unfit even for reading before the Society." (Incredibly, these words were written by John William Lubbock, who had cowritten the Royal

Society's inaugural reader's report with William Whewell.) Rayleigh thought the episode "probably retarded the development of the subject by ten or fifteen years," though he was not willing to impugn the referees nor the system itself. He drew the moral that because scientific societies were necessarily conservative institutions, "highly speculative investigations, especially by an unknown author, are best brought before the world through some other channel." He even suggested that young authors avoid ambitious topics altogether and first "secure the favourable recognition of the scientific world by work whose scope is limited, and whose value is easily judged."[122]

Many bristled at Rayleigh's apologia on behalf of referees. A series of responses ran in the *Chemical News* under the heading "An Obstacle to Scientific Progress." One chemist reflected that "the 'referee system' which prevails in some of the 'learned societies' has broken down." Another warned that the rejection of researches "because they are original" was "a great obstacle to scientific and national progress." Scientific societies can only "legislate for average truth," for the "non-existent individual born of statistics and convention."[123] The referee had become a subject of attacks on elite science already in the 1840s, but by the 1890s questioning these systems was becoming fashionable in mainstream scientific publications such as *Nature* and the *Chemical News*. The Waterston controversy was followed by several more controversies and revelations over the next decade. One physicist saw referee systems, just over a half century old at that point, as "antiquated forms" whose time had passed, although he had little hope that in Britain, "this stick-in-the-mud country," modernization was likely to happen.[124]

In 1903, the Geological Society of London launched a general inquiry into referee systems. The feedback they received suggested that while referee systems were widespread, there was still a great deal of variation in the systems in use, from the use of a relatively large pool of anonymous readers (as at the Royal Society) to vesting most of the power in the secretary and Council (as at the Linnean Society). One respected geologist and editor, Charles Davies Sherborn, came down hard against the anonymous referee: "The referee system should be abolished," he wrote.

> The editor could decide on 99% [90 penciled in] of papers whether they are rot or not — and this is the <u>sole business of a referee</u> — and if he thinks the paper nonsense in any form he [is] to refer the matter for consideration to the

committee who shall ask some competent person to read it, writing his name clearly on the author's MS or on a separate paper to be given to the author.

Sherborn thought that anonymity was simply a license to abuse authors, and that committees "composed of people well up to all shades of modern work" were a far more legitimate method of judgment. When all was said and done, the secretary compiling all these opinions found the "referee" to be such a controversial figure that his draft recommendations urged that in the new regulations "the word referee should not be used."[125]

The Royal Society's 1893 inquiry into reforming British scientific publishing had been centrally concerned with the problem of who should have the power to decide on the acceptance of papers and how this should be done. In this sense it was the precursor to not only the Geological Society's later inquiry on referee systems but to countless twentieth-century attempts to control and bureaucratize scientific judgment. Faced with opposition on several different fronts, however, Foster and his allies largely gave up the hope of creating a publishing cartel to control scientific print. They shored up their rules for the acceptance of papers and use of specialized committees and referees, but they channeled their efforts in two other directions. First, they adopted bylaws meant to resolve ambiguities surrounding the Society's role as a meeting place for savants and as a scientific publisher and gatekeeper. Informal opportunities to socialize and discuss scientific topics were to be expanded and encouraged. Society meetings would be unburdened of the formal reading of long papers. Some would be designated "Meetings for Discussion," in which papers would be distributed beforehand as a basis for dialogue. Other meetings would include shorter discussions about a select group of papers that were summarized either informally by their author or in a descriptive abstract. Most papers submitted would no longer be read out at all and would instead go straight to the relevant specialized committees and referees. Second, the Society took another approach to the problem of access to the scientific literature.[126] In 1894, it sent out a circular calling on all nations that took an interest in science to participate in the founding of a massive international organization for producing and maintaining a subject-classified index of the world's scientific production. The response was overwhelming, but they soon found that they were far from the only ones to have hit upon this idea.

A World System

In 1895, rival access fantasies clashed head-on as various groups rushed to build organizations to deliver scholarly information to an international polity of knowledge workers. At the Royal Society, Henry Armstrong and Michael Foster teamed up to lead an ambitious project for the creation of a permanent intergovernmental organization to index the scientific literature of the globe. Through diplomatic channels, they quickly sought collaborators among representative scientific bodies in other nations—including the Academy of Sciences in France, the Smithsonian Institution in the United States, and the several state academies in Germany—suggesting that they pool their collective authority to tame the wild landscape of scientific print.

At a moment when European powers were competing not simply for colonies but for leadership of international coordination efforts that included standardizing weights and measures and time, the Royal Society was keen to be at the forefront of a project that might become a flagship example of international scientific cooperation.[127] Through the power that it would come to hold over scientific authors and editors in search of readers, the architects of this scheme hoped to create a world system of scientific information, all of which would be set in motion and coordinated from a central office in London. The Society had already developed a reputation in the business of scientific cataloguing. If the *Philosophical Transactions* was emblematic of all that was wrong with the Royal Society, the *Catalogue of Scientific Papers* was a beacon of light, praised by many of the Councils of the foreign academies they contacted.[128]

They were only dimly aware that in its basic outline their scheme had been preceded by a number of smaller-scale scientific information services on the continent. Many of these information entrepreneurs were discovering their shared interests. In Paris, some of them had come together at the Association française, where Charles Richet was leading an effort to improve access to scientific publications. Then, starting in 1895, nearly all of them became affiliated with a group of socialist internationalists in Brussels who had come to believe that the key to a new internationalist order—not only within the sciences but in politics too—lay in a utopia of standardized information exchange.

The leaders of the Institut International de Bibliographie of Brussels were Paul Otlet and Henry La Fontaine. Both were lawyers by training. Otlet was a young, affluent devotee of Herbert Spencer and other positivist philosophers, looking to make his mark in the world in some grand way. La Fontaine was establishing a reputation in the field of international arbitration. Their first moment of inspiration came from France, when in 1891 Otlet stumbled on an article about Henri Poincaré's new repertoire of mathematical publications. Otlet found the grandeur of Poincaré's scheme to organize mathematics stunning. He realized that what had made the natural sciences a progressive enterprise unlike any other was their high level of *documentation*: "Each of their discoveries, every new contribution to the advancement of their science appears immediately to be recorded and to become the point of departure of new research for all. Thus an admirable unity of work exists among chemists, physicists, and biologists . . ."[129] He laid down plans to found a science of society through bibliographical organization.

Otlet's second moment of inspiration came from the United States. In 1894 he learned of the existence of a peculiar classification system taking America's public libraries by storm. Unlike the systems invented by philosophers, this one was produced by an obscure American librarian whose aims were entirely practical. It had the advantage that it was made of numbers, and so could be understood by anyone. When he finally acquired a copy of Melvil Dewey's Decimal Classification in early 1895, Otlet immediately perceived possibilities that he suspected American librarians, perhaps even Dewey himself, had missed. Far more than a bureaucratic scheme for standardizing library organization, Otlet saw in these flat numerical tables the foundations of a system for organizing the world's information. On the lookout for international collaborators, Otlet and La Fontaine found several of them in France especially. Many of the *hommes à fiches*—Charles Richet, Charles Limousin, Marcel Baudouin, Charles Gariel (the secretary of the Association Française), and Herbert Haviland Field—had already struck out on their own to build classification systems, but Otlet convinced them of the moral imperative of coordinated international action.

Setting up the International Catalogue was as vast an undertaking as its projectors imagined. Armstrong, Foster, and their team at the Royal Society held a series of international conferences in London that brought savants and catalogers from Europe, North America, and several British colonies to London, and prompted worldwide attention in the press. In the United States, *Science* predicted that these summits would go far toward "converting Tennyson's 'parliament of man' and 'federation of the world' from a poetical vision into a beneficent reality."[130]

Because the literature of science had become by the end of the century so closely identified with what scientists knew, these conferences took on the character of summits in which scientists across several fields and from dozens of nations came together to decide just what that enterprise was. What were its bounds? How did it change over time? And in what sense could it be said to be a unified enterprise at all? Answers to these questions were sought largely through interrogating the landscape of scientific periodicals and how best to classify their contents.

Otlet and La Fontaine were permitted to attend the conference by special arrangement, but it was clear that the Royal Society and many of the delegates were moving in a different direction. All parties agreed that the enterprise's success depended on building an infrastructure that could maintain a detailed classification system that would evolve along with the progress of knowledge. But they disagreed about nearly everything else. To set up such a system required deciding on its bounds, and no issue was more divisive than this. The Royal Society's plan called for focusing on what they deemed to be the natural sciences, putting aside all other branches of knowledge even if it meant cutting fields such as psychology or anthropology in two. Otlet and his allies rejected both the logic and the politics of such divisions: "a government has no right to concern itself with the natural sciences more than with sociology, philosophy, history."[131] Other delegates pointed out that if such an arrangement allowed the natural sciences to "drain all the resources that each nation had amassed for bibliographical projects," then all the better.[132]

It also turned out that the Belgians had embraced a different vision of the kind of public they had in mind. The International Catalogue of the Royal Society was aimed squarely at knowledge producers: "Individual scientific workers all the world over would be kept constantly informed of the progress of the particular Sciences in which they were interested, and would be able at a small cost, and with little difficulty or labour, to maintain a 'catalogue current' available for reference at any moment."[133] The restriction to original contributions was crucial for capturing knowledge in the making, they thought. Part of the ambition driving both Foster and Armstrong was that such a system would itself have the effect of training authors to write papers in specific ways, making them easy to index and find. But for those who saw the goal as embracing education and broad cooperation, this limitation missed the whole point of the enterprise. Otlet was incensed by the idea that the "literature" of each field of knowledge was to be defined in terms of "original contributions to knowledge." His French ally, the engineer Hippolyte Sebert, argued that the Royal Society was defining the scientific literature in a way that put it "out of the reach of most readers":

> An international bibliographical classification for a particular science must be established and directed in such a way as to assure a precise classification of all subjects that constitute the <u>literature</u> of that science: that is, not only that which forms the current work of promoters of that Science, but also that which is the foundation of classical education and that which makes up the basic literature of that science in diverse nations.[134]

But arguments such as these did not go very far with most of the Royal Society's collaborators.

Instead of combining an overarching vision of encyclopedic unity with a plan for a loose structure of coordinated institutions, as Otlet recommended, the International Catalogue was planned as a complex, hierarchical international bureaucracy with regulations that laid out the powers and responsibilities on all matters relevant to the collection of data, maintenance of the classification, and publication. High-level decisions were "entrusted to a representative body, hereinafter called the International Council," whose members were appointed by the respective national governments on a rotating basis. Each science was represented by an "International Committee of Referees" to be appointed by the Council but which was always supposed to maintain geographical diversity. Referees were to be consulted regarding ambiguous cases of inclusion and exclusion, and also of classification.[135]

After protracted negotiations, the International Catalogue began operation in 1901. It consisted of Regional Bureaus scattered in thirty-two countries and colonies, a Central Bureau in London, and a mandate to classify and catalog every original contribution to scientific knowledge that scientific men gave to the world each year. It was the most ambitious intergovernmental organization dedicated to science ever put together. When the first volumes appeared, Michael Foster informed the British public with a celebratory introduction to the new international organization. Although of great importance as a model of international cooperation, he warned that the catalog itself would be "absolutely unintelligible to the general reader." The volumes were not meant to be read at all, but simply consulted by specialists and clerks. In order to explain what kind of object it was, and why it had become necessary to invest public money into this massive international enterprise, he looked to the history of science.[136]

Centuries ago, Foster explained, the man of science first communicated new discoveries "to his more immediate friends by word of mouth or by letter." Later, they would find their way into a book. Eventually learned societies were established, and these sometimes facilitated authors in publishing and distributing such books. One of these, the Royal Society of London, did something new when it "authorised its secretaries to publish, under the name of 'Philosophical Transactions,' brief accounts of important communications made to the society." Other societies picked up on this idea, and these scientific journals spread. In the late eighteenth century, new independent journals were founded, first focusing on general science, but eventually becoming restricted to particular branches of science. Independent journals were useful, he explained, because scientific societies were "occasionally led to reject as rubbish what turned out in the end to be a pearl of great price." Periodicals continued to multiply, and became as numerous as the "sands of the sea."[137] From then on, the future of science would depend on finding means of producing and ordering these grains into grand edifices of knowledge.

After surviving into the second decade of the new century, the International Catalogue of Scientific Literature crumbled, as did many international organizations, in the aftermath of the First World War. But heroic stories about the relationship of the scientific journal to the rise of modern science proved more resilient. The very belief in a unified journal format that these stories entailed brought a certain idea of stability to the scientific literature, even if that term still referred to a heterogeneous and always-changing set of formats and practices of judgment that varied across disciplines and geographical settings. Even as scientific authors, editors, and publishers continued to experiment with new media and formats, historical tales such as Foster's continued to be passed down and retold right through that century and into the present one.

CONCLUSION

Impact Stories

In the years after the mathematician Grigori Perelman announced his proof of the Poincaré Conjecture in 2002, observers described him as having "stunned the mathematical world by posting his proof to the Internet, flouting tradition by declining to submit his papers to a peer reviewed mathematics journal."[1] Sylvia Nasar and David Gruber explained in the *New Yorker* that "judgments about the accuracy of a proof are mediated by peer-reviewed journals . . . Publication implies that a proof is complete, correct, and original."[2] Journals were also the means by which authors established their priority and credit; it was therefore suggested to be evidence of Perelman's eccentricity and odd ethical convictions — his possibly being mad — that he refused to publish in a regular journal. Viewed from up close, however, the story didn't quite hold up.[3] Observers who met him noted that "Perelman was much more normal than they expected." Moreover, as Nasar and Gruber soon found out, Perelman was perfectly willing to defend his priority when challenged.[4] In posting his paper to arXiv.org, a preprint database widely embraced by physicists and mathematicians, he was simply following an already well-established trend in his field.

The shock that these journalists expressed just a few years ago now seems quaint. Wherever we look, we find signs that whatever dominance the journal may have possessed is fast unraveling. Not only have authors in large segments of mathematics and physics shifted much of what they publish to the arXiv, but other fields are also experimenting with similar systems.[5] Elsewhere, activists and editors are challenging conventions surrounding submission, prepublication review, and priority adjudication. Entrepreneurs are testing new platforms for communicating discoveries, gathering and collating feedback, and locating relevant research. Many fields are struggling with the meaning and consequences of open data policies. Since 2014, the US National Science Foundation has required principal investigators to list their "Products" rather than "Publications" on grant applications.

But institutions are not easily dismantled, and those that replace them often find uses for many of the pieces. It is not only a matter of nostalgia for, or financial interests in, maintaining the status quo, but there certainly is that. Among some publishing groups, the narrative that starts with Henry Oldenburg as heroic inventor has been mobilized to argue that science can ill afford to undergo a media transformation: "all subsequent journals, even those published electronically in the 21st century, have conformed to Oldenburg's model. All modern journals carry out the same functions as Oldenburg's and all journal publishers are Oldenburg's heirs."[6] The implication is that if we mess with the scientific literature, we are messing with the very social structure that made modern science possible.

Even among many who have dedicated themselves to challenging the system, the epistemic virtues and bounds of expert legitimacy associated with the scientific literature remain as operative ideals. Many of the challenges to the dominance of the big academic publishers have focused on revolutionizing the economics of producing and of accessing the literature, but have left the definition of the literature — peer-reviewed research papers collected in periodicals, written for an audience of other specialists — relatively untouched. Even preprint databases, which dispense with periodicity entirely, have maintained the genre of the scientific paper as the preeminent form of knowledge expression in science.[7]

Professing veneration for an unrealistic image of the apparatus of scientific publishing has been a strategy taken up by critics of science looking to cast doubt on whole fields of study. Skeptics of climate science research, for example, have invoked a fairy-tale image of scientific publishing as the bedrock of legitimate consensus only to profess outrage when it turns out not to live up to this fantasy. This reaction was exemplified by the reaction to the leak of thousands of emails and documents from the Climate Research Unit at the University of East Anglia in November 2009. The emails seemed to re-

Figure 7.1 "Manufacturing a Climate Consensus." A depiction of climate scientists doctoring the scientific literature that accompanied an article by climate change skeptic Patrick J. Michaels in the *Wall Street Journal*, 17 December 2009. Illustration by Martin Kozlowski.

veal climate scientists engaging in secretive behavior and politicking with peer review. Evidence that scientific life behind the printed pages of journals was not a precise reflection of its better-behaved public face was seized upon by commentators to argue that the bottom had fallen out. One described this as tampering with the Bible: "The bible I'm referring to, of course, is the refereed scientific literature. It's our canon, and it's all we have really had to go on in climate science"[8] (see Figure 7.1). Claims such as these trade on a caricatured view of the social life of science, one in which the dynamics of trust and consensus in science have been reduced to and made to hinge solely on the reliability of the scientific literature.

More generally, contemporary worries that the scientific enterprise itself might be breaking down, exemplified by the replication crisis, are closely bound up with a mismatch between expectations that the scientific literature ought to be a repository of carefully vetted claims and a rather less tidy reality.[9] Some believe that the collapse of the journal system and of prepublication referee systems is a foregone conclusion, soon to be replaced by a variety of Internet platforms and data analysis that will make science increasingly open, efficient, and secure, from knowledge claims to lab notebooks. As one popular argument has it, the increasing scale of scientific exchange is making earlier technologies and practices of judgment obsolete, and there is simply no alternative but to replace them by more efficient and objective algorithmic measures of merit and credibility.[10]

Detailed critiques of many of these claims for "Open Science" and "Science 2.0" are beginning to emerge. Some tap into a familiar but still important line of argument that dismisses the technological utopianism or "solutionism" embedded in any claim that an information technology can, by the force of its own logic and construction, solve complex problems of knowledge and social order. Others argue that despite appearances, much of what passes for the revolution in scientific communication is in fact part and parcel of the increasing commercialization of scientific life as it further succumbs to neoliberal orthodoxy. Indeed, examples abound of start-ups whose business model depends on capitalizing on a claim to democratize knowledge or to enhance the objectivity of scientific judgments by algorithmic means. Many current endeavors to open up the research process can also be seen as attempts to bring expert communities under even more detailed algorithmic surveillance, whether for purposes of controlling them, profiting from them, or both.[11]

These are powerful criticisms. But so far they have either not offered constructive alternatives, or else they have implied that the conventions of scientific life that are being challenged once did exist in relative isolation from the influence of commerce and politics. In fact, however, much of what we are now observing is the latest in a pattern of appropriation, adaptation, and dissociation that cannot be adequately captured by blanket statements assigning praise or blame.

Throughout this book, many of the actors who sought to transform scientific communications and judgment—to make it more efficient or more public—adapted technologies, genres, and even values from commercial media formats. For those who aimed to shield science from the corrupting influence of the market or political influence, this inevitably meant borrowing from the very domains that they wished to exclude. Early nineteenth-century promoters of scientific journals often framed their critique of elite science by appeals to a public opinion constituted by the emerging marketplace of reading and taste. In adopting the journal format themselves, academies and societies made a claim to be opening themselves to the marketplace of public opinion while reasserting their own special status as forums for scientific judgment. In Britain, referee systems evolved as a new infrastructure of discrimination to differentiate legitimate discovery claims from pretenders, but in doing so they adapted to wider cultural confidence in anonymous criticism. In France and elsewhere, attempts to formulate a doctrine of property rights in science were indebted to modern regimes of patent law even as they were designed to create a separate space for recognizing discovery as distinct from marketable inventions. What emerged was a medium of scientific exchange that carved out a special domain of the press but that retained traces of its commercial origins. Later in the century, as new information systems such as newspaper clipping services and public card catalog systems spread, these too were swiftly embraced by savants and scientific editors as a solution to problems of scientific coordination and cohesion. These subsequently became part of renewed debates over what constituted the bounds of authoritative scientific expression.

All the while, epistemic commitments regarding legitimate forms of collaboration could not, and did not, remain stable. The credit that came from publishing scientific papers privileged certain kinds of scientific activity. Individual authorship of short texts, themselves vetted by individual readers, became standardized products of scientific life. Late nineteenth-century positivist representations of the growth of knowledge as the serial accretion of small but established facts were undergirded by the image of an archive of papers, each of which might be conceived as corresponding to a single discovery and author. Collaboration happened behind the scenes, prior to discovery, or else through the impersonal consolidation of knowledge in print over time. In a sense, the image of science as a machine for producing papers contributed to a specialized notion of scientific productivity that at once insulated scientific life from commercial criteria of success but in doing so borrowed a powerful market metaphor that lent itself especially well to quantification. Among the ironic consequences of this shift was the legitimation of brief notes as equal to or even preferable to longer memoirs and books. Short papers, especially when they were stripped of their introductory materials, historical framing, and detailed methodological information that often accompanied longer memoirs, were of use only to the most informed inner circle of readers, an enigma to everyone else.

Since then, when controversies have arisen regarding the politics of scientific knowledge, claims and proposals for reforming scientific publishing have not been far off. Beginning in the 1930s, battles over whether scientific progress might be subject to central planning broke out between socialist scientists such as the British crystallographer John Desmond Bernal and proponents of scientific freedom. Bernal's reflections on the social functions of science included ambitious schemes to dismantle the system of scientific journals and replace it with a central clearinghouse of scientific information. In the new system, the submission, selection, organization, and distribution of papers would all be handled by a set of coordinated offices.[12] When Bernal's ideas reached the public in 1948, headlines warned of "Truth in Danger" and of a coming "political domination of scientific thought" by what amounted to a recreation of the "Nazi scientific information service."[13] His idea was dismissed by opponents in the free science movement as a brazen plot to stifle scientific freedom by hijacking the critical institution through which scientific power was distributed.

The conception of the scientific literature as a total source of scientific information contributed to and was bolstered by the idea that modern society entered a new epoch in the twentieth century, the information age. Scientific papers — public, countable, and seemingly able to retain their truth value far from their context of production — were among the most exalted instantiations of the morselized character of this abstract sense of information that was purported to pervade society. Like the scientific journal itself, invocations of information often fused together several different, sometimes contradictory ideas, including facts about particular phenomena, a substance that was transferred in any act of communication, and a mathematical theory of signal and noise. In the concept of the "informed public," this conceptual flexibility made it possible to suppose that the gathering up of objective facts in documents might itself be a precondition for the functioning of democratic society and of the open exchange of ideas. Indeed, this vision of an open society based on untethered flows of information has for one of its most potent archetypes a scientific literature imagined as accessible and free to all.[14]

In the 1960s, the scientific literature came more commonly to stand for objective judgment and expert consensus in science. Especially in the United States, scientific spokespeople worked to justify the massive government funding that had flowed into science after World War II but also to preserve their autonomy to pursue basic research. A key approach to making this case was to highlight the notion that researchers constituted a "scientific community," one that was accountable to the public because it had means of self-regulation.[15] Although scientific journals remained independent products controlled by a wide array of scholarly bodies and commercial publishers, the publishing apparatus became crucial to this argument. Elaborate processes of "peer review" of grants were especially central to these claims, but the same term was adopted to describe editorial oversight as well. As the physicist and science writer John Ziman proclaimed, it was scientific papers that made up "the consensus of public knowledge," and the referee was "the lynchpin about which the whole business of Science is pivoted."[16]

It is common now to diagnose the current crises in science as arising from a mismatch between the reward structure of science—with its publication lists, impact factors, and other metrics—and the true goals of the scientific enterprise. We speak of unintended consequences, Goodhart's Law, and the distortion of research agendas wrought by quantification. But the very framing of the problem is itself a Cold War legacy, when observers and sociologists came to describe science as an autonomous, self-regulating social system with the scientific literature and its gatekeepers at its core. But this view depends on our supposing that there is wide public agreement on the nature and goals of the scientific enterprise, or at least that questions about the role of scientific knowledge in governance can be kept separate from how science works on an everyday basis. But there is no warrant for this separation of the politics of knowledge from practical epistemology, and as long as we continue to insist on it, we will continue to spin our wheels.

The point is this: of course it has always been a delusion to suppose that any information technology can solve complex problems of knowledge. But it is also naïve to assume that the epistemic virtues that govern scientific life have ever developed in isolation from the economic interests, technological visions, or political commitments of the individuals and groups that have shaped them. Both defenders of the heroic narrative of the scientific journal and those who now wish to throw the format aside as moribund must attend to the hybrid role that the scientific literature has played as a site linking scientific, political, and market ecologies.[17]

This has practical consequences for both historians and philosophers of knowledge and those focused on the future of knowledge expression and scholarly publishing. The evolution of scholarly publishing has not only depended on establishing trust and transparency within those communities, but it has also been linked to the shifting bases on which scientific researchers claim legitimacy as experts on matters of public concern. We cannot study scientific communication and practices of judgment without also considering the broader histories of media and of politics in which and through which science has always taken place.

Similarly, reimagining knowledge communication requires not simply changing the ways in which scientific claims are vetted, circulated, and recorded within research communities, but also reconsidering the nature of the relationships between those groups and other constituencies. Whether it is faith in the journals system as a necessary bulwark keeping scientific fields from epistemic collapse or zealous predictions that networked platforms will revolutionize science, both rely on impoverished conceptions of the dynamics of scientific life.

The vision of serial knowledge accretion that is a legacy of the rise of the scientific paper has certainly become increasingly untenable, as the bylines of papers become crowded with names and confidence diminishes that scientific papers were intended for reading at all. But as we experiment with new genres, formats, and platforms for bringing varied scientific publics together, we should certainly remember that these genres and formats have been more fluid than they might seem. And let us also strive to hold together our examinations of practical epistemology, on the one hand, and of the nature of the legitimacy of expertise, on the other. The scientific literature has long contained within itself contradictions, and it has only awkwardly answered to the many roles it has been asked to perform. But it has nevertheless been a juncture where the cultures of trust that still undergird much of scientific life and the cultures of accountability that dominate much of public life have come together.

ACKNOWLEDGMENTS

This book has evolved through so many stages that the end result bears little resemblance to what I thought it was going to be about when I began it. I do think the changes have been for the better, and most would never have happened if it weren't for the inspiration, criticism, and support of colleagues, mentors, and friends.

I owe a great deal to the patience, encouragement, and intellectual engagement of my many mentors. During my dissertation, Peter Galison was steadfast in his intellectual support, and he has remained so as a colleague and friend. My many conversations with Mario Biagioli over the years have been both a fount of good ideas and a cherished intellectual pleasure. Steven Shapin taught me a lot about how to think about the social history of science, and he also taught me a lot about how to write about it. The genesis of this project owes a particular debt to James Secord, who subtly but persistently encouraged me not to take the scientific journal for granted. For being my guide in all things book history, many thanks are due to Ann Blair, whose unparalleled intelligence and generosity have been a model to which I aspire as an academic.

The book has benefited from conversations with many colleagues, friends, and mentors over the years, including Etienne Benson, Carin Berkowitz, Ann Blair, Janet Browne, Henry Cowles, Paul Erickson, Robert Fox, Anna Gielas, Lisa Gitelman, Daniela Helbig, Evan Hepler-Smith, David Kaiser, Robert Kohler, Markus Krajewski, Daniel Margocsy, Everett Mendelsohn, Projit Mukharji, Ted Porter, Gregory Radick, Chitra Ramalingam, Bill Rankin, Boyd Rayward, Lukas Rieppel, Simon Schaffer, Sam Schweber, Hanna Rose Shell, Geert Somsen, Alistair Sponsel, Simon Stern, David Unger, Scott Walter, Alex

Wellerstein, and Elizabeth Yale. Friends and colleagues have offered generous feedback on draft chapters at various stages of completion, including Bruno Belhoste, Aileen Fyfe, Michael Gordin, Adrian Johns, James Secord, Megan Shields Formato, Jonathan Topham, John Tresch, and Iain Watts. Special thanks to Thomas Broman for providing especially detailed feedback and criticism on the entire manuscript, and to Nasser Zakariya, who as both a friend and lively critic played a key role in the development of my thinking about this book and many other things. In addition, I received valuable feedback on particular chapters from the Boston-Area French History Group, the History of Physical Sciences Working Group of the Philadelphia-Area Consortium on the History of Science, the writers' workshop at the Chemical Heritage Foundation, and several working groups at Harvard University.

Much research for this book was accomplished while I was an ACLS Early Career Postdoctoral Fellow in 2010–11. Since then, I have had the great boon of wonderfully supportive colleagues in the Department of the History of Science at Harvard University. Under the steadfast and caring leadership of Janet Browne, this has been a very congenial space to teach, write, and scheme. This book has also been profoundly shaped through my teaching; students—both undergraduate and graduate—have helped me think broadly about many of the themes that became central to this project. I have also had the benefit of highly productive periods of leave. At the Max Planck Institute for the History of Science in Berlin, I owe special thanks to the warm hospitality of Lorraine Daston and Fernando Vidal. At the Chemical Heritage Foundation, where most of this book was finally written, I thank especially Carin Berkowitz and Ron Brashear.

I have received kind assistance from many archivists and librarians while carrying out the research for this project. The staff at the library of the Royal Society of London and at the archives of the Academy of Sciences in Paris in particular have put up with near-annual visits from me for a decade. I owe thanks for generous assistance to the librarians and archivists of the American Philosophical Society, the Bibliothèque Inguimbertine, the Bibliothèque nationale de France, the Bodleian Library at Oxford, the Chemical Society of London, the Archives de l'École polytechnique, the Geological Society of London, the Harry Ransom Center at the University of Texas, the Archives Henri Poincaré, the Bibliothèque de l'Institut de France, the Linnean Society of London, the Mundaneum, the Bibliothèque centrale du Muséum d'histoire naturelle in Paris, the National Library of Scotland, the Natural History Museum in London, the Royal Astronomical Society, St. Bride Library, Trinity College at the University of Cambridge, the University Museum of Zoology at Cambridge, the Special Collections of the University of St. Andrews, the Archives départementales du Val-de-Marne, and the Zoological Society of London. I also owe special thanks to the librarians at my home institution, in particular Houghton Library and the Ernst Mary Library of the Museum of Comparative Zoology.

It has been a pleasure to work with Karen Merikangas Darling at the University of Chicago Press. Thanks as well to Evan White, Marianne Tatom, and Kelly Finefrock-Creed, and to everyone else at the press who helped transform this manuscript into a book.

Portions of this book appear in earlier versions in "Seriality and the Search for Order: Scientific Print and Its Problems during the Late Nineteenth Century," *History of Science* 48 (2010): 399–434; "Objectivities in Print," in Flavia Padovani, Alan Richardson, and Jonathan Y. Tsou, eds., *Objectivity in Science: New Perspectives from Science and Technology Studies* (New York: Springer, 2015), 145–169; "Peer Review: Troubled from the Start," *Nature* 532 (2016): 306–308; and "How Lives Became Lists and Scientific Papers Became Data: Cataloguing Authorship during the Nineteenth Century," *British Journal for the History of Science* 50 (2017): 23–60.

Most of this book was written after Emily Dolan came into my life. Her intelligence, creativity, and generosity have since then been a constant inspiration. I owe her loving thanks for her support and for many other things besides.

I have dedicated this book to my mother, Ibolya, whose love has never flagged, and to the memory of my father, Attila, who, despite having scant interest in the topic, would without a doubt have been its most enthusiastic reader.

ARCHIVES AND ABBREVIATIONS

Britain

BL   British Library, London, UK

    CB   Charles Babbage Papers, MS Add 37182–37205

    JB   Joseph Banks Papers, MS Add 33981

    RS   Royal Society Papers, MS Add 4441

GGP   George Greenough Papers, Special Collections, University College London, UK

GSL   Archives of the Geological Society of London, London, UK

    CM1   Council Minutes

HES   Hugh Edwin Strickland Papers, University Museum of Zoology, Cambridge, UK.

JFOR   James David Forbes Papers, University of St. Andrews Special Collections, St. Andrews, UK

LSL   Library of the Linnean Society of London, London, UK

NLS   National Library of Scotland, Edinburgh, UK

    JM   John Murray Papers

    RSE   Papers of the Royal Society of Edinburgh, Acc.10000

OU   Bodleian Library, Oxford University

BAAS   Papers of the British Association for the Advancement of Science, Dep. B.A.A.S.

RAS   Archives of the Royal Astronomical Society, London, UK

    CM   Council Minutes

RSL   Library of the Royal Society of London, London, UK

    CMB   Committee Minute Books

    CMP   Council Minutes Printed

    CMO   Council Minutes Original

    DM   Domestic Manuscripts

    HP   Herschel Papers

    JWL   John William Lubbock Papers

    MC   Miscellaneous Correspondence

    MM   Miscellaneous Manuscripts

TFP   Richard Taylor Papers / Taylor & Francis Papers, St. Bride Library, London, UK

WJP   William Jardine Papers, Royal Scottish Museum, Edinburgh, UK

WWP   William Whewell Papers, Trinity College, University of Cambridge, UK

ZSL   Zoological Society of London, London, UK

France

AAS   Les Archives de l'Académie des sciences, Paris, FR

    AMA   Personal papers of André-Marie Ampère

    CA   Commission Administrative

    CS   Comité Secret

    PS   Pochettes de Séances

    R   Registres de l'Académie des sciences

AHP   Archives Henri Poincaré, Nancy, FR

BIF   Bibliothèque de l'Institut de France, Paris, FR

    HR   Huzard

BNF   Bibliothèque nationale de France, Département des Manuscrits, Paris, FR

    GL   Papers of Guglielmo Libri, NAF 3254–3284, 22114–22115

FVR   François-Vincent Raspail Papers, Bibliothèque Inguimbertine, Carpentras, FR

MNHN   Archives du Muséum national d'histoire naturelle, Paris, FR

Other Archives

APS   American Philosophical Society, Philadelphia, PA, USA

HPT   Herschel Family Papers, Harry Ransom Center, University of Texas, Austin, Texas, USA

MUN   Archives of the Mundaneum, Mons, Belgium

Published Collections

CCD   Frederick Burkhardt et al., eds., *The Correspondence of Charles Darwin* (Cambridge: Cambridge University Press, 1985–2013).

CGA   L. de Launay, ed., *Correspondance du Grand Ampère* (Paris: Gauthier-Villars, 1936–1943). 3 vols.

CMF   Frank A. J. L. James, ed., *The Correspondence of Michael Faraday* (London: Institution of Electrical Engineers, 1991–2012). 6 vols.

PVAS   *Procès-verbaux des séances de l'Académie: tenues depuis la fondation de l'Institut jusqu'au mois d'août 1835* (Hendaye: Basses-Pyrénées, 1795–1835). 10 vols.

SCJB   *The Scientific Correspondence of Sir Joseph Banks, 1765–1820* (London: Pickering & Chatto, 2007). 6 vols.

NOTES

Introduction

1. On dissemination, see Committee on Science, Engineering, and Public Policy, *On Being a Scientist: Responsible Conduct in Research*, 2nd ed. (Washington, DC: National Academy Press, 1995), 10–11. (The statement has been softened somewhat in the 2009 edition.) On sewage, see Michael Foster, "On the Organisation of Science," *Nature* 49 (1894): 563–564. For contemporary watchdogs, see, for example, the website Retraction Watch http://retractionwatch.com.

2. J. D. Bernal, "Provisional Scheme for Central Distribution of Scientific Publications," in *The Royal Society Scientific Information Conference, 21 June–2 July 1948. Report and Papers Submitted* (London: Royal Society, 1948), 253–258; Aaron Swartz, "Guerilla Open Access Manifesto" (July 2008), https://archive.org/details/GuerillaOpenAccessManifesto, accessed 12 July 2016; "Sci-Hub . . . to remove all barriers in the way of science," sci-hub.cc, accessed 12 August 2016.

3. In fact, most scientific practitioners work in private industry or government institutes where they may not be expected to publish at all. By some estimates, less than 20 percent of working scientists have ever published an article in a refereed journal. See Donald W. King and Carol Tenopir, "An Evidence-Based Assessment of the 'Author Pays' Model," *Nature Web Focus: Access to the Literature* (2004), http://www.nature.com/nature/focus/accessdebate/26.html, accessed 4 January 2017. On the neglect of non-university -based science in traditional sociology of science, see Steven Shapin, *The Scientific Life: A Moral History of a Late Modern Vocation* (Chicago: University of Chicago Press, 2008), 93–127.

4. Among the many reflections on current transformations in scholarly publishing, see,

for example, Christine L. Borgman, *Scholarship in the Digital Age: Information, Infrastructure, and the Internet* (Cambridge, MA: MIT Press, 2007); Kathleen Fitzpatrick, *Planned Obsolescence: Publishing, Technology, and the Future of the Academy* (New York: New York University Press, 2011); Michael Nielsen, *Reinventing Discovery: The New Era of Networked Science* (Princeton, NJ: Princeton University Press, 2012); and Paul N. Edwards et al., *Knowledge Infrastructures: Intellectual Frameworks and Research Challenges* (Ann Arbor: Deep Blue, 2013), http://hdl.handle.net/2027.42/97552, accessed 10 August 2016.

5. On politics and paperwork, see Ben Kafka, *The Demon of Writing: Powers and Failures of Paperwork* (New York: Zone Books, 2012); Jacob Soll, *The Information Master: Jean-Baptiste Colbert's Secret State Intelligence System* (Ann Arbor: University of Michigan Press, 2009); and Matthew S. Hull, *Government of Paper: The Materiality of Bureaucracy in Urban Pakistan* (Berkeley: University of California Press, 2012).

6. I am guided here by the extensive literature in science and technology studies concerned with the relationship between scientific expertise and modern democracies. See, for example, Yaron Ezrahi, *The Descent of Icarus: Science and the Transformation of Contemporary Democracy* (Cambridge, MA: Harvard University Press, 1990); Bruno Latour, *Politics of Nature: How to Bring the Sciences into Democracy* (Cambridge, MA: Harvard University Press, 2004); and Sheila Jasanoff, *Designs on Nature: Science and Democracy in Europe and the United States* (Princeton, NJ: Princeton University Press, 2005).

7. Some journals might focus on review essays, and many might also include news or editorials, but gradually publishers became careful to segregate these different kinds of content from original research papers.

8. In some respects, the early history of the scientific journal recounted in this book roughly parallels the early history of scholarly journals more generally. Perhaps as a means of bolstering their perceived legitimacy as knowledge-producing fields, other academic disciplines have routinely taken up institutional forms and customs associated with the sciences. But the assimilation has often only been partial. Indeed, reflecting on the rough-and-ready distinction between book-fields and journal-fields suggests some of the ways in which the strong emphasis on publishing papers rather than books (which are usually required to be more synthetic and where the potential general sale affects publishing decisions) has given a particular cast to the criteria that determine valuable work in the sciences.

9. See James A. Secord, *Visions of Science: Books and Readers at the Dawn of the Victorian Age* (Oxford: Oxford University Press, 2014), viii; and John Tresch, *The Romantic Machine: Utopian Science and Technology After Napoleon* (Chicago: University of Chicago Press, 2012).

10. "Advertisement," *Journal of Science and the Arts* 1 (1816): i–iv, on iv.

11. Jean-Baptiste Biot, [Review of] *Mémoires de la classe des sciences mathématiques et physique, Journal des savants* (1817): 143–151, on 144. All translations, unless otherwise indicated, are my own.

12. Jean-Baptiste Biot, [Review of] *Comptes rendus hebdomadaires, Journal des savants* (November 1842): 641–661, on 651.

13. For early modern versions of the trope, see Klaas van Berkel and Arjo Vanderjagt, eds., *The Book of Nature in Early Modern and Modern History* (Leuven: Peeters, 2006).

14. Maxwell, "Essays at Cambridge—1853 to 1856," in *The Life of James Clerk Max-*

well (London: Macmillan, 1882), 223–246, on 243; Helmholtz, "Über das Verhältniss der Naturwissenschaften zur Gesammtheit der Wissenschaft" [1862], in *Populäre wissenschaftliche Vorträge*, vol. 1 (Braunschweig: Friedrich Vieweg und Sohn, 1865), 1–30, on 12; and Lord Rayleigh, "Presidential Address," in *Report of the Fifty-Fourth Meeting of the British Association for . . . 1884* (1885), 3–23, on 20.

15. "Autobiography of a Physiologist" [on Karl Ernst von Baer], *Quarterly Review* 122 (1867): 335–47, on 343.

16. "A Conspectus of Science," *Quarterly Review* 197 (1903): 139–160, on 147–148.

17. On scientific authorship, see, for example, Mario Biagioli and Peter Galison, eds., *Scientific Authorship: Credit and Intellectual Property in Science* (New York: Routledge, 2003); David Pontille, "Qu'est-ce qu'un auteur scientifique?," *Science de la société* 67 (2006): 77–93; David Pontille, *Signer ensemble: contribution et évaluation en sciences* (Paris: Economica, 2016); James A. Secord, *Victorian Sensation: The Extraordinary Publication, Reception, and Secret Authorship of Vestiges of the Natural History of Creation* (Chicago: University of Chicago Press, 2000); and Mario Biagioli, "Documents of Documents: Scientists' Names and Scientific Claims," in Annelise Riles, ed., *Documents: Artifacts of Modern Knowledge* (Ann Arbor: University of Michigan Press, 2006), 127–157.

18. *Putting Scientific Information to Work* (Philadelphia: Institute for Scientific Information, 1967).

19. F. J. Cole and Nellie B. Eales, "The History of Comparative Anatomy," *Science Progress* 11 (1917): 578–596; John Ziman, "Information, Communication, Knowledge," *Nature* 224 (1969): 318–324, on 318.

20. Harriet Zuckerman and Robert K. Merton, "Patterns of Evaluation in Science: Institutionalisation, Structure and Functions of the Referee System," *Minerva* 9 (1971): 66–100. For broader considerations on the emergence of the "scientific community" idea in the 1960s United States, see David A. Hollinger, "Free Enterprise and Free Inquiry: The Emergence of Laissez-Faire Communitarianism in the Ideology of Science in the United States," *New Literary History* 221 (1990): 897–919.

21. Zuckerman and Merton, "Patterns of Evaluation in Science," 69. Merton and Eisenstein shared a lively correspondence on the topic of printing and its relationship to science during the 1960s. During this period Eisenstein published "Some Conjectures about the Impact of Printing on Western Society and Thought: A Preliminary Report," *Journal of Modern History* 40 (1968): 1–56. The linking of the invention of printing to the rise of modern thought, which parallels to some degree the linking of the invention of the scientific journal to the rise of modern science, has been subjected to critique in Adrian Johns, *The Nature of the Book* (Chicago: University of Chicago Press, 1998).

22. For the later persistence of the heroic narrative in academic history, see, for example, Mary Boas Hall, *Henry Oldenburg: Shaping the Royal Society* (Oxford: Oxford University Press, 2002), 84.

23. *On Being a Scientist* (1995 ed.), 9–10.

24. Derek de Solla Price, "Society's Needs in Scientific and Technical Information," in Boris Pregel, Harold Dwight Lasswell, and John McHale, eds., *World Priorities* (New Brunswick: Transaction Books, 1975), 126–136, on 129.

25. H. M. Collins, "The Seven Sexes: A Study in the Sociology of a Phenomenon, or the Replication of an Experiment in Physics," *Sociology* 9 (1975): 205–224. For early ethnographies of scientific spaces, see Michael Lynch, *Art and Artifact in Laboratory Science: A Study of Shop Work and Shop Talk in a Research Laboratory* (London: Routledge & Kegan Paul, 1985); Bruno Latour and Steve Woolgar, *Laboratory Life: The Social Construction of Scientific Facts* (Beverly Hills: Sage Publications, 1979); and Karin D. Knorr-Cetina, *The Manufacture of Knowledge: An Essay on the Constructivist and Contextual Nature of Science* (Oxford: Pergamon Press, 1981).

26. Steven Shapin, *A Social History of Truth: Civility and Science in Seventeenth-Century England* (Chicago: University of Chicago Press, 1994); Steven Shapin and Simon Schaffer, *Leviathan and the Air-Pump: Hobbes, Boyle, and the Experimental Life* (Princeton, NJ: Princeton University Press, 1985); and Steven Shapin, "Discipline and Bounding: The History and Sociology of Science as Seen Through the Externalism-Internalism Debate," *History of Science* 30 (1992): 333–369.

27. Jean Dhombres, ed., "Regards sur la science: le journal scientifique," special issue, *Sciences et techniques en perspective* 28 (1994); Andreas W. Daum, *Wissenschaftspopularisierung im 19. Jahrhundert: bürgerliche Kultur, naturwissenschaftliche Bildung und die deutsche Öffentlichkeit, 1848–1914* (München: R. Oldenbourg, 1998); Secord, *Victorian Sensation*; Bernadette Bensaude-Vincent, *La science contre l'opinion: Histoire d'un divorce* (Paris: Seuil, 2003); Aileen Fyfe, *Science and Salvation: Evangelical Popular Science Publishing in Victorian Britain* (Chicago: University of Chicago Press, 2004); Geoffrey Cantor et al., eds., *Science in the Nineteenth-Century Periodical: Reading the Magazine of Nature* (Cambridge: Cambridge University Press, 2004); Geoffrey Cantor and Sally Shuttleworth, eds., *Science Serialized: Representations of the Sciences in Nineteenth-Century Periodicals* (Cambridge, MA: MIT Press, 2004); Bernard Lightman, *Victorian Popularizers of Science: Designing Nature for New Audiences* (Chicago: University of Chicago Press, 2007); Aileen Fyfe and Bernard Lightman, eds., *Science in the Marketplace: Nineteenth-Century Sites and Experiences* (Chicago: University of Chicago Press, 2007); and Martha Cecilia Bustamante, ed., "Oralités et documents scientifiques écrits: Nouvelles perspectives en histoire des sciences," special issue, *Archives internationales d'histoire des sciences* 65, no. 175 (2016).

28. On these points, see *Leviathan and the Air-Pump*, 342; and Steven Shapin, "Science and the Public," in R. C. Olby et al., eds., *A Companion to the History of Modern Science* (London: Routledge, 1990), 990–1007.

29. Shapin and Schaffer themselves did pay close attention to the role that generic conventions of writing played in the politics of early modern natural philosophy. The concept of "literary technology," which they deployed to describe the import of Robert Boyle's writing and strategies, has been especially influential. Following this, and Shapin's subsequent work on science and civility, Adrian Johns's *The Nature of the Book* demonstrated the full potential of bringing book historical concerns to the study of the history of science. My study of nineteenth-century scientific publishing owes a great deal to this work.

30. This neglect of the history of elite scientific publishing is beginning now to change. For some earlier exceptions, see, for example, A. J. Meadows, ed., *Development of Science Publishing in Europe* (New York: Elsevier Science Publishers, 1980); and Jean Dhombres,

ed., "Le journal professionnel au XIXe siècle," special issue, *Rivista di storia della scienza* 2, no. 2 (1994).

31. See Aileen Fyfe and Bernard Lightman, "An Introduction," in *Science in the Marketplace*, 1–19, on 2.

32. This modifies a phrase used by Steven Shapin (which he attributes in turn to Bruno Latour), for whom the key word is "society." See "Discipline and Bounding," 355.

33. The social and discursive approaches to historicizing the public sphere were in tension in Jürgen Habermas's *Strukturwandel der Öffentlichkeit* (Berlin: Hermann Luchterhand, 1962), translated as *Structural Transformation of the Public Sphere* (Cambridge, MA: MIT Press, 1989). Historians of eighteenth-century political culture have explored the category in both these senses. For the "public" as a discursive category that represented a major "abstract source of legitimacy" of political bodies and judgments in the eighteenth century, see Keith Michael Baker, *Inventing the French Revolution* (Cambridge: Cambridge University Press, 1990), 167–199. For a discussion of this historiography, see Harold Mah, "Phantasies of the Public Sphere: Rethinking the Habermas of Historians," *Journal of Modern History* 72 (2000): 153–182. The place a material history of texts might play in modifying this binary is raised in a joint interview of Baker and Roger Chartier, "Dialogue sur l'espace public," *Politix* 26 (1994): 5–22.

34. Thomas Broman, "The Habermasian Public Sphere and 'Science in the Enlightenment,'" *History of Science* 36 (1998): 123–149.

35. Indeed, in one recent special issue on the topic, the editors flirt with the prospect that no generalization about scholarly journals is possible at all, and that "each and every journal represents a stand-alone story and model." See Jeanne Peiffer, Maria Conforti, and Patrizia Delpiano, eds., "Scholarly Journals in Early Modern Europe: Communication and the Construction of Knowledge," special issue, *Archives internationales d'histoires des sciences* 63, nos. 170–171 (June 2013): 6.

36. David M. Bickerton, *Marc-Auguste and Charles Pictet, the Bibliothèque britannique (1796–1815), and the Dissemination of British Literature and Science on the Continent* (Geneva: Slatkine Reprints, 1986); Heinz Sarkowski, *Springer-Verlag: History of a Scientific Publishing House*, 2 vols. (New York: Springer, 1996); W. H. Brock and A. J. Meadows, *The Lamp of Learning: Two Centuries of Publishing at Taylor & Francis* (London: Taylor & Francis, 1998); Frank Holl, *Produktion und Distribution wissenschaftlicher Literatur: Der Physiker Max Born und sein Verleger Ferdinand Springer 1913–1970* (Frankfurt am Main: Buchhändler-Vereinigung, 1996); Norbert Verdier, "Le Journal de Liouville et la presse de son temps: une entreprise d'édition et de circulation des mathématiques au XIXe siècle" (PhD diss., Paris, 2009); Melinda Baldwin, *Making Nature: The History of a Scientific Journal* (Chicago: University of Chicago Press, 2015); and Nathalie Montel, *Ecrire et publier des savoirs au XIXe siècle: une revue en construction: les Annales des ponts et chaussées, 1831–1866* (Rennes: Presses universitaires de Rennes, 2015).

37. D. A. Kronick, *A History of Scientific and Technical Periodicals: The Origins and Development of the Scientific and Technical Press, 1665–1790* (Metuchen, NJ: Scarecrow Press, 1976); Susan Sheets-Pyenson, "Low Scientific Culture in London and Paris, 1820–1875" (PhD diss., University of Pennsylvania, 1976); and Volker R. Remmert and Ute Schneider,

Eine Disziplin und ihre Verleger: Disziplinenkultur und Publikationswesen der Mathematik in Deutschland, 1871–1949 (Bielefeld: transcript, 2010).

38. "Sag, was du weißt, kurz und bestimmt." Some harmonization between German publishing practices and those in place elsewhere did occur in the 1930s, in part by the intervention of a Nazi government concerned about losing face internationally. For some of the critiques of German scientific publishing, see Wilfrid Bonser, "The Cost of German Biological Periodicals," *Library Association Record* 6 (December 1928): 252–256; Charles Harvey Brown, "Hazard to Research: The Danger to Research Through the Increasing Cost of Scientific Publications," *Journal of Higher Education* 2 (1931): 420–426; and Léon Bultingare, "Le prix des périodiques allemands et sa répercussion sur le budget de nos bibliothèques," *Revue scientifique* (1933): 301–305. Springer's writer's manual, *Anleitung zur Niederschrift und Veröffentlichung medizinischer Arbeiten* (Berlin: Springer, 1929), was an adaptation of G. H. Simmons and M. Fishbein, *The Art and Practice of Medical Writing* (Chicago: American Medical Association, 1925). For secondary accounts, see H. Edelman, "Precursor to the Serials Crisis: German Science Publishing in the 1930s," *Journal of Scholarly Publishing* 25 (1994): 171–178; and Michael Knoche, "Wissenschaftliche Zeitschriften im nationalsozialistischen Deutschland," in Monika Estermann und Michael Knoche, eds., *Von Göschen bis Rowohlt: Beiträge zur Geschichte des deutschen Verlagswesens* (Wiesbaden: O. Harrassowitz, 1990): 260–281.

39. William Ramsay, "Molecular Weights," *Chemical News* 69 (1894): 51–54, on 51.

40. I have adopted the term "knowledge expression" from Geoffrey C. Bowker, "Emerging Configurations of Knowledge Expression," in Tarleton Gillespie, Pablo J. Boczkowski, and Kirsten A. Foot, eds., *Media Technologies: Essays on Communication, Materiality, and Society* (Cambridge, MA: MIT Press, 2014), 99–118.

Chapter 1

1. "Prospectus: *Annales du Muséum national d'histoire naturelle*," *Annales de chimie* 44 (1802): 100–112.

2. "Advertisement," *Annals of Philosophy* 1 (1813): iii–iv.

3. Derek J. de Solla Price, *Science Since Babylon* (New Haven, CT: Yale University Press, 1961).

4. "Prospectus: *Annales du Muséum national d'histoire naturelle*," 103–104.

5. "L'imprimeur au lecteur," *Journal des sçavans* 1 (1665). See Jean-Pierre Vittu, "De la Res publica literaria à la République des Lettres, les correspondances scientifiques autour du *Journal des savants*," in *La plume et la toile*, ed. Pierre-Yves Beaurepaire (Arras: Artois presses université, 2002), 225–252; and idem, "La formation d'une institution scientifique: le *Journal des savants* de 1665 à 1714," *Journal des savants* (January 2002): 179–203 and (July 2002): 349–377.

6. The literature on early modern learned journals is immense. See, for example, the recent special issue edited by Jeanne Peiffer, Maria Conforti, and Patrizia Delpiano, "Scholarly Journals in Early Modern Europe: Communication and the Construction of Knowledge"; Thomas Broman, "Criticism and the Circulation of News: the Scholarly Press in the Late Seventeenth Century," *History of Science* 51 (2013): 125–150; Otto Dann, "Vom Jour-

nal des sçavants zur wissenschaftlichen Zeitschrift," in *Gelehrte Bücher vom Humanismus bis zur Gegenwart*, ed. Bernhard Fabian et al. (Wiesbaden, 1983), 64–80; Jeanne Peiffer and Jean-Pierre Vittu, "Les journaux savants, formes de la communication et agents de la construction des savoirs (17e–18e siècles)," *Dix-huitième siècle* 40 (2008): 281–300; Joachim Kirchner, *Das deutsche Zeitschriftenwesen: seine Geschichte und seine Probleme* (Wisebaden: O. Harrassowitz, 1958–1962); and Frank Donoghue, *The Fame Machine: Book Reviewing and Eighteenth-Century Literary Careers* (Stanford, CA: Stanford University Press, 1996).

7. Scipione Maffei, *Giornale de' letterati d'Italia* (1710), quoted and translated in Luigi Balsamo, *Bibliography; History of a Tradition* (Berkeley, CA: Rosenthal, 1984), 97. See also Denis François D. F. Camusat, *Histoire critique des journaux*, vol. 1 (Amsterdam: J. F. Bernard, 1734), 5–6; and the entry for "Journal" in *Encyclopédie, ou dictionnaire raisonné des sciences, des arts et des métiers*, vol. 8 (Neufchastel: Samuel Faulche, 1865), 896–897.

8. On the *Journal des sçavans* as a current bibliography, see Balsamo, *Bibliography*, 95–97.

9. Oldenburg presented the first issue of the *Journal des sçavans* to the Society on 11 January 1665. R. K. Bluhm, "Henry Oldenburg, F.R.S.," *Notes and Records of the Royal Society of London* 15 (1960): 183–197, on 190.

10. Undated MS in Oldenburg's hand, quoted in E. N. da C. Andrade, "The Birth and Early Days of the Philosophical Transactions," *Notes and Records of the Royal Society of London* 20 (1965): 9–27, on 11–12. On the early *Transactions*, see especially Adrian Johns, "Miscellaneous Methods: Authors, Societies and Journals in Early Modern England," *British Journal for the History of Science* 33 (2000): 159–186.

11. For the ambivalent role of print in the early Royal Society, see Johns, *The Nature of the Book*, 444–542. The account that follows owes a great deal to this account.

12. Johns, *Nature of the Book*, 475–499. See also Shapin, *Social History of Truth*, 303.

13. Thomas Sprat, *The History of the Royal-Society of London* (London: J. Martyn, 1867), 115; and Lorraine Daston and Katharine Park, *Wonders and the Order of Nature, 1150–1750* (New York: Zone Books, 1998), 215–254.

14. See Joad Raymond, *The Invention of the Newspaper: English Newsbooks, 1641–1649* (Oxford: Clarendon Press, 2005); on periodicity, C. John Sommerville, *The News Revolution in England: Cultural Dynamics of Daily Information* (New York: Oxford University Press, 1996); for a comparative perspective with France, see Bob Harris, *Politics and the Rise of the Press: Britain and France, 1620–1800* (London: Routledge, 1996).

15. "Advertisement," *Philosophical Transactions* 1, no. 12 (7 May 1666): 214.

16. Johns, *The Nature of the Book*, 514–521.

17. On the adjudicative functions of the Academy, see Chapter 1 of Roger Hahn, *The Anatomy of a Scientific Institution: The Paris Academy of Sciences, 1666–1803* (Berkeley: University of California Press, 1971). On the Academy's various publishing strategies, and the mediation of collective and individual rights, see Alice Stroup, *A Company of Scientists* (Berkeley: University of California Press, 1990), 204–218.

18. On the Academy's early experiment in collective research and publishing, see Hahn, *Anatomy*, 1–34; Stroup, *A Company of Scientists*, 207–209; and Frederic L. Holmes, "Argument and Narrative in Scientific Writing," in *The Literary Structure of Scientific Argument:*

Historical Studies, ed. Peter Dear (Philadelphia: University of Pennsylvania Press, 1991), 164–181.

19. The project of the periodical was outlined on 19 December 1691. Each issue consisted of sixteen pages, and it ran for two years. See Jean-Pierre Vittu, "La formation d'une institution scientifique: le *Journal des savants* de 1665 à 1714. 2. L'instrument central de la République des Lettres," *Journal des savants* (July–December 2002): 349–377, on 371.

20. On the *Histoire et mémoires*, see James McClellan, "Les Mémoires de l'Académie royale des sciences, 1699–1790: bilan public et processus privés," in *Règlement, usages et science dans la France de l'absolutisme*, ed. Christiane Demeulenaere-Doyère and Éric Brian (Paris: Tec & Doc Lavoisier, 1999), 453–68; and *Les publications de l'Académie royale des sciences de Paris (1666–1793)*, 2 vols. (Turnhout: Brepols, 2001).

21. On the publications of learned societies more generally, see James E. McClellan, *Science Reorganized: Scientific Societies in the Eighteenth Century* (New York: Columbia University Press, 1985).

22. The Academy originally intended that the annual volumes be issued shortly after the completion of each year. But the volumes appear as many as six years after the fact, suggesting that promptness was ultimately subordinate to refinement.

23. See for example Kronick, *A History of Scientific and Technical Periodicals*; and McClellan, *Science Reorganized*.

24. Statute 31 stated that "the Academy will examine, if the King orders it, all machines for which a privilege has been requested from His Majesty. It will certify whether it is new and useful . . . ." Quoted in Bernard Fontenelle, *Histoire du renouvellement de l'Académie royale des sciences en MDCXCIX* (Paris: Jean Boudot et Fils, 1708), 52. Prior to 1699, the state had already been employing the Academy to evaluate inventions on an ad hoc basis.

25. For an account of the Academy's procedures of evaluation in the eighteenth century, see Jean Dhombres, "Formes publiques de la 'veille' académique au siècle des Lumières," in *Règlement, usages et science dans la France de l'absolutisme*, 265–291; and Patrice Bret, "La prise de décision académique: pratiques et procédures de choix et d'expertise à l'Académie royale des sciences," in *Règlement, usages et science dans la France de l'absolutisme*, 321–362. The state-sponsored rapport was a well-established genre in a number of other areas of the business of the state.

26. Lavoisier, in a report on 28 May 1788, AAS-R/107, 134v. Also translated by Hahn, *Anatomy*, 70.

27. Statute 27, in Fontenelle, *Histoire du renouvellement de l'Académie royale des sciences*, 37.

28. The decision to publish outsiders' work was announced in 1748, but the Academy did not begin to do so until 1752, claiming that it took that long to accumulate strong enough memoirs. See "Avertissement," *Histoire de l'Académie royale des sciences, année 1744* (Paris, 1748), 63–64.

29. In 1793, Antoine Lavoisier insisted on the importance of the distinction between the opinions of individual reporters and the judgment of the Academy as a whole. It was the latter that mattered in terms of public judgment. See Antoine Lavoisier, "Mémoire sur les

rapports académiques" (n.d.), reprinted in *Œuvres de Lavoisier*, vol. 6 (Paris: Imprimerie nationale, 1893), 48–50.

30. Eustache Marcot, letter to the Academy (read 29 May 1715), quoted in Daston and Park, *Wonders and the Order of Nature*, 152.

31. On the ways in which the world of the journalists and the censors overlapped, see Anne Goldgar, "The Absolutism of Taste: Journalists as Censors in 18th-Century Paris," in *Censorship & the Control of Print*, ed. Robin Myers and Michael Harris (Winchester: St Paul's Bibliographies, 1992), 87–110; and Raymond Birn, *Royal Censorship of Books in Eighteenth-Century France* (Stanford, CA: Stanford University Press, 2012). For the specific case of the Academy of Sciences, see Jean Perkins, "Censorship and the Académie des Sciences: A Case Study of Bonnet's *Considérations sur les corps organisés*," *Studies on Voltaire and the Eighteenth Century* 199 (1981): 251–262.

32. This argument is developed in Mario Biagioli, "Etiquette, Interdependence, and Sociability in Seventeenth-Century Science," *Critical Inquiry* 22 (1996): 193–238.

33. Hahn, *Anatomy*, 74.

34. "Preface," *Mémoires de mathématique et de physique rédigés a l'Observatoire de Marseille* 1 (1755), iii–ix, on iii.

35. Maupertuis to Johann Bernoulli, late October 1730, Basler Edition der Bernoulli-Briefwechsel, available at http://www.ub.unibas.ch/bernoulli/index.php/1730-10-12_TP_Maupertuis_Pierre_Louis_Moreau_de-Bernoulli_Johann_I, accessed 14 July 2014.

36. Maupertuis to Samuel Formey, 12 May 1747, translated in Mary Terrall, *The Man Who Flattened the Earth: Maupertuis and the Sciences in the Enlightenment* (Chicago: University of Chicago Press, 2002), 245.

37. Malesherbes to unknown, 11 December 1771, AAS, Dossier Malesherbes; Lavoisier to Louis Robinet, 21 July 1787; and Lavoisier to [Baron de Breteuil], 1 September 1787, in *Œuvres de Lavoisier: Correspondance*, vol. 5 (Paris: Albin Michel, 1993), 59, 66.

38. Maupertuis to Bernoulli, late October 1730.

39. On the potential disadvantages of this arrangement, see Stroup, *A Company of Scientists*, 216–217.

40. In 1730, the Academy used its connections to threaten the *Journal de Trévoux* after it published a review of its *Histoire et mémoires* that members thought was too critical. The journal posted an apology and was pressured to fire the writer, Louis Bertrand Castel. See AAS-R, 1 April 1730, p. 63.

41. Other French academies that published such volumes included the Académie royale des inscriptions et belles-lettres (1710), the Académie royale du chirurgie (1743), and the Académie des belles lettres de Caen (1754).

42. Maupertuis to Formey, 12 May 1747. My account is based on Terrall, *The Man Who Flattened the Earth*.

43. Leonard Euler, "Lettre de M. Euler à M. Merian," *Histoire de l'Académie royale des sciences et belles lettres à Berlin*, Année 1750 (1752): 520–532, on 520, 523. On the relationship between the Prussian Academy's publication and the press, see also Jens Häseler, "Journaux savants et l'Académie de Berlin: Deux acteurs sur le marché de l'information scientifique en Prusse," *Archives internationales d'histoire des sciences* 63 (2013): 199–214.

44. When in 1742 the Royal Society reformed the manner in which it dealt with the papers and letters it received, the question of the *Transactions* hardly came up, except as a hindrance to keeping things in order because its publisher sometimes held on to manuscripts for very long. See Meeting Minutes, 12 July 1714, RSL-CMO/3.

45. On the trusting epistemology of the early Royal Society, see Daston and Park, *Wonders and the Order of Nature*, 231–253.

46. *The Monthly Review* 2 (1750): 470–474. This article is attributed to Hill by Kevin J. Fraser, "John Hill and the Royal Society in the Eighteenth Century," *Notes and Records of the Royal Society of London* 48 (1994): 43–67. See also George Rousseau, *The Notorious Sir John Hill: The Man Destroyed by Ambition in the Era of Celebrity* (Bethlehem: Lehigh University Press, 2012), 65–81.

47. John Hill, *Lucina Sine Concubitu. A Letter Humbly addres'd to the Royal Society: In Which Is proved by most Incontestible Evidence, drawn from Reason and Practice, that a Woman may conceive and be brought to Bed without any Commerce with Man* (London: M. Cooper, 1750).

48. John Hill, *A Dissertation on Royal Societies* (London: John Doughty, 1750), 8, 9, 29.

49. "A Dissertation on Royal Societies," *British Magazine*, March 1750, 104–113, on 112. (This is a wholly distinct work from the pamphlet quoted above.) Hill also published a full volume, *A Review of the Works of the Royal Society of London* (London: R. Griffiths, 1751), dissecting what he took to be the far-fetched claims that had appeared in the *Transactions* over the years.

50. *A Dissertation on Royal Societies*, 38, 39.

51. Draft of a prefatory note proposed to be added to volumes of the *Philosophical Transactions*, BL-RS f. 21. The revised version was somewhat less emphatic.

52. The original recommendations were made by Lord Macclesfield on 5 February 1752. The results of the historical inquiry were presented on 20 February. New statutes were proposed on 27 February, revised and passed on 19 March. Financial arrangements were worked out in Council Meetings between 27 February and 28 May 1752, RSL-CMO/4, 66–88.

53. Incredibly, it was at this moment that the Paris Academy also took on the responsibility of publishing memoirs written by non-academicians when it launched the *Mémoires des savants étrangers*.

54. Mean article lengths increased more dramatically, but these are somewhat misleading because the Society began occasionally to include very long memoirs, sometimes greater than 100 pages.

55. In January 1773 the Council resolved that the *Transactions* would keep to a regular schedule of two parts per year.

56. These changes balanced themselves out so that the number of pages printed per year stayed relatively constant.

57. Council Minutes, 9 July 1789, RSL-CMO/7.

58. Peter C. G. Isaac, "William Bulmer, 1757–1830: An Introductory Essay," *The Library* 13 (1958): 37–50. On Banks's motive, see Banks to John Nichols, 28 December 1791, printed in Nichols, *Illustrations of the Literary History of the Eighteenth Century* (London:

Nichols, Son, and Bentley, 1822), vol. 4, 697. This also contains Nichols's description of "Fine Printing." The Royal Society Council approved Bulmer as printer 22 December 1791, RSL-CMO/8.

59. Banks to Josiah Wedgwood, 28 December 1791, SCJB vol. 4, 94–95. The Society tended to print editions on two sizes of paper, but the paper size used for the larger edition was increased somewhat. See May F. Katzen, "The Changing Appearance of Research Journals in Science and Technology," in A. J. Meadows, ed., *Development of Science Publishing in Europe* (New York: Elsevier Science Publishers, 1980), 177–214, on 189–199. In 1808, the Council put an end to the smaller edition (RSL-CMO, 24 March 1808). Note that the new layout only increased the margins without significantly increasing the space taken up by the printed text. This made it possible for binders to trim the new version to maintain consistency with the previous one. Eventually enlarged illustrations forced a change, however, with the result that the shift in bound size in surviving runs of the *Transactions* occurs at varying times in the 1790s.

60. The same was later true of the Horticultural Society of London, the Geological Society of London, and the Astronomical Society of London, though the latter titled its volumes *Memoirs* instead, likely in direct homage to Paris.

61. Broman, "Criticism and the Circulation of News."

62. On Haller's journal, see Martin Stuber, "Journal and Letter: The Interaction Between Two Communications Media in the Correspondence of Albrecht von Haller," in *Enlightenment, Revolution and the Periodical Press*, ed. Hans-Jurgen Lusebrink and Jeremy D. Popkin (Oxford: Voltaire Foundation, 2004), 114–141.

63. On Rozier and his journal, see J. E. McClellan, "The Scientific Press in Transition: Rozier's Journal and the Scientific Societies in the 1770s," *Annals of Science* 36 (1979): 425–449, 430; and Nicolas-François Cochard, *Notice historique sur l'abbé Rozier* (Lyon: D. L. Ayné, 1832). On the contested status of the transfer of the privilege, see Anne-Marie Chouillet, "Observations sur l'histoire naturelle, sur la physique et sur la peinture (1752–1757, 1771–1793)," *Dictionnaire des journaux, 1600–1789*, vol. 2 (Paris: Universitas, 1991), 995–998.

64. On the Academy's idea, see Keith M. Baker, "Les débuts de Condorcet au secrétariat de l'Académie royale des sciences (1773–1776)," *Revue d'histoire des sciences* 20 (1967): 229–280, on 263.

65. "Lettre VI," *Année littéraire*, 1772 (vol. 1), 125–132.

66. "Avis," *Observations sur la physique* 1 (1773): iii, v.

67. "Nouveautés," *Observations sur la physique* 14 (1779): 88; and "Avis essentiels," *Observations sur la physique* 10 (1777), front matter. Its price (24 livres for an annual subscription in Paris) was even greater than a volume of the Academy's *Histoire et mémoires* (18 livres for a bound copy in 1780).

68. McClellan, "The Scientific Press in Transition," 435.

69. On the Academy's attempts to change during the late ancien régime, see Hahn, *Anatomy*, 116–158; and Janis Langins, "Un discours prérévolutionnaire à l'Académie des sciences: l'exemple de Montalembert," *Annales historiques de la Révolution française* 320 (2000): 159–171. For Condorcet's reform project, see Baker, "Les débuts de Condorcet."

70. Carla Hesse, *Publishing and Cultural Politics in Revolutionary Paris, 1789–1810* (Berkeley: University of California Press, 1991), 33–45.

71. Before Rozier's journal, for example, weekly journals such as the *Avant-Coureur* could be used for publishing scientific news in France.

72. Thomas H. Broman, "J. C. Reil and the Journalization of Physiology," in *Literary Structure of Scientific Argument*, 13–42; idem, "Expertise Without Experts: Anonymity and Medical Authority in Johann August Unzer's Der Arzt," *Archives internationales d'histoire des sciences* 63 (2013): 29–47; and G. Mann, "Ernst Gottfried Baldinger und sein Magazin für Aerzte," *Sudhoffs Archiv* 42 (1958): 312–318.

73. On Crell's journal, see Karl Hufbauer, *The Formation of the German Chemical Community (1720–1795)* (Berkeley: University of California Press, 1982), 62–95.

74. Crell to Banks, 19 April 1789, SCJB vol. 4, 481. For more examples, see Hufbauer, *Formation*, 77–78.

75. Crell to Hatchett, 14 December 1801, RSL-MS/859/2/7; see also Crell to Banks, 18 April 1788 and 19 April 1789, SCJB vol. 4, 385–386, 481.

76. On the relationship between correspondence and learned journals during the eighteenth century, see Anne Goldgar, *Impolite Learning: Conduct and Community in the Republic of Letters, 1680–1750* (New Haven, CT: Yale University Press, 1995); Françoise Waquet, "De la lettre érudite au périodique savant: les faux semblants d'une mutation intellectuelle," *XVIIe siècle* 140 (1983): 347–359; Vittu, "De la Res publica literaria"; and Stuber, "Journal and Letter."

77. "Avertissement," *Annales de chimie* 1 (1789): 1–4; *Prospectus* announcing reissue of volume 1 of the *Annales de chimie* (1791), 1. On the origins of the *Annales*, see Patrice Bret, "Les origines et l'organisation éditoriale des *Annales de chimie* (1787–1791)," in *Œuvres de Lavoisier: Correspondance*, vol. 6, 415–426; and Maurice Crosland, *In the Shadow of Lavoisier* (British Society for the History of Science, 1994).

78. A vivid description of an editorial meeting of the *Annales* is given by M. A. Pictet when he visited Paris from Geneva in 1798. See Bickerton, *Marc-Auguste and Charles Pictet*, 233–234.

79. On the politics of print during and following the Revolution, see Hesse, *Publishing and Cultural Politics*; Bernard Vouillot, "La Révolution et l'empire: une nouvelle réglementation," in *Histoire de l'édition française*, vol. 2, ed. Henri-Jean Martin and Roger Chartier (Paris: Promodis, 1984), 526–535; and Jeremy D. Popkin, *Revolutionary News: The Press in France, 1789–1799* (Durham, NC: Duke University Press, 1990).

80. The Administration of the Book Trade was disbanded in August 1790.

81. *Bulletin des sciences*, no. 12 (June 1792), quoted in Jonathan Mandelbaum, "La Société philomathique de Paris de 1788 à 1835" (PhD diss., École des hautes études en sciences sociales de Paris, 1980), 55.

82. Marcel Dorigny, "Le Cercle social ou les écrivains au cirque," in *La Carmagnole des muses*, ed. Jean-Claude Bonnet (Paris: A. Colin, 1988), 49–66. The same was true of the Société national des Neuf Sœurs; see Jean-Luc Chappey, "Sociabilités intellectuelles et librairie révolutionnaire," *Revue de synthèse* 128, no. 6 (2007): 71–96, on 76.

83. Ginguené, "Projet de rapport à présenter au ministre de l'intérieur" [1795], BN/

NAF/9192/102. For an account of the work of Ginguené's Bureau, see Hesse, *Publishing and Cultural Practices*, 159.

84. On the intellectual project associated with republican encyclopedism, see Jean-Luc Chappey, "Usages et enjeux politiques d'une métaphorisation de l'espace savant en Révolution 'l'Encyclopédie vivante' de la République thermidorienne à l'Empire," *Politix* 48 (1999): 37–70. For cases of specific journals, see the essays by Yasmine Marcil, Isabelle Laboulais, and Pierre-Yves Lacour in a special issue titled "La Presse" in *La Révolution française* (2012, no. 2).

85. On the first wave of British scientific journals, see Jonathan R. Topham, "Anthologizing the Book of Nature: The Circulation of Knowledge and the Origins of the Scientific Journal in Late Georgian Britain," in *The Circulation of Knowledge between Britain, India, and China*, ed. Bernard Lightman and Gordon McOuat (Boston: Brill, 2013), 119–152. More than one British speculator attempted to produce a translated English edition of the *Annales de chimie*, although none lasted.

86. William Nicholson, "Preface," *A Journal of Natural Philosophy, Chemistry and the Arts* 1 (1797): iii–iv; Alexander Tilloch, "Prospectus of a new periodical work, intitled The Philosophical Magazine," *Star* 3069 (29 June 1798); epigraph translated in Brock and Meadows, *The Lamp of Learning*, 262, which also contains more on the history of the *Philosophical Magazine*.

87. Verso of covers of monthly issues beginning in 1804. A list of "Communications of men of the first eminence" is included in a separately printed advertisement for the journal printed in 1799.

88. Tilloch, "Prospectus of a new periodical work"; and Nicholson, "Preface," iv.

89. See Broman, "J. C. Reil and the Journalization of Physiology."

90. Nicholson's *Journal* was absorbed into Tilloch's in 1813.

91. Verso of paper wrapper of issues of Nicholson's *Journal* beginning in at least 1803.

92. Iain Watts, "'Current' Events: Galvanism and the World of Scientific Information, 1790–1830" (PhD diss., Princeton University, 2015). On reprinting practices in periodicals, see Will Slauter, "The Paragraph as Information Technology: How News Traveled in the Eighteenth-Century Atlantic World," *Annales. Histoire, Sciences Sociales* 67 (2012): 253–278.

93. Crell to Banks, 16 January 1791, SCJB vol. 4, 32. Young's letter was quoted by Brewster in *Quarterly Review* 43 (1830): 327.

94. "Préface," in the reprinting of the first years of the *Bulletin des sciences* (published ca. 1803): iii.

95. Conscious of their precarious standing, the compilers of the *Bulletin des sciences* started it as a manuscript news sheet copied out by hand, only hiring a printer when they couldn't find reliable copyists. Note in *Bulletin des sciences* 16–17 (1792); quoted in Mandelbaum, "La Société," 560.

96. On transformations in conceptions of authorship in France, see Geoffrey Turnovsky, *The Literary Market: Authorship and Modernity in the Old Regime* (Philadelphia: University of Pennsylvania Press, 2010); Carla Hesse, "Enlightenment Epistemology and the Laws of Authorship in Revolutionary France, 1777–1793," *Representations* 30 (1990):

109–137; Christine Haynes, *Lost Illusions: The Politics of Publishing in Nineteenth-Century France* (Cambridge, MA: Harvard University Press, 2009); and Roger Chartier, *The Order of Books: Readers, Authors and Libraries in Europe Between the Fourteenth and Eighteenth Centuries* (Stanford, CA: Stanford University Press, 1994). For Britain, see Mark Rose, *Authors and Owners: The Invention of Copyright* (Cambridge, MA: Harvard University Press, 1993); Nigel Cross, *The Common Writer: Life in Nineteenth-Century Grub Street* (Cambridge: Cambridge University Press, 1985); and Adrian Johns, "The Ambivalence of Authorship in Early Modern Natural Philosophy," in *Scientific Authorship*, 67–90, on 73–80.

97. Arthur Aspinall, "The Social Status of Journalists at the Beginning of the Nineteenth Century," *Review of English Studies* 21 (1945): 216–232, on 216.

98. J. A. Roebuck, *The Stamped Press of London and Its Morality* (London: C. & W. Reynell, 1835), 3. (Partially quoted in Aspinall, "Social Status.")

99. Review of Nicholson's *Journal*, in *The Monthly Review* 29 (1799): 301–311; for Young's refusal to be known as an author on nonmedical topics, see Young to Macvey Napier, 12 February 1816, in George Peacock, *Life of Thomas Young, M.D., F.R.S.* (London: J. Murray, 1855), 253. On Nicholson and Tilloch's problems with the Royal Society, see Iain Watts, "'We Want No Authors': William Nicholson and the Contested Role of the Scientific Journal in Britain, 1797–1813," *British Journal for the History of Science* 47 (2014): 1–23, on 9.

100. In a letter to Albrecht von Haller, Crell mentioned his plan of publishing a chemical magazine in the context of building up his literary reputation. See Crell to Haller, 7 February 1777, quoted in Hufbauer, *Formation*, 68–69.

101. Crell to Banks, 16 January 1791, SCJB vol. 4, 32; and 20 January 1796, SCJB vol. 4, 412.

102. Nicholson, "Preface," *A Journal of Natural Philosophy, Chemistry and the Arts* 1 (1797): iv.

103. "Journal, by Frances Anne Butler," *London Review* 2 (1835): 194–227, on 197.

104. See William Reddy, "Condottieri of the Pen: Journalists and the Public Sphere in Postrevolutionary France (1815–1850)," *American Historical Review* 99 (1994): 1546–1570.

105. Crosland, *In the Shadow of Lavoisier*, 71–72.

106. For Brande's journal, see Brande to John Murray, 9 November 1821, NLS-JM/MS/40142/208. For the *Edinburgh Philosophical Journal*, see Brewster to Herschel, 11 February 1820, RSL-HP/4.251; for the *Magazine*, see Susan Sheets-Pyenson, "From the North to Red Lion Court: The Creation and Early Years of the *Annals of Natural History*," *Archives of Natural History* 10 (1982): 221–249. See also Jonathan R. Topham, "The Scientific, the Literary and the Popular: Commerce and the Reimagining of the Scientific Journal in Britain, 1813–1825," *Notes and Records of the Royal Society of London* 70 (2016): 305–324.

107. The *Annales de chimie* offered a per-page rate of two livres for those who were on the editorial board. Patrice Bret, "Les origines," 301–302.

108. Maboth Moseley, *Irascible Genius: A Life of Charles Babbage, Inventor* (London: Hutchinson, 1964), 60–62.

109. Babbage to J. W. F. Herschel, 20 July 1816, RSL-HP/2.65.

110. Babbage to Herschel, 15 November 1815, RSL-HP/2.47. Nearly two decades later,

Babbage published an explicit argument about the deceitful practices of review journals, which he characterized as covert advertising bureaus for publishers. See Babbage, *On the Economy of Machinery and Manufactures* (London: C. Knight, 1832), 329–333.

111. Herschel to Babbage, 20 November 1815, RSL-HP/2.48.

112. Herschel to Babbage, 14 July 1816, RSL-HP/2.64.

113. "Advertisement," *Journal of Science and the Arts* 1 (1816): i–iv, on iv; Babbage to Herschel, 20 July 1816, RSL-HP/2.65.

114. Letter from Herschel to Edward Daniel Clarke, ca. 1817, printed in Sydney Ross, "Unpublished Letters of Faraday and Others to Edward Daniel Clarke," *Bulletin for the History of Chemistry* 11 (1991): 79–86.

115. Herschel to Francis Baily, 2 June 1821, Bodleian Library, MS. Autogr. d. 14, f. 6–7.

116. David Brewster, Review of *Reflections on the Decline of Science in England*, *Quarterly Review* 43 (1830): 305–342, on 327. This is quoted, and the general question of authorship and science discussed, in James A. Secord, "Science, Technology and Mathematics," in *The Cambridge History of the Book in Britain, vol. 6: 1830–1914*, ed. David McKitterick (Cambridge: Cambridge University Press, 2009), 443–474.

117. Brewster to Forbes, 11 February 1830, JFOR 1/230.

118. Brewster to Forbes, 27 March 1830, JFOR 1/240.

119. "Analysis of Scientific Journals," *Chemist* 1 (1824): 138.

120. "Preface" to vol. 3 of the *Magazine of Natural History* 3 (1830): iii–iv.

121. Forbes's publications as "Delta" appeared between 1827 and 1829 in Brewster's *Journal*.

122. Babbage to Brewster (ca. 1824), BL-CB/37183/125. Cuvier gave a similar reason for publishing in the *Bulletin* of the Société philomathique in a letter ca. 1798. See Dorinda Outram, *Georges Cuvier: Vocation, Science and Authority in Post-Revolutionary France* (Manchester: Manchester University Press, 1984), 122.

123. Babbage to Brewster (ca. 1824).

124. "Offprint" was coined in 1885 by Walter W. Skeat. See *The Academy* 28 (22 August 1885): 121.

125. The shift was gradual rather than instantaneous. For example, most (but not all) articles in vol. 55 (1765) begin on a new page. This suggests that the shift was an evolving printer's practice rather than an editorial choice on the part of the Royal Society, and thus may have been prompted by authors requesting separate copies. (It was only in 1828, however, that the Society shifted to beginning each memoir on an odd-numbered page.)

126. The policy is explained in Banks to John Latham, 28 December 1791, SCJB vol. 4, 93–94, but it was in place by 1788. See John Hunter to Edward Jenner (1788), printed in John Baron, *The Life of Edward Jenner* (London: Henry Colburn, 1827), 78.

127. To determine which societies allowed authors to obtain separate copies ahead of publication of volumes, one can often compare the date of publication of separate copies with those of the volumes.

128. "Preface & Laws," *Memoirs of the Literary and Philosophical Society of Manchester* 1 (1790): vii. For more on the early distribution of offprints by societies and academies, see chapters 4 and 7. See also Alex Csiszar, "Seriality and the Search for Order: Scientific

Print and Its Problems during the Late Nineteenth Century," *History of Science* 48 (2010): 399–434.

129. There are many separate copies in circulation of essays from Rozier's journal from its earliest year (these can be identified by a footnote giving bibliographical information on the first page); separate copies from the *Annales de chimie* exist as well.

130. Separate copies from the *Philosophical Magazine* survive from the early nineteenth century, but they become more common in the 1810s and 1820s. Genuine separately printed papers from Nicholson's *Journal* are extremely rare. (Nearly all of those that I have seen are actually disbound copies.) By the late 1820s Richard Taylor was actively offering a limited number of separate copies to authors. See for example John Blackburn to Taylor, 9 September 1828, TFP/Author letters A–C; and Taylor to John William Lubbock, 19 February 1830, RSL-JWL/T/68.

131. The Academy of Sciences in Paris allowed twenty-five free (PVAS 8, 29 March 1824, 52); the Royal Society allowed twenty (RSL-CMO/10, 28 February 1828); the Geological Society of London allowed fifteen (GSL-CM1/2, 12 December 1825); and the Astronomical Society of London allowed twenty-five (RASL-CM/1, 8 June 1821).

132. For example, Thomas Beddoes to Banks, 3 January 1791. SCJB vol. 4, 29. Swainson letter quoted in Taylor to William Jardine, 10 February 1838, WJP/5/129.

133. Banks to Latham, 28 December 1791.

134. Separates from Rozier's journal are often only identifiable by a footnote on the first page. Separates from the *Recueil des travaux de la Société des sciences, de l'agriculture et des arts, de Lille* give no indication at all.

135. The *Mémoires littéraires, critiques, philologiques, biographiques*, for example, occasionally mentioned that it was reviewing a work that had appeared in a journal only because it had been circulated as a separate copy. See, for example, the volume for year 1775, 209.

136. See for example the list of his publications Augustus Bozzi Granville appended to his *The Royal Society in the XIXth Century* (London, 1836). Among twenty-three works, only two reference a source journal (both the *Philosophical Transactions*), although over half were published in independent journals.

137. Quote from John Latham to Joseph Banks, 21 December 1791, SCJB vol. 4, 91–92.

138. Georges Cuvier, *Recherches sur les ossemens fossiles de quadrupèdes*, vol. 1 (Paris: Deterville, 1812), i–ii.

139. Ibid.

140. He explained his method to his son in a letter dated 28 August 1832, CGA, vol. 2, 755–756. Ampère's tribulations constructing the volume can be followed through his letters to the editors of the *Bibliothèque universelle* in Geneva (Marc-Auguste Pictet and Gaspard de la Rive) between 15 May 1821 and 21 August 1823, in Marc-Auguste Pictet, *Correspondance: sciences et techniques*, vol. 2, ed. R. Sigrist (Genève, 1998).

141. *Annals of Philosophy* 16 (1820): 297.

142. The legal status of reprinting in periodicals and newspapers was ambiguous in both Britain and France. On the British case, see Will Slauter, "Upright Piracy: Understanding the Lack of Copyright for Journalism in Eighteenth-Century Britain," *Book History* 16 (2013): 34–61.

143. Quotation from draft letter to censor, 1775, labeled "Affaire Paulet." AAS-DG/31, Folder "Imprimeurs de l'Académie royale des sciences." The Academy would even hold off producing reports on manuscripts, or printing them, until it could confirm that the author had not printed elsewhere. Several cases are recorded in the Minutes of the Academy's *Comité de Librairie* (AAS). A detailed account appears in James E. McClellan, "Specialist Control: The Publications Committee of the Académie Royale des Sciences (Paris), 1700–1793," *Transactions of the American Philosophical Society* 93 (2003): 1–99.

144. Draft letter to censor; the incident is discussed in McClellan, "Specialist Control," 72–74.

145. Augustin-François de Silvestre, "Rapport des travaux de la Société philomathique," in *Rapports généraux des travaux de la Société philomathique de Paris* (Paris: Ballard, 1792), 141–212, on 144.

146. Banks to Planta, 5 April 1783, and Planta to Banks, 7 April 1793, SCJB vol. 2, 65–70. The law passed as follows: "The Question was put whether any abstracts of papers read at the society and not ordered to be printed may be allowed to be published. It was by ballot unanimously determined in the negative." RSL-CMO/7, 3 April 1783.

147. Charles Blagden to Joseph Banks, 11 September 1783 and 14 September 1783, SCJB vol. 2, 130–133.

148. Letter from Crell to Banks, 10 March 1785, SCJB vol. 3, 33.

149. In 1802 Banks claimed not to enforce the prohibition on reprinting in foreign publications (Banks to Nicholson, 12 March 1802), but in 1812 the Society revoked an order to print a paper by Count Rumford because of "information having been communicated to the Council" that he had printed it in France. RSL-CMO/9, 5 March 1812.

150. A note on the verso of the cover in issues in 1804 "begs that authors of printed academical papers will favour him with copies."

151. Watts, "'We Want No Authors,'" 20–22.

152. Nicholson to Banks, 25 April 1802, BL-JB/27–28. Nicholson also made the argument that "this Distribution [of separate copies] is the true first publication," which Banks also rejected (Nicholson to Banks, 12 March 1802, SCJB vol. 5, 147–148).

153. Banks to Nicholson, 12 March 1802, SCJB vol. 5, 148–149; 24 April 1802, BL-JB/26.

154. The resolution to include the warning to readers on the flyleaf of separate copies was passed on 22 July 1802, RSL-CMO/8. A manuscript version of it exists in the Joseph Banks papers at the British Library. When the Astronomical Society of London began publishing memoirs in the 1820s, it included the same warning on its separate copies, and the Royal Society of Edinburgh adopted such a notice on 21 January 1822, NLS-RSE/16.

155. This may be seen for example by following the notices of articles in the commercial scientific press, such as the *Philosophical Magazine*, which reported heavily on the contents of Royal Society publications. Newly received pamphlets were often cited and reported on before the issues that contained them were in circulation.

156. The submission by Marshall Hall on the oxidation of iron was rejected by the Council on 11 February 1819, RSL-CMO/9.

157. On Banks and the early Geological Society, see Martin J. S. Rudwick, "The Foun-

dation of the Geological Society of London: Its Scheme for Co-Operative Research and Its Struggle for Independence," *British Journal for the History of Science* 1 (1963): 325–355.

158. Olinthus Gregory, "A Review of some leading points in the official character and Proceedings of the late President of the Royal Society," *Philosophical Magazine* 56 (1820): 252.

159. For the several letters between the Linnean Society and the Attorney General on this topic, see the Linnean Society Council Minutes, 23 June 1823, 16 March, 22 March, 4 May, 21 May 1824, LSL/CM vol. 1.

160. Robison is quoted to himself in a letter from Brewster, 24 September 1828, NLS-RSE/352.

161. Vigors to William Kirby, 1 October 1822. Correspondence published in John Freeman, *Life of the Rev. William Kirby, M.A.* (London: Longman and Co., 1852), 373.

162. Earlier in 1788 Carey had drawn the ire of none other than Benjamin Franklin for having published excerpts of memoirs of the *Transactions* of the APS in his journal. See Franklin to Carey, 10 June 1788. Carey was at that time already upset about the expensive quartos. See Edward Carter II, "The Political Activities of Mathew Carey, Nationalist, 1760–1814" (PhD diss., Bryn Mawr College, 1962), 115–117.

163. Carey's most articulate plea was his circular "To the Members of the American Philosophical Society," published in 1824 (reprinted in *Miscellaneous Essays* [Philadelphia: Carey & Hart, 1830], 241–246). His several motions and the subsequent discussions are in the APS Council Minutes, 1815–1824 (between 5 March and 7 May). Carey was successful in convincing the Council to publish the quarto issues more frequently, which led to the formation of a standing Committee of Publication.

164. There are many interpretations of the last days of the Academy. See for example Charles Gillispie, "The Encyclopédie and the Jacobin Philosophy of Science," in *Critical Problems in the History of Science*, ed. M. Clagett (Madison: University of Wisconsin Press, 1969), 255–289; Hahn, *Anatomy*; Jean Dhombres and Nicole Dhombres, *Naissance d'un pouvoir: sciences et savants en France (1793–1824)* (Paris: Payot, 1989); and Ken Alder, *Engineering the Revolution: Arms and Enlightenment in France, 1763–1815* (Princeton, NJ: Princeton University Press, 1997), 292–318.

165. "Loi portant que les séances ordinaires et journalières de l'institut ne seront point publiques," 9 floréal an IV [28 April 1796], *Institut national de la république française* (Paris: prairial, an IV), 40–41.

166. The number of commissions named per year to analyze manuscripts and inventions was almost always greater than 100, and tended to increase toward 150 in the 1820s. The number of reports actually written and recorded hovered at about fifty per year. The bulk of the reports were copied out *in extenso* in the *Procès-verbaux* (indeed, the text of the reports makes up approximately 75 percent of the bulk of the *Procès-verbaux* between 1795 and 1835). For statistics on the reports during this period, see Hugues Chabot, "Le jugement de l'Académie: étude quantitative des commissions et rapports entre 1795 et 1835," in *Règlement, usages et science*, 363–379.

167. This decision was taken on 6 ventôse an VIII [25 February 1800]. See PVAS 2, 114.

168. The rule against reporting on printed memoirs was put forward on 26 pluviôse an IV [15 February 1796].

169. "Avant-Propos," *Journal de l'École polytechnique* 1 (an III/1795). On the transformation of the periodical, see Loïc Lamy, "Le Journal de l'École polytechnique de 1795 à 1831," *Sciences et techniques en perspective* 32 (1995).

170. "Prospectus: *Annales du Muséum national d'histoire naturelle*" (1802). The new prospectus was distributed along with the front matter and contents of the second volume.

171. Jean-Luc Chappey, "Héritages républicains et résistances à l'organisation impériale des savoirs," *Annales historiques de la Révolution française* 4 (2006): 97–120, on 105. On the creation of a new nobility by Napoleon, see Natalie Petiteau, *Elites et mobilités: la noblesse d'Empire au XIXe siècle (1808–1914)* (Paris: La Boutique de l'histoire éditions, 1997).

172. On the politics of science during the Consulate and the Empire, see Chappey, "Héritages républicains"; Outram, *Georges Cuvier*; and Dhombres and Dhombres, *Naissance d'un pouvoir*.

173. Letter from Cuvier to Kielmeyer, 30 January 1808, quoted in Outram, *Georges Cuvier*, 132.

174. See Outram, *George Cuvier*.

175. Hesse, *Publishing and Cultural Politics*, 225–226; and Bernard Vouillot, "La Révolution et l'Empire: une nouvelle réglementation," in *Histoire de l'édition française*, vol. 2, 528.

176. The report on the offer was written by Tessier and Monge, 6 ventôse an VIII [25 February 1800], PVAS 2, 113–114.

177. This development became particularly significant during the Napoleonic Wars, for the *Moniteur* was one of the most reliable means by which scientific news made it out of France.

178. Chappey, "Sociabilités intellectuelles," 90.

179. Hesse, *Publishing and Cultural Politics*, 230–239.

180. Pietro Corsi, "The Revolutions of Evolution: Geoffroy and Lamarck, 1825–1840," *Bulletin du Musée d'anthropologie préhistorique de Monaco* 51 (2011): 113–134.

181. "Rapport de la Commission chargée d'examiner les moyens d'augmenter l'activité de la Classe," 17 July 1809, PVAS 4, 227–229.

182. Ibid. The draft of this report included a further directive, subsequently crossed out, to publish "under the auspices of the class, a periodical work that will contain the most important observations made by non-members, and where members can insert a summary of their works in order to safeguard their property." AAS-PS, 17 July 1809.

183. There are relatively few outright negative reports during this period, and it is clear that rapporteurs regularly abstained from writing them. (This policy led to later infamous cases when little-known savants, such as Évariste Galois and Niels Henrik Abel, submitted manuscripts that languished.) When negative reports did get written, they sometimes noted that the author whose work was being examined insisted on it. See, for example, Bouvard's report on Demonville's "Système du monde," 17 December 1832, PVAS 10, 167–169.

184. In most such cases a report simply never appeared. Sometimes the reasoning was made explicit. See for example PVAS on 6 pluviôse an VIII [26 January 1800], 5 June 1820, 10 May 1824, and 4 February 1833.

185. From the first volumes of the *Histoire et mémoires* of the First Class, many reports were often published in full, or in excerpt as part of the *Histoire* section. The decision to do so was made on 4 prairial an VI [23 May 1798], PVAS 1, 397.

186. This was true both of volumes of academicians' memoirs and of the volumes of *Savants étrangers*, although the latter volumes remained exceedingly rare. A recommendation that a memoir be published in the *Mémoires des savants étrangers* was thus largely symbolic, as Ampère warned Michael Faraday on 14 July 1824, CMF 1, 355.

187. Jean-Baptiste Biot, "Review: *Mémoires de la classe des sciences mathématiques et physique*," *Journal des savants*, March 1817, 143–151. Biot was reacting in part to an attack by a writer in the *Philosophical Magazine*, who argued that the low state of French science meant that it took Humphry Davy traveling to Paris to publicly announce the discovery of iodine. In denying that claim, Biot also defended the Academy's *Mémoires*.

188. Ibid., 144.

189. Arago's new measures were introduced on 29 March 1824, PVAS 8, 52–53.

190. Attempts to assert control over attendance at meetings arose regularly. See PVAS on 6 nivôse an IX [27 December 1800], 9 nivôse an XIV [30 December 1805], 17 July 1809, and 4 April 1825.

Chapter 2

1. "Prospectus: Nouvelle Série," *Gazette Médicale de Paris*, January 1833, VII.

2. An important previous account of the Arago and the press during this period is Bruno Belhoste, "Arago, les journalistes et l'Académie des sciences dans les années 1830," in *La France des années 1830 et l'esprit de réforme: actes du colloque de Rennes, 6–7 octobre 2005*, ed. Harismendy (Rennes: Presses universitaires de Rennes, 2006), 253–266. See also Bensaude-Vincent, *La science contre l'opinion*; and Maurice Crosland, *Science under Control: The French Academy of Sciences, 1795–1914* (New York: Cambridge University Press, 1992), 279–299. For an outstanding reading of the divisions among physicists over the politics of representation, see Theresa Levitt, *The Shadow of Enlightenment: Optical and Political Transparency in France, 1789–1848* (New York: Oxford University Press, 2009).

3. "Parliamentary Reporting," *The Companion to the Newspaper*, 1 April 1833, 17–20, on 20; Thomas Babington Macaulay, [Review of Henry Hallam, *The Constitutional History of England*], *Edinburgh Review* 48 (1828): 96–169, on 165. On parliamentary reporting in London, see A. Aspinall, "The Reporting and Publishing of the House of Commons' Debates, 1771–1834," in *Essays Presented to Sir Lewis Namier*, ed. R. Pares and A. J. P. Taylor (London: St. Martin's Press, 1956), 227–257; Dror Wahrman, "Virtual Representation: Parliamentary Reporting and Languages of Class in the 1790s," *Past & Present* 136 (1992): 83–113; and Andrew Sparrow, *Obscure Scribblers: A History of Parliamentary Journalism* (London: Politico, 2003).

4. François Guizot, "Cours d'histoire moderne," *Journal des cours publics de jurisprudence, histoire, et belle lettres, Première Année* (Paris, 1820–1821), 83.

5. For example, Charles-Joseph Panckoucke's *Moniteur*. On political journalism in revolutionary France, see Popkin, *Revolutionary News*; Claude Labrosse and Pierre Rétat,

Naissance du journal révolutionnaire, 1789 (Lyon: Presses universitaires de Lyon, 1989); and Harvey Chisick and Ilana Zinguer, eds., *The Press in the French Revolution* (Oxford: Voltaire Foundation, 1991).

6. From 1819, floor plans were often published of the Chamber of Deputies (*Plan figuratif de la Chambre des Députés*) showing the layout and occupants of each seat. The "Tribune des journalistes" tended to increase in size with each new plan.

7. In 1824, an editor lobbied for the English parliament to adopt French custom, allowing a reporter at the Clerk's table so that he could properly hear what was said. See Aspinall, "Reporting," 254.

8. Letter from Edmond Burke to French Laurence, 16 March 1797, quoted in Aspinall, "Reporting," 243; William Windham speech recorded in *The Parliamentary Register* 7 (London, 1799), 475.

9. Quotation from Richard Carlile, *An Address to Men of Science* (London: R. Carlile, 1821), 18. On radical British journalism in the 1810s and 1820s, see Kevin Gilmartin, *Print Politics: The Press and Radical Opposition in Early Nineteenth-Century England* (Cambridge: Cambridge University Press, 1996); Jon P. Klancher, *The Making of English Reading Audiences, 1790–1832* (Madison: University of Wisconsin Press, 1987); and Iain McCalman, *Radical Underworld: Prophets, Revolutionaries and Pornographers in London, 1795–1840* (Oxford: Clarendon Press, 1993).

10. Robert Southey, "Inquiry into the Poor Laws, &c," *Quarterly Review* (1812): 319–356, on 348–349 (also quoted in Gilmartin, *Print Politics*, 26).

11. Gilmartin, *Print Politics*, 27–31.

12. On the radical origins of the cheap periodicals of the 1820s, see Jonathan Topham, "John Limbird, Thomas Byerley, and the Production of Cheap Periodicals in the 1820s," *Book History* 8 (2005): 75–106.

13. Carlile, *An Address to Men of Science*, 17, 30–31.

14. "To the Mechanics of the British Empire," *Mechanics' Magazine* 1 (1823): 16.

15. "Apology and Preface," *The Chemist* 1 (1824): vii.

16. Letter to the editor signed "Tom Telltruth," *Mechanics' Magazine* 1 (1824): 197–198.

17. *The Chemist* 2 (1824): 47.

18. Affidavit sent to Lord Elgin, [November 1825], quoted in Mary Bostetter, "The Journalism of Thomas Wakley," in *Innovators and Preachers: The Role of the Editor in Victorian England*, ed. Joel H. Wiener (Westport, 1985), 275–292. My account of Wakley is based on this and William Brock, "The Development of Commercial Science Journals in Victorian Britain," in *Development of Science Publishing in Europe*, 95–122, on 102–104.

19. For a survey of these, see James Secord, "Progress in Print," in *Books and the Sciences in History*, ed. M. Frasca-Spada and N. Jardine (Cambridge: Cambridge University Press, 2000), 369–389.

20. *Chemist* 1 (1824): 138, 275 (quoting an article in *The Scotsman*), 138.

21. *Chemist* 1 (1824): 275.

22. "The Editor's Address," *Mechanic's Oracle* 1 (1825): 1.

23. On Richard Taylor's life, see Brock and Meadows, *The Lamp of Learning*, 19–63.

24. On the Unitarian connections that helped establish his scientific printing empire,

see Jonathan R. Topham, "Technicians of Print and the Making of Natural Knowledge," *Studies in History and Philosophy of Science Part A* 35 (2004): 391–400.

25. On the influence of the early Geological and Astronomical Societies on the Royal Society, see David Philip Miller, "Method and the 'Micropolitics' of Science: The Early Years of the Geological and Astronomical Societies of London," in *The Politics and Rhetoric of Scientific Method: Historical Studies*, ed. J. Schuster and R. Yeo (Dordrecht: Reidel, 1986), 227–257.

26. See Council Minutes, 21 February 1823, RAS-CM/1 for the decision to give the *Philosophical Magazine* access to the minutes books. See RAS Letters, 1820–1829, for several letters from Gregory in 1824 on his progress in this regard; and RAS Papers/17.1 for manuscripts and proof sheets from the *Magazine*.

27. For the Royal Society's restriction on taking notes, see Walter Crum, "Sketch of the Life and Labours of Dr. Thomas Thomson," *Proceedings of the Royal Philosophical Society of Glasgow* 3 (1855): 250–264, on 256. There was, however, a very brief period in 1802 when the Royal Institution was given permission to print proceedings of the Royal Society. (These were done by Thomas Young.)

28. See Taylor to RAS in Council Minutes, 9 March 1827, RAS-CM/1, and GSL-CM1/2, 19 March 1827.

29. Taylor's account of charges to these societies, extant at St. Bride Library, makes clear that he used standing type from the *Philosophical Magazine* for both the Geological and the Astronomical Society's proceedings and charged accordingly. TFP, *Journal 1823*, 31–32. Likewise, Taylor indicated to the Royal Society of London—whose *Proceedings* he printed in the same way beginning in 1831—that the "price of the composition of such part of the Proceedings as can be used in the Philosophical Magazine & Annals—not to be charged, but only the presswork & paper." R. Taylor to P. M. Roget, 1831 RSL-DM/1/97.

30. "Report of the Council of the Society to the Eighth Annual General Meeting" (1828). For the parallel discussion at the Geological Society, see the Report of the Council at the Annual General Meeting of 15 February 1828, in *Proceedings of the Geological Society of London* 1, 48.

31. Adam Sedgwick, "President's Address at the Annual General Meeting (19 February 1830)," *Philosophical Magazine* 7 (1830): 291.

32. Taylor was confirmed as the Society's printer of the *Transactions* on 14 February 1828, and on 28 February the decision was made that "the Secretaries be authorized to communicate to the public, at their discretion, accounts of the proceedings of the Society." For Taylor's winning bid (as well as the others), see RSL-DM/1, f. 95–96.

33. The Council passed a resolution to create "a monthly publication in the cheapest form, under the name of proceedings, [to] be distributed to the Members of the Society," on 15 December 1830. (The question whether to publish transactions was postponed.) ZSL, Council Minutes 2.

34. This is not to say that the press *did* embody such opinion, but that the frequent claim that it could do so had important consequences for both politics and the history of the press. Quotation from Keith Michael Baker, *Inventing the French Revolution: Essays on French Political Culture in the Eighteenth Century* (Cambridge: Cambridge University Press,

1990), 172. On the embodiment of public opinion in the press, see Popkin, *Revolutionary News*. On "public opinion" in prerevolutionary France, see Mona Ozouf, "'Public Opinion' at the End of the Old Regime," *Journal of Modern History* 60 (1988): 3–21.

35. On the rise of voluntary sciéntific societies after the Revolution, see Hahn, *Anatomy*, 159–195, 252–286; and Jean-Luc Chappey, "Héritages républicains et résistances à l'organisation impériale des savoirs," *Annales historiques de la Révolution française* 4 (2006): 97–120; see also Chappey's introduction to *Des naturalistes en révolution: les procès-verbaux de la Société d'histoire naturelle de Paris, 1790–1798* (Paris: CTHS, 2009).

36. Starting in the 1820s, there are occasional exceptions such as the *Bulletin de la Société de géographie* of Paris, founded in 1821, which did include proceedings of some meetings.

37. On the position of the Institut under Napoleon, see Chappey, "Héritages républicains."

38. Pierre Flourens, "Analyse des travaux de l'Académie des sciences de Paris," *Annales générales des sciences physiques* 3 (1820): 243–255.

39. On both Dubois and Leroux, see Jean-Jacques Goblot, *La jeune France libérale: Le globe et son groupe littéraire, 1824–1830* (Paris: Plon, 1995), 29–34. On Dubois, see Paul Gerbod, *Paul-François Dubois, universitaire, journaliste et homme politique, 1793–1874* (Paris: C. Klincksieck, 1967). On Leroux, see Tresch, *The Romantic Machine*, 223–251. On the press in the French romantic imagination, see John Tresch, "The Order of the Prophets: Series in Early French Social Science and Socialism," *History of Science* 48 (2010): 315–342.

40. According to a law passed on 17 March 1822, new journals reporting on politics were first required to obtain authorization from the King. See "Loi relative à la police des journaux et écrits périodiques," in *Collection complète des lois, décrets, ordonnances, règlemens avis du Conseil d'état*, vol. 23 (Paris, A. Guyot et Scribe, 1838), 479. On press regulations during this period, see Gilles Feyel, *La presse en France des origines à 1944: histoire politique et matérielle* (Paris: Ellipses, 1999), 73–78.

41. *Globe* 1 (16 September 1824): 1, 2.

42. This is a retrospective view of Leroux in "D'une nouvelle typographe," *Revue indépendante* 6 (1843): 262–291, on 274.

43. *Globe* 1 (10 October 1824): 51.

44. On Villermé, see William Coleman, *Death Is a Social Disease: Public Health and Political Economy in Early Industrial France* (Madison: University of Wisconsin Press, 1982); and David S. Barnes, *The Great Stink of Paris and the Nineteenth-Century Struggle Against Filth and Germs* (Baltimore: Johns Hopkins University Press, 2006), 65–104.

45. *Globe* 1 (9 December 1824): 182.

46. Pierre Leroux, "Bertrand (Alexandre)," in *Encyclopédie nouvelle, ou dictionnaire philosophique*, ed. P. Leroux and J. Reynaud (Paris: Charles Gosselin, 1836), 641–644, on 643.

47. Over the course of 1825, the median length of *Le Globe*'s reports was over 1,000 words, while those in the *Annales de chimie* had a median length of about 160 words. The size of *Le Globe*'s reports crept up over time, and by 1829 had nearly doubled to a median of about 1,900.

48. Leroux, "Bertrand," 643.

49. PVAS, 4 April 1825, 206; 18 April 1825, 210. See "Projet de règlement ayant pour objet de désigner les personnes à qui l'Académie accorde l'autorisation d'assister à ses séances," AAS-PS, 18 April 1825.

50. A.-M. Ampère to Auguste de la Rive, 5 August 1826, CGA, vol. 2, 686.

51. Ann F. La Berge, *Mission and Method: The Early Nineteenth-Century French Public Health Movement* (Cambridge: Cambridge University Press, 2002), 105–106.

52. The regulations gave them the right to form in secret whenever they wished—it was a matter of taking advantage of the rule.

53. This account depends on Bertrand's own transcript of the debate: *Globe* 2 (8 October 1825): 875. For the analysis in the *Gazette de santé*, see 15 October 1825, 225.

54. *Globe* 2 (6 October 1825): 867. See *Journal des débats,* 15 September 1825, p. 3, for a verbatim copy of the report.

55. *Globe* 3 (17 January 1826): 60.

56. *Globe* 4 (25 November 1826): 235. On the Doctrinaires' theory of representation, see Pierre Rosanvallon, *Le moment Guizot* (Paris: Gallimard, 1985); Dario Roldán, *Charles de Rémusat* (Paris: L'Harmattan, 1999); and Aurelian Crăiuțu, *Liberalism under Siege: The Political Thought of the French Doctrinaires* (Lanham, MD: Lexington Books, 2003).

57. On the problem of representation during the Revolution itself, see Keith Baker, "Representation," in *The French Revolution and the Creation of Modern Political Culture,* vol. 1, ed. Keith Baker (Oxford: Pergamon Press, 1987), 469–492.

58. François Guizot, "Cours d'histoire moderne," *Journal des cours publics de jurisprudence, histoire, et belle lettres, Seconde Année* (Paris, 1821–1822), 134–135.

59. Ibid., 16.

60. Charles de Rémusat, "De la publicité donnée par les journaux aux délibérations de la chambre," *Globe* 7 (28 April 1829): 266–267.

61. Guizot, *Mémoires pour servir à l'histoire de mon temps,* vol. 1 (Paris: Michel Lévy Frères, 1858), 176. In 1819 Rémusat wrote that with publicity, "la société se fait spectacle à elle-même" (*De la liberté de la presse et des projets de loi présentés à la chambre des députés dans la séance du Lundi mars 1819* [Paris: Delaunay, 1819], 35). And Guizot wrote in 1819 that "a vital effect of freedom of the press is to continuously make France known to itself"; *Courier,* 1 July 1819, quoted in André-Jean Tudesq, "Guizot et la presse sous la Restauration," in *Guizot, les Doctrinaires et la presse (1820–1830)* (Le Val Richer: Fondation Guizot, 1994), 3.

62. The problem of the relationship between science and democracy, a crucial theme for science and technology studies today, would not have arisen for these liberal theorists. See Ezrahi, *The Descent of Icarus*; and Jasanoff, *Designs on Nature.*

63. By 1828, the annual subscription for many papers was eighty francs, about 420 hours of work for an unskilled worker. See Feyel, *La presse en France,* 66–67.

64. *Globe* 3 (11 February 1826): 120.

65. See for example Bertrand's self-congratulatory tone in the footnote to his report in *Globe* 3 (22 June 1826): 416.

66. *Globe* 6 (17 May 1828): 412.

67. *Globe* 4 (29 March 1827): 518.

68. *Globe* 6 (26 January 1828): 154; *Globe* 6 (17 May 1828): 412; and David Brewster, "On the Supposed Influence of the Aurora Borealis Upon the Magnetic Needle . . . ," *Edinburgh Journal of Science* 8 (April 1828): 189–201.

69. *Le Globe*'s report on the meeting appeared on 7 February 1829, the *Gazette*'s report on 6 March 1829, and *Le Globe*'s response on 11 March 1829. Villermé and Milne-Edwards's memoir appeared that fall in *Annales d'hygiène publique et de médecine légale* 2 (October 1829): 291–307. On the debate over early baptism, see V. Gourdon, "L'hygiénisme français et les dangers du baptême précoce," in *Baptiser: pratique sacramentelle, pratique sociale, XVIe–XXe siècles* (Saint-Etienne: Université de Saint-Etienne, 2009), 103–123.

70. See Isabelle de Conihout, "La Restauration: contrôle et liberté," in *Histoire de l'édition française*, vol. 2, 534–543. For the detailed debates, see the *Collection relative au Projet de loi sur la police de la presse, proposé le 29 décembre 1826* (Paris: Moutardier, 1827).

71. Less frequent periodicals such as the *Revue encyclopédique* and the *Annales de chimie* also provided regular, though shorter, reports. The only other daily that ran such reports relatively consistently at the time was the *Journal du commerce*, though these were very brief.

72. On the history of the "feuilleton," see the essays in Marie-Françoise Cachin, ed., *Au bonheur du feuilleton: naissance et mutations d'un genre, Etats-Unis, Grande-Bretagne, XVIIIe–XXe siècles* (Paris: Creaphis, 2007); Ingemar Oscarsson, "De supplément indépendant à un rez-de-chaussée sous le filet," *Annales historiques de la Révolution française* 292 (1993): 269–294; and Lise Dumasy-Queffélec, "Le feuilleton," in *La civilisation du journal*, ed. D. Kalifa et al. (Paris: Nouveau Monde, 2011), 925–936.

73. Specimen issue of *Le Temps* published with prospectus, ca. October 1829, 1.

74. There is one exception to this pattern, which in its way proves the rule: *Le Constitutionnel*, the daily with the largest circulation in continental Europe, never reported regularly on the Academy during this period. But it was also the least militant of the opposition journals in the lead-up to the Revolution, and it was the only one with real financial interests at stake, as it may have been the only one that was reliably profitable. See Daniel L. Rader, *The Journalists and the July Revolution in France* (The Hague: Nijhoff, 1973), 19.

75. Toby Appel, *The Cuvier-Geoffroy Debate: French Biology in the Decades Before Darwin* (New York: Oxford University Press, 1987), 175–201.

76. Johann Peter Eckermann, *Gespräche mit Goethe in den letzten Jahren seines Lebens*, vol. 3 (Leipzig: Brockhaus, 1899), 234.

77. Etienne Geoffroy Saint-Hilaire, *Principes de philosophie zoologique* (Paris: Pichon et Didier, 1830), 74–76. Dorinda Outram (*Georges Cuvier*, 133–134) has shown that despite Cuvier's apparent resistance to publicity at the Academy of Sciences, he engaged extensively in writing and speaking for broader publics. In the debate with Geoffroy, he had a mouthpiece in the press through Alfred Donné's reports in the *Journal des débats*.

78. The *Journal de Paris* used the feuilleton for academic reports somewhat earlier, but its reports were shorter and tended not to get first priority.

79. On Roulin, see Marguerite Combes, *Pauvre et aventureuse bourgeoisie; Roulin et ses amis, 1796–1874* (Paris: J. Peyronnet, 1928).

80. Flourens wrote short reports for the short-lived journal *Annales générales des sciences physiques* in 1820.

81. *Annales des sciences naturelles* 18 (1829): 108.

82. There is no official record of this policy being adopted, though Arago took credit for it in 1835, and it was sufficiently long-standing that one reporter could worry about losing this access at the hands of Arago by early 1834 (letter from Saigey to Raspail ca. late February 1834, MNHN/MS2388). Joseph Bertrand, Alexandre Bertrand's son, later stated that Arago instituted the practice after Cuvier passed away. See "Un article anonyme de la Revue des deux mondes," *Revue des deux mondes* 37 (1896): 277–295, on 281.

83. J. N. P. Hachette to Michael Faraday, 30 August 1833, CMF vol. 2, 147.

84. *Annales de chimie* 46 (1831): 97. The *Annales des sciences naturelles* used *Le Globe*'s reports as a source beginning in mid-1829.

85. *Le Temps* was his first source, but he also used *La France nouvelle, Le moniteur du commerce, the Journal de Paris*, and *L'Impartial*. These manuscript copies are preserved among his massive volumes of academic ephemera at the Bibliothèque de l'Institut de France.

86. These were entitled *Extrait du Temps* and featured continuous pagination, thus encouraging their being bound together. Bound copies of these for 1833 and 1834 can be found in the Bibliothèque nationale de France and the Bibliothèque de l'Institut de France. The editor of *Le Temps* began depositing copies of its procès-verbaux at the Academy's meetings in December 1832. See PVAS, 10 Décembre 1832, and letter from Jacques Coste (Director of *Le Temps*) to the Academy, 6 December 1832, in AAS-PS, 17 December 1832.

87. This is the subject, for example, of Mathias Dörries, "The Public Face of Science: François Arago," in *Actes de les V Trobades d'història de la ciència i de la tècnica: Roquetes* (Barcelona: Societat Catalana d'Història de la Ciència i de la Tècnica, 2000), 43–54; Levitt, *The Shadow of Enlightenment*; and Belhoste, "Arago, les journalistes."

88. There is no modern scholarly biography of Arago. But see François Sarda, *Les Arago: François et les autres* (Paris: Tallandier, 2002); Maurice Daumas, *Arago, 1786–1853: la jeunesse de la science* (Paris: Belin, 1987); and Levitt, *The Shadow of Enlightenment*.

89. The incident was widely reported in the press, including in *Le Globe*, 21 July 1830, and in the *Courrier française, Le Temps*, and the *Journal des débats*, all 20 July 1830.

90. Parliamentary speech of Guizot in 1837, quoted by Crăiuțu, *Liberalism under Siege*, 229.

91. See Jeremy D. Popkin's *Press, Revolution, and Social Identities in France, 1830–1835* (University Park: Pennsylvania State University Press, 2002). For specifics on government control of the press during the period, see Irene Collins, *The Government and the Newspaper Press in France, 1814–1881* (London: Oxford University Press, 1959); and Pierre Casselle, "Le régime législatif," *Histoire de l'édition française*, vol. 3, 46–55.

92. Roldán, *Charles de Rémusat*, 166.

93. Tresch, *The Romantic Machine*.

94. Raspail, *Nouveau système de chimie organique* (Paris: J. B. Baillière, 1833), 31.

95. For biographical information on Saigey, see Bernard Bru and Thierry Martin, "Comptes rendus des *Annales des sciences d'observation*," in *Œuvres complètes de A. A. Cournot*, vol. 11 (Paris: Librairie J. Vrin, 2010), 579–590; and "Jacques-Frédéric Saigey," in *Biographie universelle et portative des contemporains* (Paris, 1836), 700–701.

96. Saigey's fullest early account, with plates, of his instrument is "Lois des phénomè-nes attribués au magnétisme en mouvement," *Annales des sciences d'observation* 2 (1829): 1–16.

97. F.-V. Raspail, *Essai de chimie microscopique appliquée à la physiologie* (Paris, 1830). On Raspail's biography and his scientific work, see Dora B. Weiner, *Raspail: Scientist and Reformer* (New York: Columbia University Press, 1968); Jean Saint-Martin, *F.-V. Raspail* (Paris: E. Dentu, F. Aureau, 1877); and Yves Lemoine and Pierre Lenoël, *Les avenues de la République: souvenirs de F.-V. Raspail sur sa vie et sur son siècle, 1794–1878* (Paris: Hachette, 1984).

98. Raspail, *Nouveau système*, 78.

99. Bernard Bru and Thierry Martin, "Présentation: articles du Bulletin de Férussac," in *Œuvres complètes de A.A. Cournot*, vol. 11, 17–31.

100. Raspail received 600 francs per year, with which it would have been difficult to maintain a comfortable existence for his family. See Férussac to F.-V. Raspail, 8 October 1825, MNHN/MS2388.

101. François-Vincent Raspail, *Nouveau système de physiologie végétale et de botanique* (Paris: Baillière, 1837), ix.

102. Raspail detailed several such cases, but the case that seems to have been dear-est to his heart occurred in December 1826, when Adolphe Brongniart, the son of Alexan-dre Brongniart (at the time president of the Academy of Sciences), read a work on "The Generation and Development of an embryo in the phanerogamic vegetables," which was both nominated for a prize and seemed to be based on work that Raspail had read to sev-eral academies and societies. Several letters were presented to the Academy on this contro-versy in late 1826 and early 1827, and Raspail recounted the case, for example, in "Le temps fait justice de tous les torts, même des torts académiques," *Annales des sciences d'observation* 4 (1830): 313–319.

103. F.-V. Raspail, "Fragment d'autobiographie," FVR/MS2745/4.

104. "Examen critique et comparatif de trois rapports faits à l'Académie des sci-ences . . . ," *Annales des sciences d'observation* 1 (1829): 230–270, on 230–231.

105. Raspail, "IIe lettre à un savant de province," *Annales des sciences d'observation* 3 (1830): 469–475; and idem, *Nouveau système de chimie organique*, 2nd ed. (Paris: J. B. Baillière, 1838), L–LIX.

106. Raspail, "IIe lettre à un savant de province," 471.

107. Charles Didier, "L'institut," in *Nouveau tableau de Paris au XIXe siècle* (Paris, 1835), 335–383, on 362; and Bertrand in *Globe* 3 (10 January 1826): 55.

108. "Séance publique annuelle," *Annales des sciences d'observation* 2 (1829): 317.

109. Raspail, "Coteries scientifiques," *Annales des sciences d'observation* 3 (1830): 151–159, on 155.

110. Ibid., 156.

111. Ibid., 153.

112. Dorinda Outram has carried out a detailed study of scientific patronage in Paris in the case of Cuvier, in *Georges Cuvier*. See especially pp. 189–202.

113. Raspail, "Examen critique et comparatif de trois rapports faits à l'Académie des sci-

ences, sur les recherches relatives à la génération chez les végétaux et aux prétendus animalcules spermatiques du pollen, et des diverses circonstances qui en ont précédé ou suivi la
lecture," *Annales des sciences d'observation* 1 (1829): 230–270, on 262.

114. Raspail, "Examen critique," 252. Raspail was referring here to his priority dispute
in 1827 over the structure of organic tissues with Adolphe Brongniart, the son of Alexandre
Brongniart, then president of the Academy.

115. François-Vincent Raspail, *Nouveaux coups de fouet scientifiques* (Paris: Meilhac,
1831), 5. Indeed, academicians continued to use these reports as part of the performative
apparatus in scientific disputes, both within France and abroad. See for example Mi Gyung
Kim, "Constructing Symbolic Spaces: Chemical Molecules in the Académie des Sciences,"
Ambix 43 (1996): 1–31.

116. F. V. Raspail, "Fragment d'autobiographie," FVR/MS2745/4.

117. See Saigey's feuilleton for 30 September, 21 October, and 12 December 1833 and
30 January 1834 for examples. Saigey later made a chronicle of his outrages against the
Academy that appeared in *Le National*. See *Réformateur*, 8 April 1835.

118. Letter from Jacques-Frédéric Saigey to Raspail, [2 December 1833], MNHN/
MS2388.

119. "Prospectus: nouvelle série," *Gazette médicale de Paris* (January 1833): vii.

120. Raspail, *Lettres sur les prisons de Paris* (Paris, 1839), 327.

121. Saigey to Raspail, [2 December 1833], MNHN/MS2388.

122. Saigey to Raspail, ca. late February 1834, MNHN/MS2388.

123. Many of Raspail's *Le Réformateur* articles on social reform and political economy
were collected by him in *Réformes sociales* (Paris: Bureau des publications de M. Raspail,
1872). For a very useful exploration of Raspail's use of his cellular theory in the development of his political views in *Le Réformateur*, see Ludovic Frobert, "Théorie cellulaire, science économique et République dans l'œuvre de François-Vincent Raspail autour de 1830,"
Revue d'histoire des sciences 64 (2011): 27–58. Raspail's would not by any means be the last
cell theory of political bodies. See Paul Weindling, "Theories of the Cell State in Imperial
Germany," in *Biology, Medicine, and Society, 1840–1940*, ed. C. Webster (Cambridge: Cambridge University Press, 1981), 99–156.

124. "A quoi se réduisent les principes républicains?," *Réformateur*, 18 October 1834.

125. "Organiser l'association de la commune," *Réformateur*, 12 March 1835 (reprinted
in *Réformes sociales*, 180–184).

126. Raspail, *Réformateur*, 25 December 1831 (reprinted in *Réformes sociales*, 229–
232).

127. Raspail, "Organiser l'association de la commune."

128. *Réformateur*, 21 March 1835, reprinted in *Réformes sociales*, 194.

129. Raspail, *Nouveau système de chimie organique*, 2nd ed., xxix, xlviii.

130. On Arago's ambiguous republican credentials during the 1830s, see Jean Sagnes,
"François Arago était-il Républicain?," in *Les Arago acteurs de leur temps* (Perpignan:
Archives départementales des Pyrénées-Orientales, 2009).

131. Raspail, "Petit coup d'état de l'Académie des Sciences," *Réformateur*, 25 March
1835.

132. "Bulletin scientifique," *Réformateur*, 14 October 1834.

133. PVAS, 17 June 1833, 287.

134. Chevreul's report (sixty-eight pages) was printed in the *Nouvelles annales du Muséum national d'histoire naturelle* 3 (1834): 239–306. The discussion involving Chevreul and Biot at the Academy, on 7 July, was reported in *Journal des débats* and *Le National*, on 11 July 1834. Direct quotations are from *Le National*, which are consistent with the account in the *Journal*.

135. "Discussions de la Semaine," *Écho du monde savant*, 11 July 1834, 57.

136. "Bulletin scientifique," *Réformateur*, 14 October 1834.

137. "Coteries scientifiques," 153.

138. "Bulletin Scientifique du Réformateur," 24 November 1834.

139. The incident was reported in detail in *Le National* and in *L'Écho du monde savant* on 11 July 1834 as well as in the *Journal des débats* on the same date.

140. "Bulletin scientifique et industrielle," *Réformateur*, 18 February 1835.

141. *Charivari*, 2 October (Académicien) and 9 October 1834 (Plagiare). The Métiers de Paris series of lithographs was the work of the caricaturist Edme-Jean Pigal.

142. *Réformateur*, 3 February 1835.

143. See the feuilleton in *Le Réformateur* from 25 February to 18 March, and in the *National* from 23 February to 25 March.

144. "Feuilleton académique," *Tribune*, 26 March 1835.

145. Arago to Raspail (circular), 21 March 1835, MN11N/MS2388.

146. Arago to Raspail, 23 March 1835, MNHN/MS2388.

147. François-Vincent Raspail, "Petit coup d'état à l'Académie des sciences," *Réformateur*, 25 March 1835, 2.

148. "L'Académie des sciences et les journaux," *Gazette des hôpitaux* 9 (26 March 1835), 1.

149. Raspail, "Petit coup d'état."

150. On 23 March 1835, Arago presented his plan to the Academy, outlining "les divers motifs qui . . . doivent faire désirer que l'Académie publie chaque semaine, tous les samedis par exemple, un compte détaillé et fidèle de ses Séances hebdomadaires. Il annonce que M. Flourens et lui se chargeront volontiers de cette publication, si l'on croit qu'elle puisse contribuer au progrès des sciences." PVAS, 23 March 1835, 678.

151. Of course, British societies on the Royal Society template did not usually allow discussion at all. The exception was the Geological Society, though it was understood that these were not to be reported on. See John C. Thackray's "Introduction," in *To See the Fellows Fight: Eye Witness Accounts of Meetings of the Geological Society of London and its Club, 1822–1868* (Faringdon: British Society for the History of Science, 2003).

152. "Projet de règlement concernant les Comptes-rendus des séances de l'Académie proposé en comité secret le 13 juillet 1835," *PVAS*, 13 July 1835, 756.

153. Raspail, *Nouveau système de chimie organique* (1838), xxxvi–xxxvii.

154. Ibid., xxxvii. Though Raspail admitted that the *Comptes rendus* had succeeded in rendering its competition relatively docile, he attributed this to the new censorship laws and to the increasingly heavy financial burdens that the state laid on the commercial press

following the political crisis in 1835. (For example, far from having to pay for the right to publish, as did most journals, the state ultimately funded the *Comptes rendus* itself.)

155. The security deposit on Paris journals was raised to 100,000 francs, for example. On the September Laws and their effects, see Charles Ledré, "La presse nationale sous la restauration et la monarchie de juillet," in Claude Bellanger and Jacques Godechot, eds., *Histoire générale de la presse française*, vol. 2 (Paris: Presses universitaires de France, 1969), 111–114.

156. The Société's decision is discussed in Marcellin Berthelot, "Notice sur les origines et sur l'histoire de la Société philomathique," in *Mémoires publiés par la Société philomathique à l'occasion du centenaire de sa fondation, 1788–1888* (Paris, 1888), i–xvii.

157. *Cours de philosophie positive*, vol. 6 (1842).

158. On the changes in French feuilletons after 1836 and the rapid growth in popularity of serialized novels in France, see Maria Adamowicz-Hariasz, "From Opinion to Information: The Roman-Feuilleton and the Transformation of the Nineteenth-Century French Press," in *Making the News: Modernity and the Mass Press in Nineteenth-Century France*, ed. Dean de la Motte and Jeannene M. Przyblyski (Cambridge, MA: MIT Press, 1999), 160–184.

159. The first issue of the Prussian Academy's *Bericht über die zur Bekanntmachung geeigneten Verhandlungen* appeared in January 1836, whereas the *Bulletin Scientifique* of the Imperial Academy of Sciences of Saint-Petersburg appeared in April 1836.

160. Cauchy's publishing behavior is discussed in Bruno Belhoste, *Augustin-Louis Cauchy: A Biography* (New York: Springer, 1991), 191–194.

161. Biot, [Review of the *Comptes rendus hebdomadaires*], *Journal des savants*, November 1842, 641–661, on 642.

162. Biot, Review of *Comptes rendus*, 654.

163. Ibid., 660–661. See Chapter 1 for the struggles of the Academy of Sciences as it faced increasing competition from journals and scientific societies.

164. Biot, Review of *Comptes rendus*, 642.

165. Biot, [Sur le discours prononcé à la réunion anniversaires de la Société royale de Londres, le 30 novembre 1836], *Journal des savants*, February 1837, 74–84, on 82. Another subtle change was that the manuscript records of meetings, which usually contained other kinds of information about what occurred besides the contents of papers read, were gradually superseded and even replaced by the printed proceedings.

166. Biot, Review of *Comptes rendus*, 651.

167. Ibid., 655.

168. Ibid., 661. Before this period, the French word *article* was most commonly used to refer to the sections of a document (such as a contract, set of statutes, or even a scientific memoir). By the early nineteenth century, it was by extension used to refer to the fragments of writing making up a political journal. Savants did not normally use the word interchangeably with *mémoire*.

169. Biot, [Sur le discours prononcé], 83.

170. Biot, Review of *Comptes rendus*, 661.

171. Outram, *Georges Cuvier*, 201.

172. "La science et les journaux," *Phalange* 1 (1840): 461–462. (This was a Fourierist journal.)

173. "M. Arago," *Presse*, 10 April 1840, 2.

174. See Chapter 5.

175. "Remarques sur l'institution récente des Comptes rendus hebdomadaires de l'Académie des sciences, et sur la publicité donnée à ses séances," in *Mélanges scientifiques et littéraires* (Paris: Michel Lévy, 1858), 257–292, on 292.

176. For the correlations between the history of scientific publishing and political history during the Revolution and First Empire, see Jean-Luc Chappey, "Enjeux sociaux et politiques de la 'vulgarisation scientifique' en Révolution (1780–1810)," *Annales historiques de la Révolution française* 338 (2004): 11–51.

177. See especially Jonathan Topham's "Rethinking the History of Science Popularization/Popular Science," in *Popularizing Science and Technology in the European Periphery, 1800–2000*, ed. Faidra Papanelopoulou, Agustí Nieto-Galan, and Enrique Perdriguero (Farnham: Ashgate, 2009), 1–20. The notion of the emergence of a "low scientific culture" was developed by Sheets-Pyenson in "Low Scientific Culture in London and Paris, 1820–1875."

178. On the historiographical dangers of the "spatialized public sphere," see Mah, "Phantasies of the Public Sphere." See also Baker, *Inventing the French Revolution*, 167–199.

Chapter 3

1. William St. Clair, *The Reading Nation in the Romantic Period* (Cambridge: Cambridge University Press, 2004), 103–121.

2. For example, Arthur Burns and Joanna Innes, eds., *Rethinking the Age of Reform: Britain, 1780–1850* (Cambridge: Cambridge University Press, 2003); J. C. D. Clark, *English Society, 1660–1832: Religion, Ideology, and Politics during the Ancien Regime* (Cambridge: Cambridge University Press, 2000); and Philip Harling, *The Waning of "Old Corruption": The Politics of Economical Reform in Britain, 1779–1846* (Oxford: Clarendon Press, 1996).

3. On the rising profile of "public opinion" in 1820s Britain, see Jonathan Parry, *The Rise and Fall of Liberal Government in Victorian Britain* (New Haven, CT: Yale University Press, 1993), 26–34; and Peter Jupp, *British Politics on the Eve of Reform: The Duke of Wellington's Administration, 1828–30* (Basingstoke: Macmillan, 1998), 330–386. On the press, see Hannah Barker, *Newspapers, Politics and English Society 1695–1855* (New York: Longman, 2000).

4. Argus, "The Royal Society," *Times*, 30 January 1830.

5. See Roy M. MacLeod, "Whigs and Savants: Reflections on the Reform Movement in the Royal Society, 1830–48," in *Metropolis and Province: Science in British Culture, 1780–1850*, ed. I. Inkster and J. Morrell (London, 1983), 55–90. On the reform movement in British science, see also David Philip Miller, "The Royal Society of London, 1800–1835: A Study in the Cultural Politics of Scientific Organization" (PhD diss., University of Pennsylvania, 1981); Jack Morrell and Arnold Thackray, *Gentlemen of Science: Early Years of the British Association for the Advancement of Science* (Oxford: Clarendon, 1981); and Susan Faye

Cannon, *Science in Culture: The Early Victorian Period* (Kent: Science History Publications, 1978).

6. Gilbert to Peel, quoted in A. C. Todd, *Beyond the Blaze: A Biography of Davies Gilbert* (Truro: D. Bradford Barton, 1967), 237; Herschel to William Fitton, 18 October 1830, RSL-HP/25.1.9, quoted in Miller, "The Royal Society of London," 353.

7. The outlines of this controversy may be found in Marie Boas Hall, *The Library and Archives of the Royal Society* (London: Royal Society, 1992), 23–25.

8. F. R. S., "Royal Society," *Times*, 2 December 1829.

9. James South, *Charges Against the President and Councils of the Royal Society* (London: Fellowes, 1830), 18, 20; and John Herschel, "Sound," *Encyclopaedia Metropolitana*, vol. 4 (London: B. Fellowes, 1830), 810.

10. Charles Babbage, *Reflections on the Decline of Science in England and on Some of Its Causes* (London: B. Fellowes, 1830), 44; and Brewster, Review of *On the Decline of Science in England*. On the science reform movement and professionals in Victorian England, see T. W. Heyck, *The Transformation of Intellectual Life in Victorian England* (London: Croom Helm, 1982); and Jack Morrell, "Individualism and the Structure of British Science in 1830," *Historical Studies in the Physical Sciences* 3 (1971): 183–204.

11. "Report of the Limitations of Admissions Committee," 1827, RSL-DM/1.

12. On these changes, which radically reduced the number of papers published by the Society, see Chapter 1.

13. Babbage's statistics are in *Reflections on the Decline of Science in England*, 154–155.

14. *Science Without a Head; or the Royal Society Dissected* (London, 1830), 34–40. The other categories were Naval Officers (seven papers by twenty-seven Fellows), Army Officers (twenty-eight papers by thirty-nine Fellows), Clergymen (eight papers by seventy-four Fellows), Men of Law (twenty-eight papers by sixty-three Fellows), Physicians (sixty-six papers by seventy-nine Fellows), Surgeons (137 by twenty-one Fellows, most of them by Everard Home), and a miscellaneous category (187 papers by 286 Fellows).

15. *Science Without a Head*, 51.

16. Banks to Charlotte Seymour, Duchess of Somerset, quoted in Harold B. Carter, *Sir Joseph Banks (1743–1820)* (Winchester, 1987), 153.

17. Johns, "The Ambivalence of Authorship."

18. On Somerset's scientific activities, see *Correspondence of Two Brothers: Edward Adolphus, Eleventh Duke of Somerset, and His Brother, Lord Webb Seymour, 1800 to 1819 and After* (London: Longmans, 1906).

19. W. H. Mallock and G. Ramsden, eds., *Letters, Remains, and Memoirs of Edward Adolphus Seymour* (London: R. Bentley, 1893), 268.

20. Lord Webb to Duke of Somerset, 17 January 1825, in *Correspondence of Two Brothers*, 159.

21. *Report from the Select Committee on Medical Education*, part 2 (London: House of Commons, 1834). The *Lancet* reported extensively on these hearings, regularly revisiting the topic of Home's literary wrongdoings in subsequent years. An interesting early attempt to understand Home's actions "according to the lights of his day and his order" appeared in

"The Centenary of the Royal College of Surgeons of England," *Nature* 62 (26 July 1900): 294–296.

22. "Règlement ordonné par le Roy" [1699], *Histoire de l'Académie royale des sciences* 1 (1702): 5.

23. Lecturers needed to have produced two disputation-dissertations, extraordinary professors needed three more of these, and full ordinary professors needed three more. See Chapter 7 of William Clark's *Academic Charisma and the Origins of the Research University* (Chicago: University of Chicago Press, 2006); R. Steven Turner, "The Growth of Professorial Research in Prussia," *Historical Studies in the Physical Sciences* 3 (1971): 137–182; and Peter Josephson, "The Publication Mill: The Beginnings of Publication History as an Academic Merit in German Universities, 1750–1810," in Peter Josephson, Thomas Karlsohn, and Johan Östling, eds., *The Humboldtian Tradition: Origins and Legacies* (Leiden: Brill, 2014), 23–43.

24. Turner, "The Growth of Professorial Research in Prussia," 170.

25. Babbage, *Decline of Science*, 10.

26. Jack Stillinger, *Multiple Authorship and the Myth of Solitary Genius* (New York: Oxford University Press, 1991); and Martha Woodmansee, "The Genius and the Copyright: Economic and Legal Conditions of the Emergence of the 'Author,'" *Eighteenth-Century Studies* 17 (1984): 425–448.

27. For an example of the elevation of authorship as a measure of scientific achievement in Bavaria and the consequent marginalization of artisans, see Myles W. Jackson, "Can Artisans Be Scientific Authors? The Unique Case of Fraunhofer's Artisanal Optics and the German Republic of Letters," in *Scientific Authorship*, ed. M. Biagioli and P. Galison (New York: Routledge, 2003), 113–131.

28. Herschel to unknown correspondent, [ca. 1820s], HPT/28/11.

29. Ironically, although Caroline Herschel did publish a few catalogs, this one — though widely known and credited to her — circulated only in manuscript.

30. On the late twentieth-century crisis of scientific authorship, see Mario Biagioli, "Rights or Rewards? Changing Frameworks of Scientific Authorship," in *Scientific Authorship*, 253–279; and Peter Galison, "The Collective Author" (same volume), 325–355.

31. Nicholas Harris Nicolas, *Observations on the State of Historical Literature* (London: W. Pickering, 1830), 199.

32. "Report of the Limitations of Admissions Committee."

33. Granville, *Science Without a Head*, 52, 55, 57.

34. Ibid., 55, 54.

35. See for example Olinthus Gregory, "Vindication of the Attack on Don Joseph Rodriguez' Paper in the Philosophical Transactions," *Annals of Philosophy* 3 (1814): 282–286.

36. Granville, *Science Without a Head*, 54.

37. John Barrow, *Sketches of the Royal Society and Royal Society Club* (London: J. Murray, 1849), 37, 112; and Gregory, "Some Leading Points," 241.

38. See Kevin J. Fraser, "John Hill and the Royal Society in the Eighteenth Century,"

Notes and Records of the Royal Society of London 48 (1994): 43–67. See Chapter 1 on Hill's criticisms.

39. Banks to Matthew Boulton, 25 May 1795, SCJB vol. 4, 366. (Banks was explaining why they would not be printing a paper.) Blagden makes a similar point to Banks on 16 October 1785, SCJB vol. 3, 104.

40. See for example "Preface," *Memoirs of the Literary and Philosophical Society of Manchester* 1 (1789): vii–viii; and "Laws of the Society," Transactions of the Royal Society of Edinburgh 1 (1788): 11–15, on 14–15.

41. *The Statutes of the Royal Society of London* (London: W. Nichol, 1823), 30.

42. See the "Advertisement" that accompanied the front matter of the *Philosophical Transactions* starting with vol. 47 (1753).

43. The expert commissions of the French Academy of Sciences are discussed in Chapter 1. See also Bret, "La prise de décision académique."

44. "Expert," *Encyclopédie, ou, dictionnaire raisonné des sciences, des arts et des métiers*, vol. 6 (Paris: Briasson, 1856), 301.

45. On the concept of expertise in France, see Olivier Leclerc, *Le juge et l'expert: contribution à l'étude des rapports entre le droit et la science* (Paris: L.G.D.J., 2005), 27–49; and Alvaro Agustín Santana Acuña, "The Making of a National Cadastre (1763–1807): State Uniformization, Nature Valuation, and Organizational Change in France" (PhD diss., Harvard University, 2014).

46. *The Law of Evidence by Lord Chief Baron Gilbert, considerably enlarged by Capel Lofft*, vol. 1 (London: J. F. & C. Rivington, 1791), 301 (also quoted in Golan, 53; see next note). According to George Cornewall Lewis, the word still had a French air to it in mid-century. *An Essay on the Influence of Authority in Matters of Opinion* (London: John W. Parker, 1849), 118.

47. My account is derived largely from Tal Golan, *Laws of Men and Laws of Nature: The History of Scientific Expert Testimony in England and America* (Cambridge, MA: Harvard University Press, 2004). See also Déirdre Dwyer, *The Judicial Assessment of Expert Evidence* (Cambridge: Cambridge University Press, 2008).

48. Young to the Admiralty, 10 January 1829, quoted in Miller, "The Royal Society of London," 324. For more on the Astronomical Society's attempts to compete with these governmental bodies, and their centrality to the declinist movement, see ibid., "The Royal Society of London," 172–218, 313–329; idem, "The Revival of the Physical Sciences in Britain, 1815–1840," *Osiris* 2 (1986): 107–134; and William J. Ashworth, "The Calculating Eye: Baily, Herschel, and the Business of Astronomy," *British Journal for the History of Science* 27 (1994): 409–441.

49. The Astronomical Society instituted its system of committees to inspect manuscripts on 11 May 1821, RAS-CM/1. All decisions about papers were made by the Council itself. For an example of instant approval by Committee, see Council Minutes, 14 December 1821.

50. Leonard Horner to George Greenough, 4 April 1809, GGP/Add.7918/823. Martin Rudwick, *The Great Devonian Controversy: The Shaping of Scientific Knowledge Among Gentlemanly Specialists* (Chicago: University of Chicago Press, 1985), 20. For the early Geologi-

cal Society, see also Roy Porter, "The Industrial Revolution and the Rise of the Science of Geology," in *Changing Perspectives in the History of Science*, ed. M. Teich and R. M. Young (London, 1973), 320–343; and Martin Rudwick, "The Foundation of the Geological Society of London: Its Scheme for Co-Operative Research and Its Struggle for Independence," *British Journal for the History of Science* 1 (1963): 325–355.

51. John Farey, "Observations on the Priority of Mr. Smith's Investigations of Strata of England; on the Very Unhandsome Conduct of Certain Persons in Detracting from his Merit Therein; and the Endeavours of Others to Supplant him in the Sale of his Maps," *Philosophical Magazine* 45 (1815): 333–334; and idem, *General View of the Agriculture of Derbyshire*, vol. 3 (London: Sherwood, Neely and Jones, 1817).

52. M. Kölbl-Ebert, "George Bellas Greenough (1778–1855): A Lawyer in Geologist's Clothes," *Proceedings of the Geologists' Association* 114 (2003): 247–254.

53. Council Minutes, 23 May 1817, GSL-CM1/1. For Greenough's MS draft—and feedback from Charles Stokes, the Society's secretary—see GGP/Add.7918/1621.

54. The Geological Society's rules stated explicitly that referees were to be selected from the Council. This was not part of the Astronomical Society's rules, but implicit evidence suggests that this was usually the case, as (unnamed) referees appear often to be in attendance at Council meetings.

55. Referee reports sometimes refer to this correspondence. An exceptional case proves the rule: Council voted to revise a paper itself for publication, the author's death "having prevented any communication between the referee & the author," GSL-CM1/1, 20 June 1823.

56. Besides these two London societies, the American Philosophical Society also appointed Committees to make recommendations whether individual papers ought to be published, although this appears to have been an entirely oral process carried out at Council meetings. See APS Council Minute Books, 1799–1804, 1815–1824.

57. For a contemporary discussion of the referee in English law, see for example *A Treatise on the Law of Arbitration & Awards* (London, 1825), Chapter 13.

58. John Crisp, *The Conveyancer's Guide* (London: A. Maxwell, 1821), 105.

59. On these and other facets of the British government's turn to publishing in the 1830s, see Oz Frankel, *States of Inquiry: Social Investigations and Print Culture in Nineteenth-Century Britain and the United States* (Baltimore: Johns Hopkins University Press, 2006). See also Edward Higgs, *The Information State in England: The Central Collection of Information on Citizens Since 1500* (Houndmills: Palgrave Macmillan, 2004). On the Public Record Office, see John Cantwell, *The Public Record Office, 1838–1958* (London: HMSO, 1991).

60. Roget to William Swainson, 5 December 1830, quoted in Miller, "Royal Society of London," 383.

61. Lubbock to Babbage, [April 1830], BL-CB/37185/139.

62. Babbage to Roget, draft letter March 1831, BL-CB/37185/501. The final draft was more measured.

63. The new arrangement was worked out by early 1831, and involved the British Library selling off duplicates and transferring the proceeds to the Royal Society for books. For the negotiations and lists of books and periodicals, see RSL-DM/2/70–73. For the

"standard list" of foreign periodicals taken in by the Society, see James Hudson to H. Bail-lière, 21 July 1834, RSL-MS/425/213.

64. Charles Babbage, *The Ninth Bridgewater Treatise: A Fragment* (London: J. Murray, 1837), 53. For more on Babbage's views of the printing trade, see Adrian Johns, "The Identity Engine: Printing and Publishing at the Beginning of the Knowledge Economy," in *The Mindful Hand: Inquiry and Invention from the Late Renaissance to Early Industrialisation*, ed. Lissa Roberts, Simon Schaffer, and Peter Dear (Amsterdam: Koninklijke Nederlandse Akademie van Wetenschappen, 2007), 403–28.

65. Babbage to Lubbock, 22 December 1831, RSL-JWL/3.

66. Ibid.

67. Lubbock to Babbage, 16 January 1832, BL-CB/37186/212; 23 December 1831, BL-CB/37185/388.

68. Babbage to Lubbock, 9 April 1832, RSL-JWL/3. In his letter of 23 December 1831, Lubbock notes that the plan to publish abstracts of papers in the *Philosophical Transactions* since 1800 was "in great measure a child of yours."

69. RSL-CMP/1, 12 June 1834.

70. For their joint work on the theory of tides, see Michael S. Reidy, *Tides of History: Ocean Science and Her Majesty's Navy* (Chicago: University of Chicago Press, 2008).

71. Whewell to Roget, 22 March 1831, RSL-DM/1/30.

72. Whewell to William Vernon Harcourt, 1 September 1831, in Jack Morrell and Arnold Thackray, eds., *Gentlemen of Science: Early Correspondence of the British Association for the Advancement of Science* (London, 1984), 52.

73. Whewell to Roget, 22 March 1831.

74. Whewell to Lubbock, 12 March 1831, RSL-JWL/41.

75. Whewell to Vernon Harcourt, 1 September 1831, 54.

76. The Committee assigned to the paper also included George Peacock, but it appears he dropped out. RSL-CMB/90/C, 43.

77. Lubbock to Whewell, 20 December 1831, WWP/a/216/58.

78. Whewell to Lubbock, 19 December 1831, RSL-JWL/41.

79. Lubbock to Whewell, 9 December 1831.

80. Airy to Whewell, 13 December 1831, WWP/a/216/76.

81. George Airy to William Samuel Stratford, 3 February 1829, RSL-JWL/1. The letter containing Airy's remarks, written to the secretary of the Society, ended up with Lubbock's papers, suggesting that Stratford forwarded it to him for his information. (The Astronomical Society appears to have made no attempt to keep or record reports, or to keep them private. It appears to have not been uncommon for referee reports to be forwarded to authors.)

82. Lubbock to Whewell, 9 December 1831, WWP/a/216/63.

83. Whewell to Lubbock, 5 January 1832, RSL-JWL/41; Lubbock to Whewell, ca. December 1831, WWP/a/216/73.

84. Whewell to Lubbock, 22 December 1831, RSL-JWL/41.

85. Lubbock to Whewell, 24 December 1831, WWP/a/216/59.

86. Lubbock to Whewell, [1831], WWP/a/216/60.

87. Ibid.

88. Lubbock to Whewell, 27 January [1832], WWP/a/216/61.

89. "Abstracts of the Philosophical Transactions," *Mechanics' Magazine* 21 (1834): 204–207, on 205.

90. "Abstracts of the Philosophical Transactions," *Mechanics' Magazine* 28 (1838): 246–249, on 246. On Raspail's interpretation of the *Comptes rendus*, see Chapter 2. On Cobbett's critique, see Frankel, *States of Inquiry*, 43–44.

91. On Whewell's metascientific pursuits, see Richard Yeo, *Defining Science: William Whewell, Natural Knowledge, and Public Debate in Early Victorian Britain* (Cambridge: Cambridge University Press, 1993); and Henry Cowles, "The Age of Methods: William Whewell, Charles Peirce, and Scientific Kinds," *Isis* 107 (2016): 722–737.

92. Duke of Sussex, "Anniversary Address," *Abstracts of the Papers Printed in the Philosophical Transactions* 3 (1832): 140–155.

93. Roget to Forbes, 22 January 1836, JFOR/1/680.

94. A. B. Granville, *The Royal Society in the XIXth Century* (London, 1836), 127–129.

95. The Society faced embarrassment when one referee lost several papers in 1855, and their author, Henry Foster Baxter, came looking for them. After a search, a letter from the referee "apologized that in moving house he had mislaid these papers." Council Minutes, 20 December 1855, RSL-CMP/2.

96. See Zoological Society Publication Committee Minutes, 21 January 1833, and Royal Society of Edinburgh Council Minutes, NLS-RSE/17, 2 December 1833.

97. James David Forbes to Whewell, 21 September 1836, WWP/a/204/29.

98. Whewell to Forbes, 25 September 1836, JFOR/1/726.

99. A debate about the shifting duties of rapporteurs occurred at the Academy of Sciences on 7 July 1834. See the reports in *Journal des débats* and *National* on 11 July 1834.

100. Marshall Hall, "A Note Addressed to the Royal Society," June 1837, RSL-MC/2.

101. On anonymity in the mid-Victorian press, see Elaine Hadley, *Living Liberalism: Practical Citizenship in Mid-Victorian Britain* (Chicago, 2010); Dallas Liddle, "Salesmen, Sportsmen, Mentors: Anonymity and Mid-Victorian Theories of Journalism," *Victorian Studies* 41 (1997): 31–68; and Oscar Maurer, "Anonymity vs. Signature in Victorian Reviewing," *Studies in English* 27 (1948): 1–27.

102. See Gilles Feyel, "La querelle de l'anonymat des journalistes, entre 1836 et 1850," in *Figures de l'anonymat: médias et sociétés*, ed. Frédéric Lambert (Paris, 2001), 27–55; and Michael Palmer, "Londres, Washington, Paris, 1830–1880," also in *Figures de l'anonymat*, 57–81.

103. [Edward Bulwer-Lytton], "To Our Friends: On Preserving the Anonymous in Periodicals," *New Monthly Magazine* 35 (November 1832): 385–389; and [S. C. Hall], "On the Anonymous in Periodicals," *New Monthly Magazine* 39 (September 1833): 2–6, on 5.

104. Stokes to J. S. Bowerbank, 28 March 1863, RSL-MS/426. Bowerbank's threat to publish the reports survives in the G. G. Stokes papers at Cambridge University Library, Add.7656/RS/374.

105. "Review: Thomas Tate, *A Treatise on Factorial Analysis. . . ,*" *Mechanics' Magazine* 43 (1845): 165–169.

106. "The Royal Society of London," *Wade's London Review* 1 (1845): 565–586, on 583.

107. An accusation leveled in Marshall Hall, "A Note Addressed to the Members of the Council of the Royal Society," pamphlet dated 7 June 1837, RSL-MC/2/255. Another early public critique of the injustice of the Royal Society's referee system can be found in [William Robert Grove], "Physical Science in England," *Blackwood's Magazine* 54 (1843): 514–525, on 518.

108. "The Royal Society of London," 584.

109. J. J. Sylvester, Report on a paper by William Spottiswoode, 26 April 1855, RSL-RR/2/226.

110. "Mazzini and the Ethics of Politicians," *Westminster Review* 42 (1844): 251. On the Mazzini Affair, see F. B. Smith, "British Post Office Espionage, 1844," *Historical Studies* 14 (1970): 189–203; and Bernard Porter, *Plots and Paranoia: A History of Political Espionage in Britain, 1790–1988* (London: Unwin Hyman, 1989). On its role in the changing culture of secrecy in British politics, see David Vincent, *The Culture of Secrecy in Britain, 1832–1998* (Oxford, 1998), 1–9, 26–50.

111. "The Secret Office, at the General Post-Office," *Illustrated London News* 4 (29 June 1844): 409; "The Secret Chamber in the General Post-Office," *London Journal* 1 (15 March 1845); and *Punch* 7 (1844), quotation on 118, numerous other articles and illustrations throughout the volume. G. W. M. Reynolds, *The Mysteries of London*, vol. 1 (London, 1845), 221.

112. Jeremy Bentham, "Of Publicity," in *The Works of Jeremy Bentham*, part 8 (Edinburgh, 1839), 310–317.

113. Reynolds, *Mysteries of London*, vol. 1, 148. On Reynolds, Dickens, and paperwork, see Richard Maxwell, *The Mysteries of Paris and London* (Charlottesville: University Press of Virginia, 1992), 160–190. On the aftermath of the Mazzini Affair, see Vincent, *Culture of Secrecy*, 26–50.

114. This scandal again involved Marshall Hall. For an overview, see Diana Manuel, *Marshall Hall (1790–1857): Science and Medicine in Early Victorian Society* (Amsterdam: Rodopi, 1996), 156–232.

115. Letter from Huxley to Eliza Huxley, 5 March 1852, published in Leonard Huxley, *The Life and Letters of Thomas Henry Huxley*, vol. 1 (London: Macmillan, 1900), 97.

116. Tony Crilly, "The Cambridge Mathematical Journal and Its Descendants," *Historia Mathematica* 31 (2004): 455–497.

Chapter 4

1. De Morgan, "Invention and Discovery," in *The Supplement to the Penny Cyclopædia* (London: Charles Knight, 1846), 98–101.

2. Ibid., 99, 101.

3. See, for example, Simon Schaffer, "Scientific Discoveries and the End of Natural Philosophy," *Social Studies of Science* 16 (1986): 387–420; Michael D. Gordin, "The Textbook Case of a Priority Dispute: D. I. Mendeleev, Lothar Meyer, and the Periodic System," in *Nature Engaged: Science in Practice from the Renaissance to the Present*, ed. Mario Biagioli and Jessica Riskin (New York: Palgrave Macmillan, 2012), 59–92; and Carin Berkowitz, *Charles*

Bell and the Anatomy of Reform (Chicago: University of Chicago Press, 2015), 130–164. An important early sociological investigation of discovery is Augustine Brannigan, *The Social Basis of Scientific Discoveries* (Cambridge: Cambridge University Press, 1981); a recent investigation of discovery and invention is Jessica Silbey, *The Eureka Myth: Creators, Innovators, and Everyday Intellectual Property* (Stanford, CA: Stanford University Press, 2015). The priority dispute was also the sociological problem through which Robert K. Merton chose to announce his new empirical program in the historical sociology of science in the late 1950s. See his "Priorities of Scientific Discovery," *American Sociological Review* 22 (1957): 635–659.

4. An early precursor was an elaborate definition of invention by the medical writer Archibald Pitcairn in 1688, "Solutio Problematis de Historicis, seu, Inventoribus," which took a probabilistic approach to solving the problem that the same individual might make multiple contradictory predictions and thus take credit for several later contingencies. It also insisted on public claims rather than private ones, since the former could be kept track of. On Pitcairn, see Stephen M. Stigler, *Statistics on the Table* (Cambridge, MA: Harvard University Press, 1999), 208–212.

5. Hachette to Faraday, 30 August 1833, CMF 2, 147.

6. Cuvier to Adriaan Camper [1798], quoted in Outram, *Georges Cuvier*, 122; Gergonne, "Prospectus," *Annales de mathématiques pures et appliquées* 1 (1810): i–iv; and *Prospectus: Annales de chimie et de physique* (1815), 4.

7. "Report of the Council of the Society to the Eighth Annual General Meeting," *Monthly Notices of the Astronomical Society of London* 1 (1828): 49–56, on 49.

8. William Nicholson, "Preface," *A Journal of Natural Philosophy, Chemistry and the Arts* 1 (1797): iii; "Analysis of Scientific Journals," *Chemist* 1 (1824): 275.

9. David Brewster, "Decisions on Disputed Inventions and Discoveries," *Edinburgh Journal of Science* 2 (1825): 144.

10. Johns attributes this crucial passage to Christopher Wren. But while it does appear in the *Parentalia; or, Memoirs of the family of the Wrens*, 231 (along with many other extracts not written by any member of the Wren family), the original source for it is an anonymous report in the *Philosophical Transactions* 2, no. 28 (21 October 1667): 517–525, on 524. This report was certainly written largely by Oldenburg, for the author refers to himself as the *Publisher* (Oldenburg's name for himself in the pages of the *Transactions*) twice, and it is evident that Oldenburg is exercising his prerogative to unravel a priority dispute between French and British claimants regarding progress in transfusion in man and animals. The paper's title is "An Account of More Tryals of Transfusion, Accompanied with Some Considerations Thereon, Chiefly in Reference to Its Circumspect Practise on Man; Together with a Farther Vindication of This Invention from Usurpers."

11. Wren, op. cit., 128–129.

12. On this wider sense of "authorship" as the subject of discovery, see Schaffer, "Scientific Discoveries."

13. Justus von Liebig, "Bemerkungen der Redaction zur vorhergehenden Abhandlung," *Annalen der Pharmacie* 12 (1834): 50–54, on 50–51. See Jakob Volhard, *Justus Von Liebig*, vol. 1 (Leipzig: J. A. Barth, 1909), 340–342. George Bentham later made the same com-

plaint about the *Annales des sciences naturelles* tending to be backdated; see *Address of the President, Linnean Society* (London: Taylor & Francis, 1867), 8.

14. Michael Faraday, *Experimental Researches in Electricity* (London: Taylor, 1839), iv.

15. See Chapter 2 for newspaper reporting at the Academy.

16. Forbes to Faraday, 18 April 1832, CMF 2, 39. See Brian Gee, "Faraday's Plight and the Origins of the Magneto-Electric Spark," *Nuncius* 5 (1990): 43–69. Faraday's own account of the incident is in the *Philosophical Magazine* 11 (1832): 401–413.

17. Hachette to Faraday, 30 August 1833, CMF, vol. 2, 147.

18. William Whewell, *History of the Inductive Sciences*, vol. 3 (London: J. W. Parker, 1837), 86.

19. Berzelius's target was the French chemist Jean-Baptiste Dumas. See Jenny Beckman, "The Publication Strategies of Jöns Jacob Berzelius (1779–1848): Negotiating National and Linguistic Boundaries in Chemistry," *Annals of Science* 73 (2016): 195–207, on 206.

20. Arago made the comparison with Watt in his *Éloge historique de James Watt* (Paris: Didot, 1839), 45–46.

21. It is notable that Robert K. Merton's famous hypothesis that priority disputes force scientists into the ambivalent position of having to advocate for their property while denying that they care rested on modern historical examples that were almost exclusively British. See Merton, "Priorities in Scientific Discovery," *American Sociological Review* 22 (1957): 635–659. Lynn Nyhart has written an excellent essay tracking a historical shift in a mid-nineteenth-century community of Germany physiologists with respect to cultures of priority: "Writing Zoologically: The *Zeitschrift für wissenschafliche Zoologie* and the Zoological Community in Late Nineteenth-Century Germany," in *Literary Structure of Scientific Argument*, 43–71.

22. See Babbage, *Reflections on the Decline of Science*, 131–132; and "Minutes of British Association Committee of Recommendations," 15 August 1835, OU-BAAS/4.

23. Yates explained his reasons for resigning in a letter to Nasmyth on 12 August 1841, TFP/Box "City Life — Societies, 1808–1933."

24. The British Association published an "Addendum to the Report of the Transactions of the Sections in 1839" containing correspondence and reports in this controversy. See *Report of the Eleventh Meeting of the British Association for . . . 1841* (1842): 1–23. My account is also based on "Volume of minutes of Council meetings, 1832–42," OU-BAAS/17.

25. Council Minutes, 3 May 1827, RSL-CMO/10. In actuality, the printing of reception dates did not happen until 1833.

26. The Geological Society of London passed a regulation that date of last revision always be printed on 19 November 1834, GSL-CM1/3.

27. The Academy was constantly passing small refinements to the regulations dealing with the dating and ordering of papers. For examples, see the printed *Procès-verbaux* on the dates 1 pluviôse an V [20 January 1797], 1 vendémiaire an VII [22 September 1798], 6 fructidor an VIII [24 August 1800], and 11 germinal an XIII [1 April 1805].

28. This question had arisen in 1814 regarding a memoir of Charles Mirbel that had been read in 1810, revised for the next few years, but published only recently. The Academy

decided that "since the volume bore the date, it would be impossible for anyone to be mistaken about the time of the observations in question." PVAS 5, 22 August 1814, 390–391.

29. These figures are based on numbers derived from the *Procès-verbaux* of the Academy between 1795 and 1835, and the *Comptes rendus* from 1835 to 1850. For another analysis beginning in the eighteenth century, see Pierre Berthon, "Les plis cachetés de l'Académie des sciences," *Revue d'histoire des sciences* 39 (1986): 71–78.

30. By mid-century the Society had accumulated only about nine of them. RSL-CMP/2, 20 January 1848 and 18 May 1854. British scientific authors looking to make use of such a service occasionally sent their notes to Paris.

31. Brewster, "Decisions on Disputed Inventions and Discoveries," 143; Arago, *Éloge historique de James Watt*, 17; and De Morgan, "Invention and Discovery," 99.

32. See Michael Shortland and Richard Yeo's "Introduction," in *Telling Lives in Science: Essays on Scientific Biography*, ed. Michael Shortland and Richard Yeo (Cambridge: Cambridge University Press, 1996), 1–42; and Rebekah Higgitt, *Recreating Newton: Newtonian Biography and the Making of Nineteenth-Century History of Science* (London: Pickering & Chatto, 2007). On the relationship between discovery narratives and the rise of biography, see Schaffer, "Scientific Discoveries," 408–413. For the case of invention and the apotheosis of Watt, see Christine MacLeod, *Heroes of Invention: Technology, Liberalism and British Identity, 1750–1914* (Cambridge: Cambridge University Press, 2007).

33. James Patrick Muirhead, "Introductory Remarks," in *Correspondence of the Late James Watt on his Discovery of the Theory of the Composition of Water* (London: J. Murray, 1846), cxxi–cxxii.

34. "La science et les journaux," *Phalange* 1 (1840): 461–462; and "M. Arago," *Presse*, 10 April 1840, 2.

35. The correspondence between Libri and Pontécoulant during this period, frequently involving matters relating "M. Ar." and the Academy, is partially preserved in BNF-GL/3273/501-554.

36. A thorough modern biography of Libri is P. Alessandra Maccioni Ruju and Marco Mostert, *The Life and Times of Guglielmo Libri (1802–1869): Scientist, Patriot, Scholar, Journalist and Thief* (Hilversum: Verloren, 1995). Earlier, less well-documented accounts include Giuseppe Fumagalli, *Guglielmo Libri* (Firenze: L. S. Olschki, 1963); and A. N. L. Munby, *The Earl and the Thief* (Cambridge, MA: Houghton Library, 1968). There are countless short articles about the scandal of his thefts.

37. The left-wing press reported glowingly and extensively on his exploits. See Maccioni Ruju and Mostert, *Life and Times*, 69–70, passim.

38. Several disputes are recorded in the *Archives parlementaires*, including 30 May 1833 on the budget of the Ministry of Public Instruction (*Archives parlementaires* ser. 2, v. 84, 399–403); a new restrictive law on associations in March 1834 (*Archives parlementaires* ser. 2, vol. 87, 710–716); and on Paris fortifications (this took place over several years). On 14 June 1834, the widely read daily *Le courrier français* devoted an article to the rivalry between Guizot and Arago.

39. This debate took place after Double's report of Civiale's memoir on 5 October 1835. This gloss is based on the report of it that appeared in *L'Impartial*, 8 October 1835, 1.

40. The specific catalyst of their violent split is not known, but it was perhaps inevitable; not only did their scientific politics diverge sharply, but Libri had become increasingly intimate with the centers of power of the bourgeois monarchy, while Arago took up an oppositional stance.

41. Libri, "Lettres à un Américain sur l'état des sciences en France. 1. L'Institut," *Revue des deux mondes* 21 (1840): 789–818, on 795–796.

42. Ibid., 795, 792.

43. On Liouville, see Jesper Lützen, *Joseph Liouville, 1809–1882, Master of Pure and Applied Mathematics* (New York: Springer-Verlag, 1990); on his struggles with Libri, see 52–64, 84–91; and specifically on this controversy, Caroline Ehrhardt, "A Quarrel between Joseph Liouville and Guillaume Libri at the French Academy of Sciences in the Middle of the Nineteenth Century," *Historia Mathematica* 38 (2011): 389–414.

44. This controversy played out in volume 17 of the *Comptes rendus*. See Liouville, "Rapport sur un Mémoire de M. Hermite; relatif à la division des fonctions abéliennes ou ultra-elliptiques" (14 August 1843): 292–296 (including discussion between Liouville and Libri); continued discussion of Liouville and Libri (21 August 1843): 327–335; "Réponse de M . Libri à la Note insérée par M. Liouville, dans le Compte rendu de la séance du 21 août" (4 September 1843): 431–445 (Liouville's response, 445–449); and Libri, "Réponse de M. Libri à la Note insérée par M. Liouville, dans le Compte rendu de la séance du 4 septembre dernier" (18 September 1843), 546–555.

45. Libri, "Réponse" (4 September 1843), 432.

46. Ibid., 432.

47. One of Liouville's implicit accusations was that most of Libri's purported mathematical discoveries were plagiarized from results that he had chanced upon in manuscript sources that he had studied.

48. A version of the *Éloge* was read at the Academy in 1834, but the full history did not appear in print until 1839 as the second part of the *Annuaire pour l'an 1839 présenté au Roi par le Bureau des Longitudes* (Paris: Bachelier, 1839), 255–441. Subsequent page numbers refer to the separately published book, *Éloge historique de James Watt* (Paris: Didot, 1839).

49. The classic rebuttal of Arago's argument for Watt was given as the Presidential Address by the Rev. Vernon Harcourt at the British Association Meeting in Birmingham 1839, which was the same meeting that set off the Owen-Nasmyth dispute (*Report of the Ninth Meeting of the British Association for . . . 1839* [1840]: 3–68). Henry Brougham responded defending Watt in his *Lives of men of letters & science who flourished in the time of George III* (London: Knight, 1845–1846). Harcourt in turn responded to Brougham in an open letter in 1846. The matter by no means ended there. The progress of the "Water Controversy" has been treated extensively by David Philip Miller in *Discovering Water: James Watt, Henry Cavendish, and the Nineteenth Century "Water Controversy"* (Aldershot: Ashgate, 2004).

50. Cavendish was a primary example for Robert Merton's theses on the ambivalence of scientists in the face of their conflicting drives to establish priority and claim modesty. In fact, Cavendish almost certainly had no role in the mistaken date. See Christa Jungnickel

and Russell McCormmach, *Cavendish: The Experimental Life* (Cranbury: Bucknell, 1999), 604.

51. Arago, *Éloge historique de James Watt*, 85.

52. Adrien Huard, ed., *Répertoire de législation et de jurisprudence en matière de brevets d'invention* (Paris: Cosse et Marchal, 1863), 351–359.

53. Arago, *Éloge historique de James Watt*, 49, 50.

54. Ibid., 86.

55. Arago, *Annuaire pour l'an 1839*, 355.

56. On the new centrality of history to political culture during the Restoration, see Ceri Crossley, *French Historians and Romanticism: Thierry, Guizot, the Saint-Simonians, Quinet, Michelet* (London: Routledge, 1993); and Stanley Mellon, *The Political Uses of History: A Study of Historians in the French Restoration* (Stanford, CA: Stanford University Press, 1958).

57. See Charles-Olivier Carbonell, "Guizot, homme d'état, et le mouvement historiographique français du XIXe siècle," in *Actes du Colloque François Guizot, Paris, 22–25 octobre 1974* (Paris: Société de l'Histoire du Protestantisme Français, 1976), 219–237; and Laurent Theis, "Guizot et les institutions de mémoire," in *Les lieux de mémoire*, vol. 2, ed. Pierre Nora (Paris: Gallimard, 1992), 569–592.

58. Maccioni Ruju and Mostert, *Life and Times*, 175–179. On Libri's conflicts with the Ecole des Chartes, see Lara Moore, *Restoring Order: The Ecole des Chartes and the Organization of Archives and Libraries in France, 1820–1870* (Duluth, MN: Litwin Books, 2008).

59. Guillaume Libri, *Histoire des sciences mathématiques en Italie* (Paris: J. Renouard, 1838–1841), 4 vols. Volumes 1 and 2 are dated 1838, and the last two are dated 1841. Libri planned more volumes, but these never appeared.

60. Libri, *Histoire des sciences mathématiques en Italie*, vol. 1, ix; vol. 4, 157.

61. Ibid., 176. On Galileo's publishing practices: "Cette indifférence pour la publication de ses ouvrages, et cette libéralité de communication caractérisent Galilée," 184–185.

62. Ibid., 277–278.

63. Ibid., 183 (note 2).

64. Ibid., 189.

65. Ibid., 160 (note 1), 162–163.

66. Arago, *Analyse de la vie et des travaux de Sir William Herschel* (Paris: Bachelier, 1843), 214. Originally published in the *Annuaire pour l'an 1842* (1843).

67. Ibid., 214.

68. Ibid., 215 (note).

69. Arago's principal argument against Galileo's claim to the discovery of sunspots is given on 215–226. Arago wavers on whether "publication" means "making public via printing" or "making public" more generally. He generally implies that printing is essential, but he occasionally makes exceptions.

70. Ibid., 219–220.

71. Arago, "Sur la prise de possession des découvertes scientifiques," in *Œuvres complètes*, vol. 12 (Paris: Gide, 1859), 60–64, on 62.

72. *Œuvres complètes* 3, 296–297. This is an English translation of Arago's French trans-

lation and abridgment of Targioni Tozzetti's anecdote. The original is in Tozzetti, *Notizie degli aggrandimenti delle scienze fisiche*, vol. 1 (Firenze: Giuseppe Buchard, 1780), 124.

73. Ibid., 297. Italics in the original.

74. Arago, *Analyse de la vie de Sir William Herschel*, 214–215 (note 1).

75. Humboldt to Arago, 23 July 1842, in E. T. Hamy, ed., *Correspondance d'Alexandre de Humboldt avec François Arago (1809–1853)* (Paris: E. Guilmoto, 1907), 237.

76. *Comptes rendus* 17 (14 August 1843): 271–273; (21 August 1843): 350–352; (28 August 1843): 367–370. This account is based on the latter sources as well as two detailed newspaper accounts, one in *Univers* (23 August 1843), and the other in *Courrier français* (16 August 1843).

77. For the narrative of this controversy, see the following in volume 17 of the *Comptes rendus*: "Rapport sur un Mémoire de M. Donné" (with discussion) (25 September 1843): 585–598; discussion of a letter of Donné (2 October 1843): 686–687; "Réponse de M. Dien à M. Donné" (9 October 1843):768–769; and discussion (16 October 1843): 815–817.

78. Libri, *Comptes rendus* 17 (9 October 1843): 768–769. Libri quotes Arago verbatim.

79. *Comptes rendus* 17 (16 October 1843): 775–777. Arago is quoting his biography of *William Herschel*, 216.

80. BNF-GL/22114, ff. 82–101.

81. Ibid., 86v, 93r, 93v.

82. This appeared in a revised version of "Invention and Discovery," in *The English Cyclopaedia*, ed. Charles Knight (London: Bradbury and Evans, 1860), 942–948, on 948.

83. François Arago's description (in French, "au bout de sa plume"), *Comptes rendus* 23 (5 October 1846): 660.

84. U.-J. Le Verrier, "Sur la planète qui produit les anomalies observées dans le mouvement d'Uranus," *Comptes rendus* 23 (31 August 1846): 428–438.

85. The letter from Galle to Le Verrier was read out by Arago at the meeting of the Academy on 5 October 1846, *Comptes rendus* 23 (5 October 1846): 659–660.

86. See Nicholas Kollerstrom, "An Hiatus in History: The British Claim for Neptune's Co-Prediction, 1845–1846: Part 1," *History of Science* 44 (2006): 1–28, for an attempt to use contemporary norms of priority adjudication to decide whether the British deserve credit for Neptune.

87. Robert W. Smith, "The Cambridge Network in Action: The Discovery of Neptune," *Isis* 80 (1989): 395–422. James Secord is exploring the complex role of print media—and of newspapers in particular—in establishing the character of the Neptune discovery claim. See "A Planet in Print: Rethinking the Discovery of Neptune," talk at STS Workshop, Department of History and Philosophy of Science, University of Cambridge, 22 February 2007.

88. See John Herschel, "Le Verrier's Planet," *Athenaeum* (3 October 1846): 1019.

89. "The New Planet," *Civil Engineer and Architect's Journal* 9 (1846): 331–332, on 331.

90. Ibid., 332.

91. George Airy, "Account of some circumstances historically connected with the discovery of the planet exterior to Uranus," *Memoirs of the Royal Astronomical Society* 16 (1847): 385–414.

92. Herschel, "Le Verrier's Planet," 1019.

93. Arago, "Examen des remarques critiques et des questions de priorité que la découverte de M. Le Verrier a soulevées," *Comptes rendus* 23 (19 October 1846): 741–755, on 751.

94. Noted in Herschel's diary, 25 October 1846, and quoted in Smith, "The Cambridge Network in Action," 415.

95. On the aftermath, see Smith, "The Cambridge Network in Action," 396.

96. Babbage's papers include a timeline with copious verbatim citations in a hand that is not Babbage's, and another brief timeline in Babbage's hand, BL-CB/37203/11–33.

97. Charles Babbage, "The Planet Neptune and the Royal Astronomical Society's Medal," *Times*, 15 March 1847, 5. Letter to François Arago, 17 March 1847, BL-CB/37193.

98. Richard Sheepshanks letter to Augustus De Morgan, 22 March 1847, RAS-ADM/1; Richard Sheepshanks, *A Reply to Mr. Babbage's Letter to "The Times"* (London: G. Barclay, 1847), 10.

99. Ibid., 10.

100. Some brought up a motto that had previously been cited in such matters known as Waring's Rule. Edward Waring, Lucasian Professor at Cambridge of the late eighteenth century, had mentioned a priority principle while writing about Newton's fluxions. It said that the inventor was he "qui primus evulgaverit, vel saltem cum amicis communicaverit" (who first made it public, or at least communicated it to friends) (*Meditationes Analyticae* [Cambridge: Cambridge University Press, 1785], ii–iii). Most of those who cited it agreed that the reference to "friends" made this rule distinct from the Arago/Babbage claim, and that even if relevant it was certainly no law.

101. Airy, "The Disputed Medal of the Astronomical Society," *Athenaeum* (20 March 1847): 309.

102. Herschel to W. H. Fitton, 20 February 1847, RSL-HP/25.7.5.

103. Herschel to Richard Jones, [January 1847], RSL-HP/22.295; Herschel to Sheepshanks, 3 December 1846, RSL-HP/16.48; Herschel to Fitton, 20 February 1847. Herschel's enduring *public* narrative of the Neptune discovery appeared in his *Outlines of Astronomy* (1849), where he avoided any mention of the dispute, and spoke only of simultaneous discovery.

104. David Brewster, "Researches Respecting the New Planet Neptune," *North British Review* 7 (1847): 207–246.

105. The author with whom Brewster engaged with directly on this point was actually Jean-Baptiste Biot, who had recently written his own history of the new planet and repeated the principle of priority of his erstwhile enemy, Arago. "Sur la planète nouvelle découverte . . ." *Journal des savants* (February 1847): 66–86, on 84.

106. Brewster, "Researches Respecting Neptune," 237.

107. On Brewster and patent reform, see Adrian Johns, *Piracy*, 250–258.

108. Brewster, "Researches Respecting Neptune," 238. Brewster cited on this point the article on patents in the *Edinburgh Encyclopaedia* (which he commissioned his friend to write). The truth of the matter was rather murkier. For several different opinions on the case law in this regard, see the various opinions discussed in the *Report from the Select Committee*

on the Law Relative to Patents for Inventions, etc. (London: House of Commons, 1829), 9–13, 19–20, 70, 74–75.

109. H. I. Dutton, *The Patent System and Inventive Activity During the Industrial Revolution, 1750–1852* (Manchester: Manchester University Press, 1984); Christine MacLeod, "The Paradoxes of Patenting: Invention and Its Diffusion in 18th & 19th-Century Britain, France, & North America," *Technology and Culture* 32, no. 4 (1991): 885–910; and Christine MacLeod, *Inventing the Industrial Revolution: The English Patent System, 1660–1800* (Cambridge: Cambridge University Press, 1988).

110. Gabriel Galvez-Behar, *La République des inventeurs: propriété et organisation de l'innovation en France, 1791–1922* (Rennes: Presses universitaires de Rennes, 2008); Christiane Demeulenaere-Douyère, "Inventeurs en Révolution: la Société des inventions et découvertes," *Documents pour l'histoire des techniques* 17 (2009): 19–45.

111. Mario Biagioli, "Patent Republic: Representing Inventions, Constructing Rights and Authors," *Social Research* 73 (2006): 1129–1172, on 1143. This suggestion has been elaborated upon in a book-length study, Alain Pottage and Brad Sherman, *Figures of Invention: A History of Modern Patent Law* (Oxford: Oxford University Press, 2010). See also Mario Biagioli, "Patent Specification and Political Representation: How Patents Became Rights," in Mario Biagioli, Peter Jaszi and Martha Woodmansee, eds., *Making and Unmaking Intellectual Property: Creative Production in Legal and Cultural Perspective* (Chicago: University of Chicago Press, 2011), 25–39.

112. See testimony of expert witnesses in the *Report from the Select Committee on the Law relative to Patents for Inventions, etc.* (London: House of Commons, 1829). Testimony of John Farey, 36; and testimony of Mark Isambard Brunel, 40.

113. Ibid., testimony of Francis Abbott, 59; and testimony of Samuel Clegg, 97.

114. Babbage, *Reflections on the Decline of Science in England*, 131–132.

115. A general outline of Brewster's early position on patent reform may be found in his review of Babbage's "Decline of Science in England," in *Quarterly Review* 43 (1830): 305–342. A more detailed, earlier description and critique of the patent system appeared in Brewster's *Edinburgh Encyclopaedia*, vol. 16, written by Brewster's friend and lawyer James Simpson under Brewster's guidance. This was more detailed and careful than anything Brewster wrote himself, but he continued to refer to it as an expression of his own views. (See Brewster's testimony on patent legislation on 30 May 1851, in *House of Commons Reports from Committees* [London, 1851], 321.) The best sources for Arago's views are the transcripts of legislative debates preceding the 1844 French Patent Law, printed in Adrien Huard, ed., *Répertoire de législation et de jurisprudence en matière de brevets d'invention* (Paris: Cosse et Marchal, 1863). Arago later collected many of these opinions in an article for his collected works called "Sur les brevets d'invention," in *Œuvres complètes de François Arago*, vol. 6, 677–698.

116. Brewster, "Researches Respecting Neptune," 241.

117. Airy, "The Disputed Medal," 309

118. See for example Stathis Arapostathis and Graeme Gooday, *Patently Contestable: Electrical Technologies and Inventor Identities on Trial in Britain* (Cambridge, MA: MIT Press, 2013); and Iwan Rhys Morus, *Frankenstein's Children: Electricity, Exhibition, and*

Experiment in Early-Nineteenth-Century London (Princeton, NJ: Princeton University Press, 1998).

119. On the Academy's loss of these powers, see Hahn, *Anatomy*, 190–193.

120. Such was the state of patent legislation, that there was sometimes confusion, even among academicians, about whether an academic report on an invention would invalidate a subsequent attempt to seek a patent. The Academy certainly welcomed submission of inventions that an author planned to patent, but when presented with the question in 1827 of whether such a report would endanger a patent, a long conversation led to no definite conclusion. See "Académie des sciences, Séance du 22 janvier 1827," *Globe* 4 (25 January 1827), 374. (By the 1830s they were more likely to admit that such reports would invalidate such claims.)

121. Published posthumously in *Oeuvres complètes de Arago*, vol. 12 (Paris: Gide et J. Baudry, 1859), 60–64.

122. Brewster, "Researches Respecting Neptune," 235–236.

123. Ibid., 236.

124. Ibid., 246.

125. "Invention and Discovery" (1860 version), 948.

126. For example, in 1881 H. Fontaine quoted Arago's publication in claiming priority for applying electricity to the transportation of objects by motor force. See *Congrès international des électriciens, Paris, 1881, Comptes rendus des travaux* (Paris: Masson, 1882), 371. There are several French examples, but it was cited in other countries as well. Alexander von Humboldt cited the principle in his *Kosmos: Entwurf einer physischen Weltbeschreibung* (Stuttgart: Cotta, 1845–1862), and it was invoked in Britain during the controversy over the discovery of argon during the 1890s.

127. For similar reflections regarding the historiography of information, see Brian Larkin in Paul N. Edwards, Lisa Gitelman, Gabrielle Hecht, Adrian Johns, Brian Larkin, and Neil Safier, "AHR Conversation: Historical Perspectives on the Circulation of Information," *American Historical Review* 116 (December 2011): 1393–1435, on 1401.

128. On this point, see Brannigan, *The Social Basis of Scientific Discoveries*. For a fascinating claim about the way in which the advent of electronic journals has led to a shift in conceptions of discovery, see James Evans, "Electronic Journals and the Narrowing of Science and Scholarship," *Science* 321 (2008): 395–399.

Chapter 5

1. On financial difficulties encountered by natural history periodicals during the 1830s, see Sheets-Pyenson, "From the North to Red Lion Court"; and Susan Sheets-Pyenson, "A Measure of Success: The Publication of Natural History Journals in Early Victorian Britain," *Publishing History* 9 (1981): 21–36.

2. Testimony of Richard Taylor in *First Report from the Select Committee on Postage Together with the Minutes of Evidence* (London: House of Commons, 1838), 314–329, on 315.

3. Ibid., 321.

4. Ibid., 321.

5. A spectacular example of such ambiguity of purpose was the anonymous *Vestiges of the Natural History of Creation* (1844). See Secord, *Victorian Sensation*.

6. Touched off by Rowland Hill, *Post Office Reform: Its Importance and Practicability* (1837).

7. From Lardner's testimony in *Select Committee on Postage*, 387.

8. Ibid., 382.

9. Ibid., 385. Lardner was incorrect. A subscription to the *Comptes rendus* was substantially cheaper for those residing in Paris. Where a Parisian could subscribe for twenty francs in 1839, it cost thirty-two francs in the provinces, and forty-four francs abroad.

10. Taylor's testimony in *Select Committee on Postage*, 315, 320.

11. Ibid., 329.

12. Lardner's testimony in *Select Committee on Postage*, 384.

13. On commercial science journals and the taxes on knowledge, see W. H. Brock, "The Development of Commercial Science Journals in Victorian Britain." On the history of the excise duty on paper, see H. Dagnall, "The Taxes on Knowledge: Excise Duty on Paper," *The Library* 20 (1998): 347–363. On their repeal from the point of view of one of the mid-century abolitionists, see Collet Dobson Collet, *History of the Taxes on Knowledge: Their Origin and Repeal* (London: T. F. Unwin, 1899).

14. Rudwick, *The Great Devonian Controversy*, 36.

15. Aileen Fyfe, *Steam-Powered Knowledge: William Chambers and the Business of Publishing, 1820–1860* (Chicago: University of Chicago Press, 2012), 130–132; Simon Eliot, *Some Patterns and Trends in British Publishing, 1800–1919* (London: Bibliographical Society, 1994). For the specific case of serial publication, see Graham Law and Robert L. Patten, "The Serial Revolution," in *The Cambridge History of the Book in Britain, vol. 6: 1830–1914*, ed. David McKitterick (Cambridge: Cambridge University Press, 2009), 144–171; and Linda K. Hughes and Michael Lund, *The Victorian Serial* (Charlottesville: University Press of Virginia, 1991).

16. The recent literature on this topic is vast. The volumes connected with the SciPer project (already cited in the introduction) are a key resource for studies of science in the British periodical press. For a study of the economics of publishing *Chambers' Edinburgh Journal*, see Fyfe, *Steam-Powered Knowledge*.

17. "Full Report of the First Meeting of the Mudfog Association for the Advancement of Everything," *Bentley's Miscellany* 2 (1837): 397–413.

18. For Hooker, see A. J. Meadows, *Science and Controversy: A Biography of Sir Norman Lockyer* (Basingstoke: Macmillan, 2008), 25. For Thomson, see Thomson to Stokes, 23 August 1848; and Stokes to Thomson, 21 August and 23 August 1848, in *The Correspondence Between Sir George Gabriel Stokes and Sir William Thomson, Baron Kelvin of Largs*, vol. 1 (Cambridge: Cambridge University Press, 1990), 50–53.

19. This characterization of the scientific content of *Punch* is due to Richard Noakes, "Punch and Comic Journalism in Mid-Victorian Britain," in *Science in the Nineteenth-Century Periodical*, 91–122.

20. On this point see Pietro Corsi, "What Do You Mean by a Periodical? Forms and Functions," *Notes and Records of the Royal Society* 70 (2016): 325–341.

21. On encyclopedias serving as periodicals, see Corsi, "The Revolutions of Evolution."

22. On publication by parts in British natural history, see Gowan Dawson, "Paleontology in Parts: Richard Owen, William John Broderip, and the Serialization of Science in Early Victorian Britain," *Isis* 103 (2012): 637–667.

23. J. C. L., "Preface," *Magazine of Natural History* 9 (1836): iii.

24. See, for example, E. W. Brayley to Taylor, 4 October 1843, TFP, authors' letters A–C; and Taylor to William Jardine, 5 March 1839, WJP/5/129, where Taylor rejects a paper because "people will not buy a number filled with Latin characters of species."

25. On the spread in the 1860s of popular science periodicals in France, see for example Bernadette Bensaude-Vincent, "Un public pour la science: l'essor de la vulgarisation au XIXe siècle," *Réseaux* 58 (1993): 47–66; idem, ed., "Sciences pour tout," *Romantisme*, no. 65 (1989); Lise Andries, "Vulgarisation scientifique et naissance de la culture générale," in *La civilisation du journal*, ed. D. Kalifa et al. (Paris: Nouveau monde, 2011), 1467–1475; Marie-Laure Aurenche, "La presse de vulgarisation ou la médiation des savoirs," ibid., 383–415; and Robert Fox, *The Savant and the State*, 2012), 184–206. For a comparative survey of commercial journals in London and Paris at mid-century, see Sheets-Pyenson, "Low Scientific Culture in London and Paris, 1820–1875."

26. Marie-Laure Aurenche, *Édouard Charton et l'invention du Magasin pittoresque, 1833–1870* (Paris: H. Champion, 2002); Jean-Pierre Bacot, *La presse illustrée au XIXe siècle: une histoire oubliée* (Limoges: Presses universitaires de Limoges, 2005); and Jean Watelet, "La presse illustrée," in *Histoire de l'édition française*, ed. Henri-Jean Martin and Roger Chartier (Paris: Promodis, 1982–), 328–341.

27. "La mission du vulgarisateur," *L'ami des Sciences*, 6 May 1855, 137. On Meunier, see Catherine Glaser, "Journalisme et critique scientifiques: l'exemple de Victor Meunier," *Romantisme* 65 (1989): 27–36; and Matthias Dörries, "Visions of the Future of Science in Nineteenth-Century France: 1830–1871" (PhD diss., Freie Universität, Berlin, 1997), 166–202.

28. In some cases, support came via guaranteed subscriptions from the state, but direct grants were not out of the question. In 1871, Liouville acknowledged that his journal had been supported by the ministry since the 1830s. See Ernst Neuenschwander, *Die Edition mathematischer Zeitschriften im 19. Jahrhundert und ihr Beitrag zum wissenschaftlichen Austausch zwischen Frankreich und Deutschland* (Göttingen, 1984), 25. On Liouville's journal, see S. Duvina, "Le Journal de Mathématiques pures et appliquées sous la férule de J. Liouville (1836–1874)," *Sciences et techniques en perspective* 28 (1994): 179–217; and Verdier, "Le Journal de Liouville et la presse de son temps."

29. Moigno, "Préface," *Cosmos* 1 (1852): i–iv; Joseph Liouville, "Avertissement," *Journal de mathématiques pures et appliquées* 1 (1836): 1–4.

30. On 27 September 1841, the Academy extended the privilege of receiving twenty-five free separate copies to authors of notes in the *Comptes rendus* (AAS-CA/1, 32) although this may not have been implemented consistently.

31. Several journalists complained bitterly about this change, including Moigno (*Cosmos* 2 [1853]: 581–582), and the young astronomer Léon Foucault, who was working as a scientific journalist for the *Journal des débats*. See *Journal des débats* (29 May 1853): 1.

32. On this transition, see Corinne Saminadayar-Perrin, *Les discours du journal: rhéto-rique et médias au XIXe siècle (1836–1885)* (Saint-Etienne: Publications de l'université de Saint-Etienne, 2007).

33. On the financial difficulties inherent in keeping the *Comptes rendus* going, see the minutes of the *Commission administrative* (AAS-CA), vol. 1 (1829–1877), which dealt with this matter on a regular basis. See also Crosland, *Science under Control*, 279–299.

34. Comité secret, 13 October 1856, AAS-CS/2, 458–462.

35. Victor Meunier, "Une réforme nécessaire," *La Presse*, 30 March 1855.

36. Ibid.

37. Jardine to Taylor, 16 March 1841, TFP, Authors D–K. Jardine makes several criti-cisms along these lines in his letters to Taylor.

38. There are several such advisory letters to Richard Taylor and William Francis in the "Authors' Letters" series (TFP) by men of science such as William Jardine, William Hal-lowes Miller, and many others.

39. Edward Forbes, the secretary of the Geological Society of London, complained to Hugh Strickland that much of his time was taken up with preparing these abstracts; J. D. Forbes to H. E. Strickland, 9 October 1843, HES/E-535.

40. "Report of the Council of the Society to the Eighth Annual General Meeting" (1828), 49; and "Explanatory Notice," *Monthly Notices* 8 (1848), front matter.

41. In May 1847 the Council of the Royal Society passed a motion to have abstracts of papers sent to the printer almost immediately upon receipt so that they were already avail-able in proof at the time a paper was read. RSL-CMP/2, 20 May 1847. Also that May, the Astronomical Society considered means of maintaining the efficiency of its *Monthly Notices*, which meant foregoing the ability of Council to vet its contents. See Richard Sheepshanks, "Report to Council on Monthly Notices," 14 May 1847, RAS-CM/5.

42. Undated and unsigned report in GSL/COM/P/4/1. There are several more examples of this reasoning in extant reports for 1830–41 in GSL/COM/P/4/2. Charles Darwin, for example, often rejected papers on this basis. Similar judgments that a short ver-sion in the proceedings would be sufficient were made by referees at the Royal Society (see, for example, RSL/RR/1/27, 34, and 66), although in this case they usually had not yet seen the abstract.

43. De la Beche to George Greenough, 10 November 1834, GGP/Add.7918/44.

44. After Taylor took over as printer in 1828, saving money by reducing the margins and the font size, margins and font were further reduced in a resolution on 9 January 1834, RSL-CMP/1. Further savings on the *Transactions* were made (though more by renegotiation than format) in December 1846, RSL-CMP/2, 6–7, and RSL-CMB/86a.

45. See the agreement with Longman dated 20 December 1844, GSL-CM1/6, 8 Janu-ary 1845. Longman pulled out after a year, but the Society kept publishing the journal.

46. "Report of the Scientific Committee appointed 17 June 1847," ZSL/GB/0814/GABQ.

47. Linnean Society Council Minutes, 18 December 1855, LSL/CM/3. Joseph Hooker explained to Charles Darwin, "We are I hope really about to have a reform in the Linnean

& get a first rate Natural History Journal established" [before 7 March 1855], CCD 5, 277–278.

48. At the Astronomical Society, a Finance Committee report celebrated an increase in membership and connected it directly to the *Monthly Notices*. See Council Minutes, 9 May 1862, RAS-CM/6. For the Chemical Society, see A. W. von Hofmann, "Report of the President at the Anniversary Meeting," *Journal of the Chemical Society* 14 (1862): 492–501, on 500.

49. Reported speech of Liebig and an unnamed observer that appears in Carl Vogt's autobiography, who studied with Liebig beginning in 1833: *Aus meinem Leben* (Stuttgart: E. Nägele, 1896), 129. A different English translation of the relevant sections is provided by H. G. Good, "On the Early History of Liebig's Laboratory," *Journal of Chemical Education* 13 (1936): 557–563.

50. Ibid. On Liebig's laboratory, see Jack Morrell, "The Chemist Breeders: The Research Schools of Liebig and Thomas Thomson," *Ambix* 19 (1972): 3–46; William H. Brock, *Justus von Liebig: The Chemical Gatekeeper* (Cambridge: Cambridge University Press, 1997); and Ernst Homburg, *Van beroep "Chemiker": De opkomst van de industriele chemicus en het polytechnische onderwijs in Duitsland, 1790–1850* (Delft: Delft University Press, 1993).

51. On the spread of the Giessen Model, see Alan Rocke, "Origins and Spread of the 'Giessen Model' in University Science," *Ambix* 50 (2003): 90–115, which focuses especially on the French case; and for British examples, Gerrylynn K. Roberts, "The Establishment of the Royal College of Chemistry: An Investigation of the Social Context of Early-Victorian Chemistry," *Historical Studies in the Physical Sciences* 7 (1976): 437–485; and Gerald Geison, *Michael Foster and the Cambridge School of Physiology: The Scientific Enterprise in Late Victorian Society* (Princeton, NJ: Princeton University Press, 1978).

52. Michael Faraday to James Sheridan Muspratt, 8 May 1846: "I rank as a reformer there & do not advocate or sign any certificate except of parties who have communicated a paper to the Transactions," CMF 3, 510.

53. T. G. Bonney, *Annals of the Philosophical Club of the Royal Society* (London: Macmillan, 1919), 1.

54. Resolution passed 17 March 1853, RSL-CMP/2.

55. See George Scott to Royal Society, 31 December 1864, RSL-MC/7. Election certificates of successful candidates to the Royal Society are cataloged under the shelf-mark EC.

56. This difference between genres for establishing credentials in Britain and France is explained in a document on French medical education: "With us, a proof of credentials is a sort of dossier in which the candidate presents the works that he has produced or in which he has collaborated . . . In England, [it] is a small work containing a series of certificates of English or foreign professors that allows a jury to determine the esteem in which the candidate is held by his teachers and colleagues." *Annales d'hygiène publique et de médecine légale* 44 (1900): 198–199.

57. On this genre, see Maurice Crosland, "Scientific Credentials: Record of Publications in the Assessment of Qualifications for Election to the French Académie des Sciences," in *Studies in the Culture of Science in France and Britain Since the Enlightenment* (Brookfield, 1995), 605–632.

58. Strickland to Darwin, 31 January 1849, CCD 4, 190.

59. Brewster, "Researches Respecting Neptune," 237.

60. Gordon McOuat gives a masterful account of the origins and deliberations of the nomenclatural reform committee in "Species, Rules and Meaning: The Politics of Language and the Ends of Definitions in 19th Century Natural History," *Studies in History and Philosophy of Science* 21 (1996): 473–519. On political and theoretical factions in early Victorian Britain, see Adrian Desmond, *The Politics of Evolution: Morphology, Medicine, and Reform in Radical London* (Chicago: University of Chicago Press, 1989).

61. Hugh Strickland, "On the Arbitrary Alteration of Established Terms in Natural History," *Magazine of Natural History 8* (1835): 36–40, 39–40.

62. Gordon McOuat, "Cataloguing Power: Delineating 'Competent Naturalists' and the Meaning of Species in the British Museum," *British Journal for the History of Science* 34 (2001): 1–28.

63. Strickland, "On the Arbitrary Alteration of Established Terms," 40.

64. For example, Maximilien Spinola, "Essai d'une nouvelle classification des diplolépaires," *Annales du Muséum national d'histoire naturelle* 17 (1811): 138–152, on 145. Spinola was even cited for the principle by the entomologist J. O. Westwood in "On the Chalcididae," *Zoological Journal* 4 (1828): 3–31, on 4.

65. Strickland et al., "Report of a Committee appointed 'to consider the rules by which the Nomenclature of Zoology may be established on a uniform and permanent basis," *Report of the Twelfth Meeting of the British Association in 1842* (1843): 105–121, on 106–107.

66. Strickland et al., "Report," 109.

67. Ibid., 114.

68. Ibid., 120.

69. For example, Karl Ludwig Willdenow's *Grundriss der Kräuterkunde* (Berlin: Haude und Spener, 1798), which was translated into English as *The Principles of Botany, and of Vegetable Physiology*. However, the 1813 textbook of Augustin Pyramus de Candolle, *Théorie élémentaire de la botanique* (Paris: Déterville, 1813), 251, did specify that publication should occur via printing.

70. Charles Lucien Bonaparte was a key ally, helping orchestrate discussions of nomenclature at a series of scientific meetings in Italy in the mid-1840s. It was also translated into French in *Institut* 11 (1843): 248–251.

71. McOuat, "Species, Rules and Meaning," 509.

72. Darwin to Strickland, [4 February 1849], CCD 4, 207.

73. Hooker to Darwin, 3 February 1849, CCD 4, 204.

74. Darwin to Strickland, 29 January [1849], CCD 4, 187.

75. Solitarius, "Remarks upon Zoological Nomenclature and Systems of Classification," *Field Naturalist* 1 (1833): 521–528, on 525.

76. On Gray's cataloguing strategy, see McOuat, "Cataloguing Power."

77. Joseph Hooker to William Jardine, [June 1865], HES/N-047; George Bentham to William Jardine, 7 June 1865, HES/N-051.

78. Asa Gray, "Nomenclature," *American Journal of Science and Arts* 37 (1864): 278–281, 279.

79. "Discussion des lois de la nomenclature botanique," in *Actes du Congrès international de botanique tenu à Paris en août 1867*, ed. Eug. Fournier (Paris, 1867), 198.

80. A. R. Wallace to C. von Felder, 20 September 1865, WCP639.811, in G.W. Beccaloni, ed., *Wallace Letters Online*, http://www.nhm.ac.uk/research-curation/scientific-resources/collections/library-collections/wallace-letters-online/639/811/T/details.html, accessed 12 June 2014.

81. *Address of the President, Linnean Society* (London, 1867), 6–8. In 1852, Richard Owen, amid a dispute with Henri Milne-Edwards, also claimed that print publication was crucial to priority claims. See Evelleen Richards, "A Question of Property Rights: Richard Owen's Evolutionism Reassessed," *British Journal for the History of Science* 20 (1987): 129–171.

82. Draft rules of the Société zoologique de France, in *Bulletin de la Société zoologique de France* 14 (1889): 212–282, on 266. See also new British Association recommendations in "The Bibliography of Zoology," *Natural Science* 9 (1896): 146–147.

83. Strickland to Darwin, 31 January 1849.

84. Strickland to Charles Lucien Bonaparte, April 1843, HES/F-160.

85. Strickland to Bonaparte, 1 March 1844, HES/F-166.

86. These remarks come from an MS of an irreverent history of the formation of the Ray Society by Jardine ca. 1844 titled "Reality, A Tale of Physical Science Founded on Facts," now at the Natural History Museum, London, MSS RAY H. It has been transcribed by Peter Davis in "Sir William Jardine's Account of the Formation of the Ray Society," *Archives of Natural History* 25 (1998): 25–28.

87. JEG to Strickland, 23 January 1844, HES/R-013; Jardine, "Reality."

88. On Darwin's publishing dilemma, see CD to J. D. Hooker, 9 May [1856], CCD 6, 106–107. On the superabundance of journals, see CD to THH, 20 July [1860], CCD 8, 294.

89. On the early history of the Ray Society, see Davis, "Sir William Jardine's Account"; and Richard Curle, *The Ray Society: A Bibliographical History* (London: Ray Society 1954), 1–8.

90. Strickland to Darwin, 31 January 1849.

91. Strickland to Thomas Bell, 12 February 1851, HES/R-163.

92. H. E. Strickland, "Preface," in *Bibliographia Zoologiae et Geologiae*, vol. 1 (London: Ray society, 1848), viii.

93. Ibid., vii, ix.

94. Secord, "Science, Technology and Mathematics," 459.

95. I use the term "canon-formation" in analogy to its use in literary theory to refer to genres and texts that achieve cultural prestige through inclusion in a select list. See for example John Guillory, *Cultural Capital: The Problem of Literary Canon Formation* (Chicago: University of Chicago Press, 1993).

96. An expanded account of the *Catalogue*'s construction is given in Alex Csiszar, "How Lives Became Lists and Scientific Papers Became Data," *British Journal for the History of Science* 50 (2017): 23–60. On the fourth series, see Hannah Gay, "A Questionable Project: Herbert McLeod and the Making of the Fourth Series of the Royal Society *Catalogue of Scientific Papers*, 1901–25," *Annals of Science* 70 (2013): 149–174.

97. On the American prehistory of the *Catalogue,* see Donald deB. Beaver, "The Smithsonian Origin of the Royal Society *Catalogue of Scientific Papers," Science Studies* 2 (1972): 385–393.

98. *Report of the Twenty-Sixth Meeting of the British Association for . . . 1856* (1857), 463–464.

99. J. D. Forbes, "Catalogue of Philosophical Memoirs," *Athenaeum* (20 September 1856): 1166–1167; and Thomson to Forbes, 6 October 1856, JFOR/1/2693.

100. "Report to the Council from the Library Committee," read 14 January 1858, RSL-CMP/2.

101. The Academy's financial problems are discussed earlier in this chapter. On an abandoned 1848 plan for major London societies to band together to publish a collective "comptes rendus" to rival Paris, see Bonney, *Annals of the Philosophical Club of the Royal Society,* 27–38.

102. Council Minutes, 5 March 1857, RSL-CMP/2.

103. RSL-CMB/47C, 17 June 1858; and Edward Sabine, "Anniversary Address, 1 December 1862," *Proceedings of the Royal Society* (1863), 12, 286. For details on the progressive hiring of indexers, see RSL-CMB/47C, passim.

104. RSL-CMB/47C, 29 January 1864. The day-to-day decisions regarding the *Catalogue*'s construction are detailed in the Minutes of the Library and Catalogue Committee, 17 June 1858–1 June 1875, RSL-CMB/47C and RSL-MM/14. See also Marie Boas Hall, *The Library and Archives of the Royal Society* (London: Royal Society, 1992).

105. "Preface," *Catalogue of Scientific Papers,* vol. 1 (London: Eyre and Spottiswoode, 1867), iv.

106. "Report of the Committee to consider the formation of a Catalogue of Philosophical Memoirs," 11 June 1857, RSL-CMP/2.

107. "Preliminary Report" and "Report to the Council," read 14 January 1858.

108. Some that were explicitly rejected in this process were *Paxton's Horticultural Register* and Férussac's *Bulletin.*

109. Preliminary List of Journals, RSL-MM/14/184. Twenty percent were published in Britain and Ireland; twenty-four percent in German states, and twenty-one percent in France. These figures are relatively meaningless, however, since many periodicals on the list were short-lived. Running the same count weighted by the span of years ultimately indexed of each periodical puts France ahead of Germany.

110. The responses to the circular are preserved in RSL-MM/14/184 ff. Poggendorff's response is summarized by William H. Miller in a note written 6 May 1864 (RSL-MM/14/202); for Haidinger, 3 April 1864 (RSL-MM/14/217); Henry, 23 March 1864 (RSL-MM/14/199).

111. W. H. Miller, "Circular to Scientific Societies," January 1864, RSL-MM/14/183.

112. See De Morgan's testimony in the *Report of the Commissioners Appointed to Inquire into the Constitution and Government of the British Museum with Minutes of Evidence* (London: W. Cloves, 1850), 375–377. Quotation from Augustus De Morgan, "Libraries and catalogues," *Quarterly Review* 72 (1843): 1–25, on 14–15.

113. The end of volume 6 contains 1,398 anonymous entries (those for which the cata-

logers were able to ascertain the author also appear under that author's list). These are distributed about evenly over 1800–1863, but because the number of entries in the *Catalogue* doubles about every two decades, this represents a significant proportional reduction over time.

114. "Preface," ix.

115. Ibid.

116. Of seventy-four publications listed under Young's name, thirty-nine were originally published anonymously or using a variety of initials.

117. Irénée-Jules Bienaymé, "Rectification de listes d'articles détachés de M. Cauchy, publiées dans deux Catalogues différents, et restitution à M. Cournot de quelques-uns de ces articles," *Comptes rendus hebdomadaires* 72 (1871): 25–29.

118. RSL-CMB/47C, 28 January 1869.

119. Murchison to Strickland, July 1852, HES/E1110; and CD to H. E. Strickland, 29 January 1849, HES/N168.

120. G. D. Campbell to White, 13 April 1876, RSL-MS/769; Brewster's responses were sent 27 May and 5 June 1867, RSL-MC/8; Herschel and Forbes requested copies of proofs. See Herschel to RS, 1 March 1869, RSL-MM/14/217; and W. White to J. D. Forbes, 13 July 1868, JFOR 1/4686.

121. "Preface," ix.

122. W. White to J. D. Forbes, 13 July 1868; and J. Herschel to the RS, 1 March 1869.

123. Coleridge, *Cabinet Edition of the Encyclopaedia Metropolitana: Prospectus* (London: Griffin & Company, 1849), 7.

124. These lists, or transcriptions of them, are preserved at the Harry Ransom Center (Austin, TX), Herschel Papers, 21.1–7. For a classified list, see 21.1 (a transcription), for a list organized by publication venue, 21.1, and for chronological, 21.6.

125. Marcellin Berthelot to RS, 19 December 1889, RSL-MS/539.

126. A thousand copies of volume 1 were printed and put on sale for 20/– (cloth binding) and 28/– (Morocco binding); RSL-CMB/47C, 24 January 1868.

127. For example, "Catalogue des brochures scientifiques," *Les Mondes* 17 (1868): 410; "Catalogue de mémoires scientifiques," *Revue des Cours Scientifiques de la France et de L'étranger* 5 (1868): 487–488; and Review in *Athenaeum* (6 June 1868): 790–791.

128. "Societies," *Athenaeum* (20 November 1869): 667; and "The Month: science and arts," *Chambers's Journal* (28 December 1867): 830.

129. *Athenaeum* (6 June 1868): 790.

130. Ibid.

131. Letter to editor signed F. R. S. in *Athenaeum* (16 January 1869): 99–100.

132. Ibid.; and *The Annual Register . . . for the year 1868* (London: Rivingtons, 1869), 348.

133. George Gore's 1878 guide, *The Art of Scientific Discovery* (London: Longmans, Green, and Co.), instructs readers in a method for using the *Catalogue* to find papers on particular subjects. Contemporary evaluations of the *Catalogue* as a research tool can be found in correspondence received by the Royal Society in 1894 in response to a survey regarding the possibility of creating a successor to the *Catalogue*. See RSL-MS/531.

134. [Review of the] *Catalogue of Scientific Papers, Athenaeum* (6 June 1868): 790.

135. *Nature* 1 (18 November 1869): 86. For a continental example, see "Zur Statistik der Naturwissenschaftlichen Litteratur," *Wiener Zeitung* (5 February 1870): 437.

136. For example, see *Athenaeum* (7 June 1879): 732.

137. "Ein Verzeichniß sämmticher naturwissenschaftlichen Abhandlungen aus den Jahren 1800 bis mit 1863," *Wiener Zeitung* (18 July 1868): 201; and *Athenaeum* (6 June 1868): 791.

138. Max Simon Nordau, *Degeneration* (London: Heinemann, 1895), 114.

139. Wilhelm Ritter von Haidinger, "Catalogue of Scientific Papers (1800–1863)," *Verhandlungen der K.K. Geologischen Reichanstalt* 4 (15 February 1870): 70–74.

140. After 1836, 391 of his 405 entries in the *Catalogue* are references to the *Comptes rendus*. To be clear, criticism of Cauchy was not the same as contemporary criticism of authors who engage in so-called "salami science" to pad their CV. There was little or no sense at that time that having published very many short articles (as opposed to longer memoirs or books) was in itself something to be rewarded.

141. J.-B. Biot, "Comptes rendus hebdomadaires," *Journal des savans* (November 1842): 641–661, on 659–660.

142. See, for example, Jöns Berzelius, *Jahres-Bericht über die Fortschritte der physischen Wissenschaften* 7 (1828): 87; and the reported speech of Felix Klein in a letter from Georges Brunel to Henri Poincaré, 7 July 1881, in "La correspondance d'Henri Poincaré avec des mathématiciens de A à H," *Cahiers du séminaire d'histoire des mathématiques* 7 (1986): 92.

143. On the transformation of this genre toward the form of a list, see Crosland, "Scientific Credentials."

144. *Athenaeum* (6 June 1868): 791.

145. This obituary appeared in *Nature* 9 (1874): 403–404. But see also *Athenaeum* (28 February 1874): 297.

146. Franz Ritter v. Hauer, "Zur Erinnerung an Wilhelm Haidinger," *Jahrbuch der Kaiserlich-Königlichen Geologischen Reichsanstalt* 21 (1871): 31–40; and Ed. Döll, *Wilhelm Ritter von Haidinger*, Vienna: Realschule, 1871.

147. "Johann Christian Poggendorff," *Proceedings of the American Academy of Arts and Sciences* 12 (1877): 331.

148. Johann Poggendorff, "Vor- und Schlusswort des Verfassers," in *Biographisch-Literarisches Handwörterbuch für Mathematik, Astronomie, Physik mit Geophysik, Chemie, Kristallographie und Verwandte Wissensgebiete*, vol. 1 (Leipzig: J.A. Barth, 1863), v.

149. Harvey C. Lehman, "Men's Creative Production Rate at Different Ages and in Different Countries," *The Scientific Monthly* 78 (1954): 321–326; and Donald deB. Beaver, *The American Scientific Community, 1800–1860: A Statistical-Historical Study* (New York: Arno Press, 1968), 1. See also Wayne Dennis, "Bibliographies of Eminent Scientists," *Scientific Monthly* 79 (1954): 180–183.

Chapter 6

1. "The Bibliography of Periodical Literature," *The Library*, viii (1896): 49–64, on 49.

2. [Norman Lockyer], "Order or Chaos?," *Nature* 48 (1893): 241; Michael Foster, "The Organisation of Science," in *Atti dell'XI Congresso medico internazionale Roma, 29 marzo–5 aprile 1894* (Rome: Ripamonti e Colombo, 1895), 247; and idem, "On the Organisation of Science," *Nature* 49 (1894): 563–564, 563.

3. For details, see for example Jeffrey Wawro, *The Franco-Prussian War: The German Conquest of France in 1870–1871* (Cambridge: Cambridge University Press, 2003), a recent history of the war.

4. For an account of the Academy's doings during the siege, see Maurice Crosland, "Science and the Franco-Prussian War," in *Studies in the Culture of Science in France and Britain Since the Enlightenment* (Brookfield, VT: Variorum, 1995), 185–214.

5. For accounts of the pigeon post, see F. F. Steenackers, *Les télégraphes et les postes pendant la guerre de 1870–1871* (Paris: G. Charpentier, 1883); and John Stirling Fisher, *Airlift 1870: The Balloon and Pigeon Post in the Siege of Paris* (London: M. Parrish, 1965).

6. Henri Sainte-Claire Deville, "De l'intervention de l'Académie dans les questions générales de l'organisation scientifique en France," *Comptes rendus* 72 (6 March 1871): 237–239.

7. Arthur Mangin, "Revue scientifique," *Correspondant* 83 (1871): 1170–1183.

8. N. Lockyer, "The Hope of France," *Nature* 3 (1871): 501–502.

9. Ernest Renan, *La Réforme intellectuelle et morale de la France* (Paris: M. Lévy frères, 1871), 153.

10. François Furet, *Revolutionary France, 1770–1880* (Oxford: Blackwell, 1992), 497–499.

11. The theme of centralization and decentralization is a classic in the historiography of nineteenth-century science. See especially Joseph Ben-David, *The Scientist's Role in Society: A Comparative Study* (Englewood Cliffs, NJ: Prentice Hall, 1971); and Rainald von Gizycki, "Centre and Periphery in the International Scientific Community: Germany, France and Great Britain in the Nineteenth Century," *Minerva* 11 (1973): 474–494.

12. Renan, *La Réforme intellectuelle*, 95.

13. Jean-Baptiste Dumas, "De l'intervention de l'Académie dans les questions générales de l'organisation scientifique en France," *Comptes rendus* 72 (6 March 1871): 265.

14. On the reform movement as it applied to the reform of higher education in particular, see George Weisz, *The Emergence of Modern Universities in France, 1863–1914* (Princeton, NJ: Princeton University Press, 1983). For the role of chemists in the 1860s in this, see Alan J. Rocke, *Nationalizing Science: Adolphe Wurtz and the Battle for French Chemistry* (Cambridge, MA: MIT Press, 2001), 269–300.

15. On the Association française, see the essays in H. Gispert, ed., *'Par la science, pour la patrie': l'Association française pour l'avancement des sciences, 1872–1914* (Rennes: Presses universitaires de Rennes, 2002).

16. Armand Quatrefages de Breau, "La science et la patrie," in *Comptes-rendus de la 1re session 1872* (Paris: AFAS, 1873), 36–41.

17. G. Darboux to J. Houël, letter ca. 1870, printed in Hélène Gispert, "La correspondance de G. Darboux avec J. Houël. Chronique d'un rédacteur (déc. 1869–nov. 1871)," *Cahiers du séminaire d'histoire des mathématiques* 8 (1987): 67–202, on 89.

18. Darboux cited Férussac in his preface to the new *Bulletin* as well as in letters to Houël ca. 1870, p. 86, 141.

19. Émile Alglave, Editorial in *Revue scientifique de la France et de l'étranger* 1 (1871): 1.

20. On Richet, see Jacqueline Carroy, "Playing with Signatures: The Young Charles Richet," in *The Mind of Modernism*, ed. M. S. Micale (Stanford, CA: Stanford University Press, 2004); Stewart Wolf, *Brain, Mind, and Medicine: Charles Richet and the Origins of Physiological Psychology* (New Brunswick: Transaction Publishers, 1993); and E. Osty, "Charles Richet," *Revue métapsychique* (1936): 1–42.

21. Charles Richet, "La Revue scientifique (1863–1888)," *Revue scientifique* 16 (1888): 721–722.

22. Ibid., 722.

23. Richet, "De la méthode à suivre dans les recherches bibliographiques," *Revue scientifique* 4 (1882): 19–21, on 20.

24. Comments of Klein were reported in secret to Poincaré by Georges Brunel, a French mathematician visiting Klein's group in Leipzig at the time. See letters from Brunel to Poincaré, June 1881 and 7 July 1881, reprinted in "La correspondance d'Henri Poincaré avec des mathématiciens de A à H," *Cahiers du Séminaire d'histoire des mathématiques* 7 (1986): 59–223, on 91–95.

25. The quotation appears in an article by Gustav Eneström, "Sur les bibliographies des sciences mathématiques," *Bibliotheca mathematica* 4 (1890): 37–42, on 39. The original circular seems not to have survived.

26. See Laisant, *À mes électeurs: Pourquoi, et comment, je suis Boulangiste* (Paris: Mayer et Cie, 1887), 31.

27. C. Laisant and E. Lemoine, "Sur l'orientation actuelle de la science et de l'Enseignement mathématique," *Revue générale des sciences pures et appliquées* 4 (1893): 719–722, on 721.

28. "Préface," *L'Intermédiaire des mathématiciens*, viii. Laisant was involved in several other social service publications for mathematicians. He cofounded *L'enseignement mathématique* (1899–), which published articles and discussions on anything related to the social and intellectual life of mathematics. In 1896 he began publishing collections of important mathematical problems—the *Recueil de problèmes de mathématiques* (1896)—for the use of teachers and for the continuing edification of mathematical amateurs.

29. Laisant borrowed this phrase from Marcellin Berthelot. See Berthelot, "La direction des sociétés humaines par la Science" [1897], in *Science et éducation* (Paris: Société française d'imprimerie et de librairie, 1901), 1–9, on 9.

30. Laisant, "Le rôle social de la science," *Enseignement mathématique* 6 (1904): 337–362, on 349–350.

31. J. E. S. Hayward, "The Official Social Philosophy of the French Third Republic: Léon Bourgeois and Solidarism," *International Review of Social History* 6 (1961): 19–48.

32. In 1898, a law came into effect giving mutual assistance societies the same rights

as trade unions and according them other subsidies; in 1893, laws were passed guaranteeing medical aid to those who could not afford it otherwise; in 1901, a bill was introduced extending pensions to several classes of workers. See Janet R. Horne, *A Social Laboratory for Modern France: The Musée Social & the Rise of the Welfare State* (Durham, NC: Duke University Press, 2002); Phillip Nord, "The Welfare State in France, 1870–1914," *French Historical Studies* 18 (1994):821–838; and Timothy B. Smith, *Creating the Welfare State in France, 1880–1940* (Montreal: McGill-Queen's University Press, 2003).

33. Baudouin, "La seconde conférence bibliographique internationale de Bruxelles en 1897," *Revue scientifique* 60 (1897): 235–239, on 236.

34. The Institut was variously known as the Institut de Bibliographie Scientifique, the Institut International de Bibliographique Scientifique, and the Institut International de Bibliographie Médicale.

35. "Le problème bibliographique," *Revue scientifique* 4 (1895): 708–715, on 708.

36. Baudouin, *Bibliographie scientifique* 2 (1896): 2.

37. Carleton B. Chapman, *Order Out of Chaos: John Shaw Billings and America's Coming of Age* (Canton, MA: Science History Publications, 1994).

38. For a recent history of index cards in libraries, see Markus Krajewski, *Paper Machines: About Cards & Catalogs, 1548–1929* (Cambridge, MA: MIT Press, 2011).

39. Baudouin, *Bibliographie scientifique* 2 (1896): 2.

40. *Bibliographie scientifique* 1 (1895): 101.

41. Circulating libraries were also founded for other disciplines. A Bibliothèque Mathématique des Travailleurs, for instance, was founded in Paris in 1895. See *Les tablettes du chercheur* 6 (1895): 212–214.

42. *Bibliographie scientifique* 3 (1897): 2.

43. *Bibliographie scientifique* 2 (1896): 18.

44. *Bulletin des sommaires des journaux scientifiques, littéraires, financiers* 1, no. 6 (1888): 44; 2, no. 13 (1889): 100; 2, no. 14 (1889): 108.

45. *Bibliographie scientifique* 3 (1897): 1.

46. *Bibliographie scientifique* 3 (1897): 2.

47. Baudouin worked for *Le Progrès médical*, *Archives provinciales de chirurgie*, and the *Gazette médicale de Paris*. In the early 1890s, he was a *préparateur* at the Faculté de Médicine in Paris.

48. Charles M. Limousin, "Coup d'œil historique sur l'Internationale," *Journal des économistes* 37 (1875): 68–87.

49. Limousin, "Causerie," *Bulletin des sommaires* 2 (1889): 100.

50. Karl Hescheler, "Dr. Phil. Herbert Haviland Field, 1868–1921" (Schaffhausen, 1921), 2. Field himself recounted that he had subsequently become fascinated with finding better means of locating papers in his well-trod subdiscipline. See *Annotationes Concilii Bibliographici* 3 (1907): 1. On Field, see also Colin Burke, *Information and Intrigue: From Index Cards to Dewey Decimals to Alger Hiss* (Cambridge, MA: MIT Press, 2014).

51. The *Société* had taken up the problem of nomenclature early in its existence, but the matter hardly progressed until Blanchard pushed the matter forcefully onto the agenda of the 1889 International Congress. See Richard V. Melville, *Towards Stability in the Names*

of Animals (London: ICZN, 1995), 15–19. See also Robert Fox, "The Early History of the Société Zoologique de France," in the *Culture of Science in France, 1700–1900* (Aldershot: Variorum, 1992), 1–16.

52. For Field's initial proposal, see "La réforme bibliographique," *Mémoires de la Société zoologique de France* (1894): 259–263.

53. E.-L. Bouvier's "Rapport sur le projet de réforme bibliographique" was followed by the nomination of a seven-member committee. The Société was among the first to put up funds for the new Bureau, which would be directed by Field.

54. On the history of the Concilium, see Colin B. Burke, *Information and Intrigue: From Index Cards to Dewey Decimals to Alger Hiss* (Cambridge, MA: MIT Press, 2014).

55. *Annotationes Concilii Bibliographici* 2 (1906): 31–37.

56. Limousin's enterprise eventually folded in 1903. Baudouin's Institut fared better; contemporary accounts suggest that it was a bustling enterprise, eventually becoming a joint-stock company, but it too folded, in 1906. The Concilium had greater longevity, and remained in operation until 1940.

57. A Free Lance, *On the Organisation of Science (Being an Essay Towards Systematisation)* (London: Williams & Norgate, 1892), 4.

58. Ibid.; and F. G. Donnan, "The Organisation of Scientific Literature," *Nature* 48 (1893): 436.

59. E.g., Foster, "The Organisation of Science," 563; T. D. A. Cockerell, "A Suggestion for the Indexing of Zoological Literature," *Nature* 46 (1892): 442; and Free Lance, *On the Organisation of Science*, 4.

60. E.g., *Review of Reviews* (British edition) 10 (1894): 194; and *Nature* 1 (1894): 539.

61. E.g., Foster, "The Organisation of Science," 564; and Armstrong, "Scientific Bibliography," *Nature* 54 (1896): 617–618, on 618.

62. Darwin to Joseph Hooker, 17 April [1865], CCD 13, 122.

63. Swinburne, "The Publication of Physical Papers," *Nature* 48 (1893): 198.

64. Henry Armstrong, "Presidential Address," *Journal of the Chemical Society (Transactions)* 65 (1894): 336–382, on 345. There were varying opinions, however, on what size format was best for reading in the train. See W. T. Thiselton Dyer, "To Accompany the Report," 6 June 1894, RSL-CMB/43: "We read the *Times* in the train, and a paper from the 'Philosophical Transactions' is at least as handy as that."

65. Foster, "The Organisation of Science," 254.

66. Free Lance, *On the Organisation of Science*, 28. A Free Lance was a pseudonym for Frank H. Perry Coste (later Perrycoste) (1864–1928). Trained as a botanist (BSc 1891; joined the Linnean Society the same year), this brief but eventful foray into science policy helped inspire him to reinvent himself as a social philosopher in the Spencerian mode. His next major book, *Towards Utopia (being speculations in social evolution)* (London: S. Sonnenschein & co., 1894), extended his advocacy of rational organization to society in general.

67. E.g., Parker, "Suggestions for Securing Greater Uniformity of Nomenclature in Biology," *Nature* 45 (1891): 68–69; Cockerell, "A Suggestion," 442; T. R. R. Stebbing, "On Random Publishing and Rules of Priority," *Natural Science* 5 (1894): 337–344; and F. A. Bather, "Zoological Bibliography and Publication," in *Report of the Sixty-Seventh Meeting*

of the British Association for . . . 1897 (1898): 359–362. See also Harriet Ritvo, "Zoological Nomenclature and the Empire of Victorian Science," in *Victorian Science in Context*, ed. B. Lightman (Chicago: Chicago University Press, 1997), 334–353, on 338.

68. Lord Walsingham and John Hartley Durrant, *Rules for Regulating Nomenclature: With a View to Secure a Strict Application of the Law of Priority in Entomological Work* (London: Longmans, Green, 1896), 3. See also T. R. R. Stebbing, "From Buffon to Darwin," *Zoologist* 1 (1897): 312–324, on 323; David Sharp, *The Object and Method of Zoological Nomenclature* (London: E. W. Janson, 1873), 17; and Foster, "The Organisation of Science."

69. Stebbing, "On Random Publishing," 341.

70. F. Jeffrey Bell, "The 'Claims of Priority,' and What they are Sometimes Worth," 19 (December 1896): 476–477; "Zoological Nomenclature: A Proposal," *Natural Science* 8 (1896): 218–220; and Review of *History of Crustacea* by T. R. R. Stebbing, *Athenaeum* (24 June 1893): 800.

71. Stebbing, "On Random Publishing," 338.

72. Sharp, *The Object and Method*, 30.

73. "The Bibliography of Zoology," *Natural Science* 9 (1896): 146–147; and "Zoological Nomenclature: A Proposal."

74. L. H. Petit, *Essais de bibliographie médicale* (Paris: G. Masson, 1887), 126.

75. J. A. Allen, "A Reprehensible Method of Determining Priority of Publication," *Science* 4 (1896): 691–693. Allen, an American ornithologist, and the paleontologist E. D. Cope had a public dispute over dating practices in *Science* 4 (1896): 760–761, 838–839, 878–879. Many concerns about dating were also raised by Charles Davies Sherborn, who was in the process of producing a massive index of animal names, the *Index Animalium*. See, for example, "The Dating of Books," *Natural Science* 9 (1896): 280.

76. "The Publication of Papers by Societies," *Natural Science* 6 (1895): 294–296.

77. "Suggestions by Professor Armstrong."

78. Swinburne, "The Publication of Physical Papers," 197; and "Suggestions by Professor Armstrong."

79. [Lockyer], "Order or Chaos?," 241.

80. Oliver Lodge, "The Publication of Physical Papers," *Nature* 48 (1893): 292–293, on 292.

81. "The Diffusion of Scientific Memoirs," *Nature* 29 (1883): 171.

82. Swinburne, "The Publication of Physical Papers," 197.

83. J. Y. Buchanan, "Publication of Scientific Papers," *Nature* 48 (1893): 340–341.

84. A. P. Trotter, "The Publication of Physical Papers," *Nature* 48 (1893): 412; and Swinburne, "The Publication of Physical Papers," 197.

85. Free Lance, *On the Organisation of Science*, 23.

86. Foster, "The Organisation of Science," 254.

87. Ramsay, "Molecular Weights," *Chemical News* 69 (2 February 1894): 51.

88. Lodge, "The Publication of Physical Papers," 292.

89. Buchanan, "Publication of Scientific Papers," 340.

90. Gustav Eneström to Poincaré, 24 May 1885, AHP.

91. A. B. Basset, "The Organisation of Scientific Literature," *Nature* 48 (1893): 436.

92. AFAS, *Rapport sur la réforme de la bibliographie scientifique* (Paris, AFAS, 1895), 10.

93. R., "Diffusion of Scientific Memoirs," *Nature* 29 (1883): 261.

94. A local bookseller advertised copies of Free Lance's *On the Organisation of Science* for those in attendance. *Nature* 48 (7 September 1893).

95. "Physics at the British Association," *Nature* 48 (1893): 525–529, on 529.

96. Hofmann, "Report of the President at the Anniversary Meeting," 500.

97. Armstrong, "Presidential Address," *Journal of the Chemical Society (Transactions)* 65 (1894): 336–382, on 343.

98. Ibid., 340–341.

99. See *Jubilee of the Chemical Society of London* (London, 1896), 201.

100. Armstrong, "Autobiography 1848–79," *The Central* 35 (1938): 3–12, on 10.

101. Armstrong, "Presidential Address."

102. Armstrong, "Scientific Bibliography," 617–618; and idem, "Presidential Address," 350.

103. Ibid., 346, 355.

104. A Free Lance, *On the Organisation of Science*, 9. This argument for the unintended consequences of the credit authors obtain for publishing their findings presages an argument often made by Robert K. Merton to the same effect beginning in the 1960s. See for example Merton, "Priorities in Scientific Discovery," 655.

105. Swinburne, "The Publication of Physical Papers," 197.

106. Armstrong, "Presidential Address," 345. See also Trotter, "The Publication of Physical Papers," 412; and A Free Lance, *On the Organisation of Science*, 8.

107. [Lockyer], "Order or Chaos?," 241.

108. Free Lance, *Towards Utopia*, 114.

109. Armstrong to Royal Society, 17 January 1892, RSL-MC/15/253; and idem, "Presidential Address," 344.

110. Armstrong, "Further Suggestions," RSL-CMB/43.

111. [Lockyer], "Order or Chaos?," 242.

112. "Wanted: A Science Reform," *Chemist and Druggist* 41 (15 October 1892): 583–584, on 583.

113. "Annual Report," *Quarterly Journal of the Chemical Society* 5 (1853): 153–165, on 163.

114. Foster, Letter to the President presented at the Council Meeting of 5 July 1893. A copy of this letter is pasted into the Procedure Committee Minutes Book, RSL-CMB/43.

115. See the Procedure Committee Minutes; see also the Royal Society Council Minutes, 5 July 1893. RSL-CMP/7.

116. The Committee was chaired by the President (William Thomson), and its members included the secretaries Michael Foster and Lord Rayleigh, as well as Henry Armstrong, Sir Benjamin Baker, William Thomas Blanford, Lauder Brunton, Sir John Evans, A. R. Forsyth, Archibald Geikie, Richard Glazebrook, Frederick DuCane Godman, John Hopkinson, Ray Lankester, Joseph Lister, Norman Lockyer, Arthur Rücker, Philip Sclater, W. T. Thiselton-Dyer, William Tilden, and Sydney Howard Vines.

117. "Suggestions by Professor Armstrong (Received 9 November 1893)," RSL-CMB/ 43.

118. Armstrong, "Further Suggestions."

119. H. B. Woodward, "President's Address," *Transactions of the Norfolk and Norwich Naturalists' Society* 5 (1889–1894): 333–363.

120. Review of *The Organisation of Science* by A Free Lance, *Lancet* (18 June 1892): 1378; and "The Organisation of Science," *Natural Science* 3 (1892): 241–243, on 243.

121. "Physics at the British Association," 529; and "The Organisation of Science," *Natural Science*, 241.

122. Rayleigh, "Introduction: On the Physics of Media that are Composed of Free and Perfectly Elastic Molecules in a State of Motion," *Philosophical Transactions Part A* 183 (1892): 1–5.

123. See the series of letters titled "An Obstacle to Scientific Progress" in *The Chemical News* 46 (1892): 39, 61. The passages cited here were published 29 July 1892 and attributed to A. Irving and X.

124. Oliver Heaviside, "The French Academy," *Nature* 69 (1904): 317. On Heaviside's troubles with Royal Society referees, see Bruce J. Hunt, "Rigorous Discipline: Oliver Heaviside Versus the Mathematicians," in *Literary Structure of Scientific Argument*, 72–95.

125. The papers of the inquiry, including Sherborn's letter (dated 28 May 1903) and summary proposals for new bylaws by Horace Monckton, are in GSL/COM/SP/4/7. They had earlier launched a smaller-scale inquiry on referee systems in 1898 as well.

126. Standing Orders of Council, in RSL-CMB/43.

127. On internationalism in science, see Elisabeth Crawford, *Nationalism and Internationalism in Science, 1880–1939* (Cambridge: Cambridge University Press, 1992); Anne Rasmussen, "L'internationale scientifique (1890–1914)" (PhD diss., École des hautes études en sciences sociales de Paris, 1995); Elisabeth Crawford, Terry Shinn, and Sverker Sörlin, eds., *Denationalizing Science: The Contexts of International Scientific Practice* (Dordrecht: Kluwer, 1993); and Brigitte Schröder-Gudehus, "Caractéristiques des relations scientifiques internationales, 1870–1914," *Cahiers d'histoire mondiale* 10 (1966–1967): 161–177.

128. The responses to the Royal Society's circular are preserved in RSL-MS/531.

129. Paul Otlet, "Un peu de bibliographie," *Palais, organe des conférences du jeune barreau de Belgique* (1891–1892): 254–271.

130. Horatio Hale, "An International Scientific Catalogue and Congress," *Science* 1 (1895): 324–326.

131. *Report of the Proceedings at the International Conference on a Catalogue of Scientific Literature, Held in London, July 14–17, 1896* (London: Harrison, 1896), 22.

132. Ibid.

133. Henry Armstrong, Circular sent with draft Index Schemes, 1 April 1896, printed in *Bulletin de l'Institut International de Bibliographie* 1 (1896): 182–188, on 187.

134. Sebert to Poincaré, 19 February 1900, MUN, ICSL Folder.

135. *Scheme for the Publication of an International Catalogue* (London: Harrison, 1900), 17.

136. Michael Foster, "A Conspectus of Science," *Quarterly Review* 197 (1903): 139–160.

137. Ibid., 144–145.

Conclusion

1. John F. McGowan, "An Unreasonable Man" [review of Masha Gessen, *Perfect Rigor* and Donal O'Shea, *The Poincaré Conjecture*], Math-Blog, 31 January 2010, https://mathblog.com/an-unreasonable-man/, accessed 25 July 2010.

2. Sylvia Nasar and David Gruber, "Manifold Destiny," *New Yorker*, 28 August 2006, 44–57.

3. On Perelman and the theme of madness, see Masha Gessen, *Perfect Rigor: A Genius and the Mathematical Breakthrough of the Century* (Boston: Houghton Mifflin Harcourt, 2009), 170–199.

4. Nasar and Gruber, "Manifold Destiny," 54–57.

5. On the arXiv, see Alessandro Delfanti, "Beams of Particles and Papers: How Digital Preprint Archives Shape Authorship and Credit," *Social Studies of Science* 46 (2016): 629–645.

6. UK Publishers Association, "Memorandum from the Publishers Association," in *Scientific Publications: Free for All? Vol. 2, Oral and Written Evidence* (London: Stationery Office, 2004), 95–105. See also Michael A. Mabe, "Scholarly Publishing," *European Review* 17 (February 2009): 3–22.

7. Bowker, "Emerging Configurations of Knowledge Expression."

8. Patrick J. Michaels, "How to Manufacture a Climate Consensus," *Wall Street Journal*, 17 December 2009.

9. John P. Ioannidis, "Why Most Published Research Findings Are False," *PLoS Medicine* 2 (2005): 696–701; and Daniel Sarewitz, "Saving Science," *New Atlantis* 49 (2016 Spring/Summer): 4–40.

10. Jason Priem, "Scholarship: Beyond the Paper," *Nature* 495 (28 March 2013); and J. Priem, D. Taraborelli, P. Groth, and C. Neylon, "Altmetrics: A Manifesto," 26 October 2010, http://altmetrics.org/manifesto.

11. Philip Mirowski, "Die offene Wissenschaft und ihre Freunde," *Frankfurter Allgemeine Zeitung*, 29 March 2014; Evgeny Morozov, *To Save Everything, Click Here: The Folly of Technological Solutionism* (New York: PublicAffairs, 2013); and Rebecca Lave, Philip Mirowski, and Samuel Randalls, "STS and Neoliberal Science," *Social Studies of Science* 40 (2010): 659–675.

12. J. D. Bernal, "Provisional Scheme for Central Distribution of Scientific Publications." On this incident see Harry East, "Professor Bernal's 'Insidious and Cavalier Proposals: The Royal Society Scientific Information Conference of 1948," *Journal of Documentation* 54 (1998): 293–302.

13. H. J. Fleure, "Scientific Papers: Professor Bernal's Scheme," *The Times*, 23 June 1948, 5; Webster Plass, "Records of Research," *The Times*, 29 June 1948, 5; and John R. Baker and A. G. Tansley, "Publications on Science: Threat to Journals," *The Times*, 21 June 1948, 5.

14. On the changing senses of information and their cultural consequences, see espe-

cially Geoffrey Nunberg, "Farewell to the Information Age," in *The Future of the Book*, ed. Geoffrey Nunberg (Berkeley: University of California Press, 1996), 103–137; Geoffrey C. Bowker, "Information Mythology: The World of/as Information," in *Information Acumen: The Understanding and Use of Knowledge in Modern Business*, ed. L. Bud-Frierman (London: Routledge, 1994), 231–247; and Paul N. Edwards, Lisa Gitelman, Gabrielle Hecht, Adrian Johns, Brian Larkin, and Neil Safier, "AHR Conversation: Historical Perspectives on the Circulation of Information," *American Historical Review* 116 (December 2011): 1393–1435.

15. David A. Hollinger, "Free Enterprise and Free Inquiry: The Emergence of Laissez-Faire Communitarianism in the Ideology of Science in the United States," *New Literary History* 221 (1990): 897–919.

16. John Ziman, *Public Knowledge: An Essay Concerning the Social Dimension of Science* (London: Cambridge University Press, 1968), 109, 111.

17. I rely here on Andrew Abbott, "Linked Ecologies," *Sociological Theory* 23 (2005): 245–274.

INDEX